Collision Processes Involving Highly Excited Atoms and Neutral Particles

I0028320

CAMBRIDGE SCIENTIFIC PUBLISHERS

COLLISION PROCESSES INVOLVING HIGHLY EXCITED ATOMS AND NEUTRAL PARTICLES

Contents

Physics Reviews 2004, Vol 21, pp 1-304
Reprints available directly from publisher
Printed in UK
Photocopying permitted by license only

COLLISION PROCESSES INVOLVING HIGHLY EXCITED ATOMS AND NEUTRAL PARTICLES

V.S. Lebedev

P.N. Lebedev Physical Institute,
Russian Academy of Sciences, Moscow [1]

ABSTRACT

The review presents an up-to-date summary of knowledge on elementary collision processes of highly excited atoms with neutral particles. Its principal goal is to provide the basic physical approaches and theoretical techniques for the description of a broad class of observed collision phenomena in gases and low-temperature plasmas involving Rydberg atoms. Emphasis is given to the consideration of different mechanisms of inelastic and quasielastic collisions with a change in the principal and orbital quantum numbers, transitions between the fine-structure components, direct and associative ionization, ion-pair formation and electron–ion recombination processes. It is shown that there is an important role of the effects associated with the ion core–projectile interaction in collisions with large energy transfer. Considerable attention is put to discussion of resonant phenomena in the excitation and quenching of highly excited states resulting from the long-range part of the electron–projectile interaction. The theory is applied for comparison with available experimental data on thermal collisions of Rydberg atoms with the ground-state rare gas and alkali-metal atoms, polar, non-polar and electron-attaching molecules. The material is illustrated by many figures, which demonstrate the dependencies of cross sections and rate constants on atomic quantum numbers, energy defects of reactions, relative velocities of colliding partners, and the gas temperatures.

[1]P.N. Lebedev Physical Institute, Russian Academy of Sciences, 53 Leninskii Prospect, Moscow 117924, Russia. E-mail: vlebedev@mail1.lebedev.ru

1

1 Introduction

The past two decades have witnessed great advances in different areas of the physics of Rydberg (highly excited) atoms and its applications to high-resolution atomic spectroscopy, kinetics and diagnostics of gases and low-temperature plasmas, astrophysics and radio astronomy. The present review concentrates on elementary collision processes of Rydberg atoms with neutral atomic and molecular particles. It includes a systematic description of basic physical approaches, theoretical methods and modern techniques for the calculation of a broad class of quasielastic, inelastic and ionizing collisions involving the Rydberg atoms. The main emphasis is given to quantal and semiclassical methods, which have found wide application only in recent years. These methods provide a basis for understanding a number of new effects in the physics of atomic and molecular collisions. Particular attention is paid to a comparison of theoretical results with available experimental data.

The observed physical phenomena in collisions involving Rydberg atoms and specific effects of their interactions with neutral targets are primarily determined by the peculiar properties of highly excited states. The properties distinguishing a Rydberg atom from an atom in the ground or low excited states are mainly due to the large magnitude of the principal quantum number n for one of its electrons. As a result, a Rydberg atom is characterized by a very large dimension and a period of orbital electron motion around the parent ion core, very small ionization potential and average velocity of a highly excited electron on the atomic scale. In accord with the basic correspondence principle, a Rydberg atom with high value of $n \gg 1$ has simultaneously both pure quantal and classical properties. That is why various versions of semiclassical methods have found wide applications in the theory of Rydberg atom-neutral particle collisions.

On the other hand, the physical mechanisms of excitation, quenching and ionization processes involving Rydberg atoms turn out to be quite different for collisions with different types of atomic and molecular projectiles. In particular, for a given value of n and the energy defect of reaction, the corresponding values of the cross sections may differ by several orders of magnitude for collisions with the ground-state rare gas and alkali-metal atoms as well as for collisions with atoms and molecules having small electron affinities. Significant differences in magnitudes of cross sections take also place for collisions

3

of highly excited atoms with strongly polar and non-polar molecules.

In spite of the specific structure and energy spectrum of a complex Rydberg atom which appreciably affect the values of collision cross sections for inelastic transitions from selectively excited levels, the basic features of any atom with a high principal quantum number are similar to the hydrogen atom. This is because of the interaction between the Rydberg electron and parent ion core is primarily determined by their Coulomb interaction at large distances. Thus, an important feature of the Rydberg atom is a great number of closely spaced energy levels. The radiation lifetime of a Rydberg atom increases drastically with increasing principal quantum number. The cross sections for collisions of Rydberg atoms with neutral and charged targets are much greater than those for collisions involving atoms in the ground or low excited states. The very large range of $n \sim 10 - 1000$ achieved in modern laboratory experiments leads to extremely large variations in the values characterizing the physical properties of highly excited atoms.

It is important to stress that the growth of research activity in the physics of atomic collisions involving Rydberg atoms over the last 10-15 years was stimulated by an extremely rapid development of laser spectroscopy, methods of selective excitation and detection of highly excited states and techniques of atomic beams. This has facilitated significant progress in different directions of the physics of Rydberg atoms which is of fundamental and practical importance. The physical principles and experimental techniques for the production and detection of Rydberg atoms, methods of their photoexcitation in electric fields and pulsed field ionization, microwave excitations and ionization, effects of interaction of Rydberg atoms with the blackbody radiation were reflected in the comprehensive books [1, 2]. General theoretical methods in the physics of highly excited atoms and ions were the subject of the book [3]. The recent monograph [4] and review articles [5–7] are devoted to the basic methods in the one– and multi-channel Rydberg spectroscopy and quantum defect theory. In addition to the formulation of basic aspects of the theory, these works contain an extensive material on the structure and spectra of Rydberg atoms, including the problems of autoionizing states, interseries interaction and electron correlations for complex atomic systems.

Over the past few years evident advances were achieved in studies of the Rydberg atoms in electric and magnetic fields, the effects of their interactions with electromagnetic radiation in the microwave and far infrared regions. The new topics in the Stark and Zeeman effects in a hydrogen atom were reflected in the monographs [8, 9]. The recent theoretical and experimental data on Rydberg atoms in electric and magnetic fields can be found, e.g., in [10–15]. There are many original papers and several reviews devoted to radiative properties of Rydberg states in resonant cavities [16] (see also [17]), to high-resolution Rydberg spectroscopy [18–21], and to interaction between the highly excited atoms and electromagnetic fields [22, 23] including some close problems of classical and quantum chaos [24, 25]. Bibliographies on the dynamics of the wave packets from highly excited atomic states are given in [26–30]. The most important results on the radio astronomy and astrophysics of Rydberg atoms are discussed in [31–33]. There are also several special reviews (see [1] and [34–36]), which contain some results on collisions involving Rydberg atoms obtained primarily before the middle of nineties.

The principal objective of the present review is to familiarize the reader with recent achievements in the physics of atomic and molecular collisions involving highly excited states and to reflect an up–to–date summary of knowledge in this field. It includes seven sections devoted to different aspects of the theory and its applications to studies of excitation, quenching, ionization and recombination processes. The text of each section is provided by some preliminary remarks which outline its specific goals and topics of interest, an historical sketch and key ideas of theoretical methods. The structure of hydrogen-like and complex atoms with high principal quantum number are discussed in Sect. 2. The aim of this section is to introduce the basic physical parameters of an isolated Rydberg atom and to describe its interaction with perturbing atoms and molecules. Another goal is to discuss qualitatively the general features and the main physical mechanisms of elementary processes involving Rydberg atoms and neutral targets. Here we also consider the effective range theory for slow electron scattering by neutral targets and discuss different forms of their interaction with Rydberg atom.

Sections 3 and 4 are devoted to the formulation and detailed description of basic theoretical methods and physical approaches to collisions involving Rydberg atoms, which take into account the peculiar properties of highly excited states. Here our attention is primarily focused on the various versions of perturbation theory, the close coupling method, semiclassical impact–parameter approach combined with the Fermi pseudopotential model, binary–encounter theory and quantal impulse approximation. The theory of quasielastic and inelastic collisions induced by the Rydberg-electron–projectile interaction and its application to different types of the bound–bound and bound–free transitions are presented in sections 5 and 6. The main emphasis is on the detailed description of the orbital and total angular momentum transfer processes, transitions with a change in the principal quantum number and ionization in thermal collisions with the target atoms or molecules. These processes were the subject of intensive experimental and theoretical research for more than 15 years. A great number of available results of calculations are compared here with the recent experimental data on the excitation, quenching and ionization of Rydberg states by the ground–state rare gas and alkali–metal atoms as well as by the atoms with low electron affinities such as Ca and Yb. Special attention is paid to the discussion of various possible mechanisms of depopulation and ionization of highly excited atoms by the polar, nonpolar and electron-attaching molecules.

Finally, in section 7 we continue to study the excitation, quenching, direct and associative ionization of highly excited atoms induced by collisions with neutral particles. However, our attention is focused on the analysis of the alternative physical mechanisms accounted for by the scattering of perturbing neutral particle on the parent core of the Rydberg atom. The most efficient mechanism for such processes is due to the resonant energy exchange between the highly excited electron and inner electrons of quasimolecule $BA^+ + e$ temporarily formed during the scattering of atom B on the ion core A^+. We also consider the dipole–induced n-changing transitions and ionization due to the mechanism of energy exchange between the Rydberg electron and translational motion of colliding atoms. The theoretical description of the identified processes with small energy defects is given on the basis of the shake–up model of Rydberg electron in the perturber–core scattering, while the ionization and inelastic transitions with large energy transfer are considered within the framework of quasimolecular approach. The results on the resonant inelastic transitions between Rydberg states are applied for a description of an efficient mechanism of the three-body electron–ion recombination in a noble gas mixture. We present the kinetic model of recombination processes involving Rydberg atomic states, taking into account inelastic collisions with the plasma free-electrons and neutral atoms of the buffer gas. At the end of this section we discuss the method of classical distribution and partition functions for diatomic systems.

2 Formulation of Problem and General Relations

In this section we shall consider the structure and the most important physical properties of atoms in highly excited states. One of its major objectives is to introduce the basic parameters of Rydberg atoms such as their orbital radius and geometrical area, the ionization potential, transition frequencies and wavelengths, period and the mean velocity of electron motion, radiation lifetime, quantum defect etc. We shall analyze their dependencies on the principal quantum number n and shall present the characteristic scales of identified physical parameters for typical values of n achieved in modern laboratory experiments and obtained from astrophysical observations. Another goal is to discuss the basic features of collisions involving Rydberg atoms and to outline the key idea of the quasifree electron model. Here we also consider some typical forms of interaction between the Rydberg atom and neutral targets and present the effective range theory for slow electron scattering by the weakly and strongly polarizable atoms.

2.1 Hydrogen-like Atom

Let us discuss the basic physical characteristics of Rydberg atoms in the hydrogen-like states. Characteristic magnitudes of the principal quantum number n of highly excited atoms observed in laboratory and cosmic space vary within a large range. Usually, in the rarefied gas cells and in the beams of the Rydberg atoms, n is of the order of $10-100$. At the same time, in recent years Rydberg atoms with very high magnitudes of the principal quantum number $n \sim 500-1000$ are also becoming available in laboratories. In the interstellar medium the principal quantum numbers of atoms in the Rydberg states are observed up to ~ 1000, whereas the typical magnitudes are about $100-300$. Although the subject of the review is the neutral Rydberg atom $(Z = 1)$, all simple relations presented below will contain the dependencies on the charge Z of the ionic parent core in explicit form.

In accord with the correspondence principle (see [37]) many properties of an atom (ion) with a large value of the principal quantum number $n \gg 1$ may be reasonably described in terms of classical mechanics. Thus, we start our consideration from introducing pure classical parameters of electron orbital motion. The classical electron motion in the central Coulomb field

$$U(|\mathbf{r}|) = -\frac{Ze^2}{r}, \qquad U_{\text{eff}} = -\frac{Ze^2}{r} + \frac{\mathcal{L}^2}{2mr^2} \qquad (1)$$

is confined to a plane perpendicular to a total angular momentum vector $\mathcal{L}$. Here U_{eff} is the effective potential energy including the centrifugal potential, m and e are the mass and charge of electron. In the polar coordinate frame (r, ϕ, ζ) with the origin 0 placed to the charge centre Ze and the ζ-axis directed along the $\mathcal{L}$-vector general solution of the Kepler problem may be directly derived from the conservation laws of energy and angular momentum (see, for example, [38]).

The relation between the polar angle ϕ and separation r of electron from the charge centre can be written as

$$\frac{\mathsf{p}}{r} = 1 + \epsilon \cos\phi, \quad \mathsf{p} = \frac{\mathcal{L}^2}{Ze^2m}, \quad \epsilon = \sqrt{1 + \frac{2E}{m}\left(\frac{\mathcal{L}}{Ze^2}\right)^2}, \qquad (2)$$

where the quantities p and ϵ are the parameter and eccentricity of the orbit, respectively. This expression determines the trajectory form of electron motion in the Coulomb field. The point with $\phi = 0$ in (2) corresponds to the nearest separation r_1 (perihelion) of the electron from the Coulomb centre (i.e. focus of the orbit).

For finite motion with the energy $E < 0$ (when the eccentricity $\epsilon < 1$), expression (2) directly demonstrates the elliptic trajectory of the electron orbit. Then, the minimal r_1 (perihelion, $\phi = 0$) and maximal r_2 (aphelion, $\phi = \pi$) possible separations of electron from the centre are given by

$$r_1 = \frac{\mathsf{p}}{1 + \epsilon}, \qquad r_2 = \frac{\mathsf{p}}{1 - \epsilon}.$$

The parameters of the electron orbit can be also determined by the values of the semimajor axis a and the semiminor axis b of the ellipse

$$a = \frac{\mathsf{p}}{1 - \epsilon^2} = \frac{Ze^2}{2|E|}, \qquad b = \frac{\mathsf{p}}{\sqrt{1 - \epsilon^2}} = \frac{\mathcal{L}}{\sqrt{2m|E|}}. \qquad (3)$$

Since the semimajor axis is dependent only on the binding energy of electron $|E|$ it is convenient to express the minimal and maximal electron separations from the Coulomb centre in the form

$$r_{\min} \equiv r_1 = a\left(1 - \epsilon\right), \qquad r_{\max} \equiv r_2 = a\left(1 + \epsilon\right). \tag{4}$$

As is apparent from (4), for a fixed value of the binding energy $|E|$, the form of the electron orbit is close to a circle if the value of the angular momentum $\mathcal{L}$ is large, i.e. when $\epsilon \ll 1$ and, hence, $r_1 \approx r_2 \approx a$. In the other limiting case of small angular momenta, for which the eccentricity of the orbit is close to unity $\epsilon \to 1$, we have $r_1 \to 0$ and $r_2 \to 2a$, so that the elliptic orbit becomes strongly drawn out. On the other hand, for a given value of the angular momentum $\mathcal{L}$ the minimal possible magnitude $E_{\min}$ of the electron energy is

$$E_{\min} = (U_{\mathrm{eff}})_{\min} = -\frac{m}{2}\left(\frac{Ze^2}{\mathcal{L}}\right)^2. \tag{5}$$

This value coincides with the minimal magnitude of the effective potential energy (1) at $r = (\mathcal{L}/Ze^2)^2/m$. In this case the eccentricity $\epsilon = 0$, so that the elliptic orbit degenerates into a circle.

The period T of the classical electron motion and corresponding angular frequency $\omega_{\mathrm{cl}} = 2\pi/T$ are independent of the angular momentum value $\mathcal{L}$ and are expressed through the binding energy $|E|$ (or through the value a of semi-major axis of the ellipse) by the relation

$$\omega_{\mathrm{cl}} = \sqrt{\frac{Ze^2}{ma^3}} = \frac{(2\,|\,E\,|)^{3/2}}{Ze^2\sqrt{m}}. \tag{6}$$

In accord with (3) and (6), the averaged velocity v_a in its orbital electron motion is given by

$$v_a = \frac{2\pi a}{T} = \sqrt{\frac{2E}{m}}. \tag{7}$$

The non-relativistic Schrödinger equation with the Coulomb interaction (1) leads to the simple expression for the electronic energy of Hydrogen atom $(Z = 1)$ or hydrogen-like ion $(Z > 1)$ [37]

$$E_n = -\frac{Z^2 Ry}{n^2}, \qquad Ry = \frac{me^4}{2\hbar^2}, \tag{8}$$

where Ry is the Rydberg constant $(2Ry = 27.212$ eV is the atomic unit of energy), and $\hbar$ is the Planck constant. The energy levels with the given magnitude of the principal quantum number n are

degenerated over the orbital angular momentum l ($l = 0, 1, 2, ..., n -$
1), its z-projection m, and the z-projection σ of the electron spin
($s = 1/2$) provided only electrostatic interaction between the electron
and nucleus with charge Ze is taken into account. The total statistical
weight g_{tot} of the given n-level is given by

$$g_{tot} = g_n g_s = 2 \sum_{l=0}^{n-1} g_l = 2n^2 , \qquad (9)$$

where $g_l = 2l + 1$ and $g_s = 2$ are the statistical weights of the l- and
s-sublevels, respectively.

The use of expression (8) and the basic relation $\mathcal{L} = \hbar l$ between
the total angular momentum $\mathcal{L}$ and orbital quantum number l allows
us to express the aforementioned parameters of the classical electron
motion in terms of the quantum numbers n and l

$$a = \frac{n^2 a_0}{Z} , \qquad b = \frac{n l a_0}{Z} , \qquad \epsilon = \sqrt{1 - \frac{l^2}{n^2}} . \qquad (10)$$

Here $a_0 = \hbar^2/me^2$ is the Bohr radius ($a_0 = 0.529 \cdot 10^{-8}$ cm).

Thus the characteristic radius $r_n \equiv a$ and geometrical area $S_n = \pi r_n^2$ of the Rydberg atom or ion as well as the mean orbital velocity
v_n and the period T_n of classical orbital motion of the highly ex-
cited electron around the parent ionic core are given by the following
expressions

$$r_n = \frac{n^2 a_0}{Z} , \qquad S_n = \frac{\pi a_0^2 n^4}{Z^2} , \qquad (11)$$

$$v_n = \frac{Z v_0}{n} , \qquad T_n = \frac{2\pi a_0 n^3}{v_0 Z^2} , \qquad (12)$$

where $v_0 = e^2/\hbar$ is the atomic unit of velocity ($v_0 = 2.188 \cdot 10^8$
cm $\cdot$ s^{-1}). As is apparent from (11) and (12), the values of r_n, S_n, and
T_n increase dramatically as n grows. The characteristic dimensions
of Rydberg atoms correspond to macroscopic sizes (> 1 μm) for
$n > 100$. For the principal quantum number $n \sim 1000$ observed now
both in laboratory and cosmic conditions the radius r_n of the Rydberg
atom turns out to be of the order of 0.1 mm. This value exceeds
the characteristic radius of the ground-state atom by more than 10^6
times (see Fig. 1). At the same time the mean orbital velocity of a

highly excited electron v_n is small compared with the characteristic velocities of atomic electrons in the ground or low excited states.

An important specific feature of the Rydberg atom with high principal quantum number n is a great number of closely spaced energy levels. This means that the density of states per unit energy interval

$$\rho(E_n) = g_s g_n \left| \frac{dE_n}{dn} \right|^{-1} = \frac{n^5}{Z^2 Ry} \qquad (13)$$

grows rapidly as n increases. Substitution of (8) and (9) into (13), yields

$$\rho(E_n) = \frac{2m^{3/2} Z^3 e^6}{\hbar^3 (2|E_n|)^{5/2}} . \qquad (14)$$

This expression can be derived using the classical microcanonical distribution function in the Coulomb field (see [3, 39] for more details)

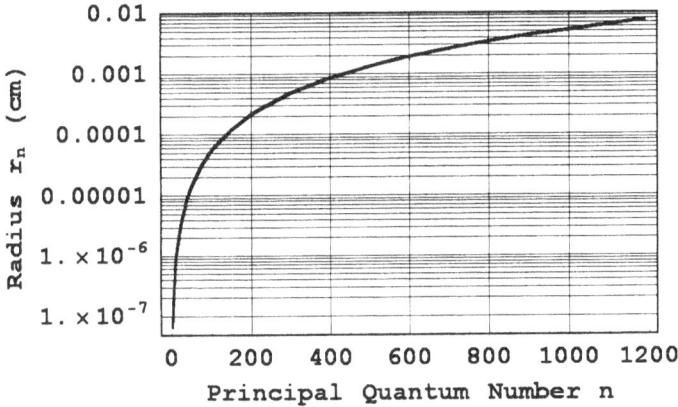

Figure 1. Radius r_n of highly excited H(n) atom ($Z = 1$) as a function of the principal quantum number n.

The line spectra corresponding to transitions from the highly excited state to the ground (or low excited) state of a neutral atom lie in the visible or ultraviolet region. At the same time, the transition frequencies between two closely spaced Rydberg states are small. For example, the transition frequency $\omega_{n,n\pm1} = |E_n - E_{n\pm1}|/\hbar$ between two levels with neighboring magnitudes of the principal quantum numbers n and $n \pm 1$ can be written as

$$\omega_{n,n\pm1} = \frac{Z^2 Ry}{\hbar} \left(\frac{1}{n^2} - \frac{1}{(n \pm 1)^2} \right) \approx \frac{2Z^2 Ry}{\hbar n^3} . \qquad (15)$$

This is the Kepler frequency of the Rydberg electron, which corresponds to the classical period $T_n = 2\pi/\omega_{n,n\pm1}$ of its orbital motion (12). The corresponding wavelength $\lambda_{n,n\pm1} = 2\pi c/\omega_{n,n\pm1}$ increases drastically with an increase in n. It is embedded in the micron-region for $n \sim 10$ and the centimeter region for $n \sim 100$. Radioastronomy observations of states with $n \sim 300$ and more are usually carried out in the meter region of wavelength. Figure 2 illustrates the typical scales of basic energy parameters of neutral hydrogen atom $(Z = 1)$ in a wide range of the principal quantum number $5 \leq n \leq 1000$.

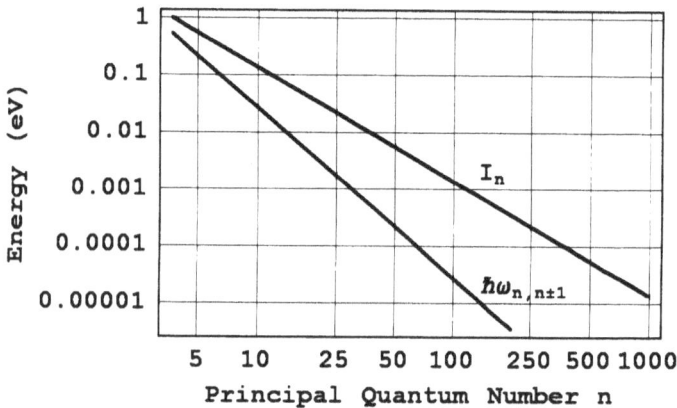

Figure 2. The n-dependencies of the ionization potential $I_n = |E_n|$ (8) and transition energy defect $\hbar\omega_{n,n\pm1}$ (15) between neighboring levels of H(n) atom.

An account of the spin–orbit interaction of highly excited electron leads to the appearance of the relativistic correction $E_{nJ}^{(1)}$ to the energy of Rydberg levels [40]

$$E_{nJ} = -\frac{Z^2 Ry}{n^2} + E_{nJ}^{(1)} , \qquad (16)$$

$$E_{nJ}^{(1)} = -\frac{\alpha^2 Z^4 Ry}{n^3}\left(\frac{1}{J+1/2} - \frac{3}{4n}\right) \tag{17}$$

and to the fine-structure splitting of states with different magnitudes $J = l + 1/2$ and $J' = |l - 1/2|$ of the total angular momentum of the electron $(\mathbf{J} = \mathbf{l} + \mathbf{s})$ but with the same n and l values

$$\omega_{J'J} = \frac{\Delta E_{J'J}}{\hbar} = \frac{\alpha^2 Z^4 Ry}{\hbar n^3 l(l+1)}. \tag{18}$$

Here $\alpha = e^2/\hbar c = 1/137.036$ is the fine structure constant. One can see that the relativistic correction decreases rapidly with n increasing and turns out to be particularly important for multi-charged ions with large Z. The second term in (16) for $E_{nJ}^{(1)}$ is negligible for Rydberg states with $n \gg 1$. It should also be noted that energies of states with equal magnitudes of the total angular momentum J, but different orbital momenta l are the same. This is a specific feature of a hydrogen and hydrogen-like ions. The difference of these energies is connected with radiatiative corrections and is usually negligible for Rydberg states. The detailed theory of the energy levels and precise data of two-photon spectroscopy of atomic Hydrogen are presented in the review article [41].

The radiation lifetime τ_n of hydrogen-like atom (ion) in the highly excited n-state $(n \gg 1)$ averaged over the lm-sublevels

$$\tau_n = \left[\sum_{l=0}^{n-1} \frac{(2l+1)}{n^2}\frac{1}{\tau_{nl}}\right]^{-1} \tag{19}$$

can be expressed in terms of the total probability A_n of spontaneous radiation from the n-level

$$\tau_n = 1/A_n, \qquad A_n = \frac{1}{n^2}\sum_{n'}\sum_{ll'}(2l+1)\,A_{n'l',nl}. \tag{20}$$

The τ_{nl} quantity in Eq. (19) corresponds to the radiation lifetime of the Rydberg nl-level with definite magnitudes of both the principal and orbital quantum numbers. The value of A_n may be successfully evaluated in a wide range of n by using a simple expression

$$A_n \approx \frac{8\alpha^3 Z^4\,(v_0/a_0)}{\pi\sqrt{3}}\frac{\ln(n/1.1)}{n^5}. \tag{21}$$

It takes into account the total contribution of all dipole $nl \to n', l \pm 1$ transitions. Here A_n is the total probability of spontaneous radiation from the n-level, which is mainly determined by transitions to the final levels with small n'-magnitudes ($n' = 1 - 2$) and to the neighboring level with $n' = n - 1$. Radiation transitions with intermediate n'-values between the given n-level and the ground-state turns out to be less effective.

As follows from Eqs. (20) and (21) the radiation lifetime increases drastically with increasing n and becomes greater than 10 s for states with $n > 100$. Comparison of characteristic values of the averaged radiation lifetime τ_n with corresponding values of the classical period T_n for highly excited electron motion around the Coulomb centre with $Z = 1$ is shown in Fig. 3.

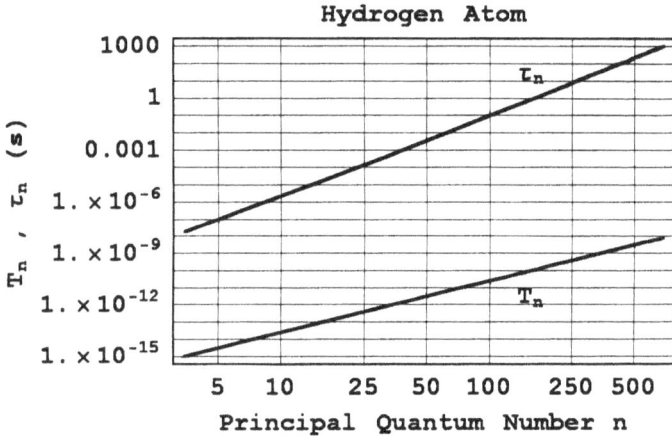

Figure 3. Averaged radiation lifetime τ_n (s) of Rydberg atom and classical period T_n (s) of electron motion as functions of the principal quantum number n.

As noted in Ref. [42], the n-dependence of the averaged radiation lifetime τ_n can also be described by the following simple law $\tau_n \propto n^{4.5}$. For the nl-states of a hydrogen atom with given values of the principal n and orbital l quantum numbers the n-dependence of the radiation lifetime is $\tau_{nl} \propto n^3$. As follows from the detailed analysis [43] of available experimental works and theoretical calculations, such behavior takes place for all hydrogen l-series ($l \geq 1$) starting from small magnitudes of n. According to [43] the radiation lifetime of the p-, d-, f-, g-, h-,...,-series of neutral Hydrogen atom satisfies the power-law $\tau_{nl} = \tau_l n^\beta$ with the β constants, whose magnitudes are very close to 3. The values of the τ_l and β coefficients for the first few p-, d-, f-, g-, h-series of Hydrogen are given in Table 1.

Table 1. The τ_l (ns) and β coefficients for the first few s-, p-, d-, f-, g-, h-, i-, and k-series of the H(nl) atom fitted to experimental data in Ref. [43].

Series	p	d	f	g	h	i	k
β	2.948	2.932	2.941	2.942	2.945	2.946	2.954
τ_l	0.209	0.627	1.235	2.068	3.113	4.372	5.766

At the same time the n-dependence of the radiation lifetime for the ns-states of the hydrogen atom exhibits an anomalous behavior. For the ns-states the lifetime τ_{ns} is not changed as n^3 if the principal quantum number is not too large ($n \leq 12$). In this range of n the value of β is monotonically changed from 1.42 at $n = 3$, 4 till 2.81 at $n = 11$, 12. The effect is somewhat similar to that observed for the ns-states of complex atoms, for which there is a perturbation induced by interaction between the excited electron and electrons of the ion core of the Rydberg atom. Since this interaction is absent for the hydrogen atom the identified effect can be attributed to the electron–proton interaction in accordance with comments in Ref. [43].

As is evident from Table 1, the values of τ_l grow monotonically with an increase of the orbital quantum number l. This fact is in agreement with the results of quasiclassical calculations according to which the radiation lifetime τ_{nl} of Rydberg atom is given by

$$\tau_{nl} \propto (l + 1/2)(l + 1) n^3 . \qquad (22)$$

This relation is formally applicable at $n \gg 1$ and $l \gg 1$. It clearly demonstrates the asymptotic behavior of radiation lifetime at large magnitudes of $l \sim n - 1$, when $\tau_{nl} \propto n^5$.

2.2   Alkali-Metal Rydberg Atoms

The structure of energy levels and spectra of non-hydrogenic Rydberg atoms have a number of important features due to the presence of the short-range part of interaction between the highly excited electron and parent ion core as well as to the long-range polarization interaction and other correction terms to pure Coulomb interaction at large distances. Let us briefly discuss this problem for the simplest case of neutral atoms ($Z = 1$) having one valence electron in the open shell (i.e. alkali-metal atoms). For the Rydberg series of the alkali-metal atoms there is particularly extensive experimental material on the energy levels and transition wavelengths as well as a large number of corresponding theoretical calculations.

It is well known that the term value T_{nl} of such an atom in a highly excited state with the given magnitudes of the principal n and orbital l quantum numbers and the energy E_{nl} measured from the series limit T_∞ (or the ionization limit) are described by the Rydberg formula

$$T_{nl} = T_\infty + E_{nl} , \qquad E_{nl} = -Ry/n_*^2 , \qquad n_* = n - \delta_l . \quad (23)$$

Here n_* is the effective principal quantum number and δ_l is the quantum defect of the nl-state. This expression differs from the case of a hydrogen atom only by the presence of the quantum defect δ_l which characterizes the non-Coulomb part of the electron–core interaction for non-hydrogenic Rydberg atoms.

The values of the quantum defect are determined by the specific type of Rydberg atom. The dependence of the quantum defect on the principal quantum number n is weak enough and can be neglected, to a first approximation. At the same time, the δ_l values strongly fall with an increase of the orbital angular momentum l (see Table 2).

Table 2. The quantum defects δ_l of the Rydberg nl-levels of alkali-metal atoms.

Atom	δ_s	δ_p	δ_d	δ_f	δ_g
Li	0.399	0.053	0.002		
Na	1.347	0.854	0.0145	0.0016	
K	2.178	1.712	0.267	0.010	
Rb	3.135	2.65	1.34	0.0164	
Cs	4.057	3.58	2.47	0.033	0.00704

It is due to the strong decrease of the probability for the Rydberg electron placing in the short-range distances near the parent ion core as well as to decrease of the long-range polarization parts of the electron–core interaction as l grows. Thus, a few Rydberg nS-, nP-, and nD-states (as well as nF-states for the heavy atoms) are usually significantly separated from the practically hydrogen-like quasidegenerate manifold of the nl-states with $2-3 \le l \le n-1$, for which $\delta_l \approx 0$. This fact is illustrated by Fig. 4, where the quantum defects of Li, Na, K, Rb, and Cs atoms are shown as functions of the orbital quantum number l ($l = 0, 1, 2, 3$).

Alkali Metal Atoms

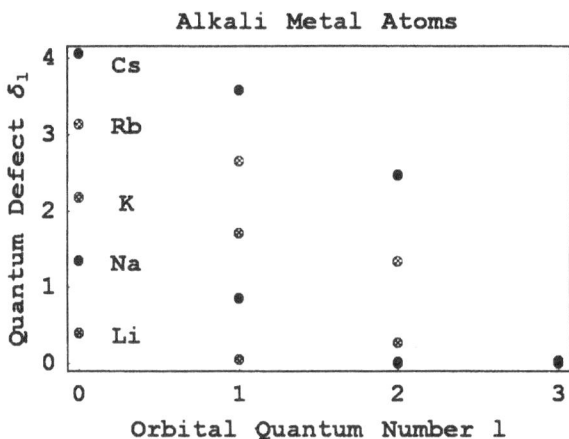

Figure 4. The quantum defects δ_l of Li, Na, K, Rb, and Cs atoms against the orbital quantum number l.

As is evident from the figure, the quantum defect values grow rapidly with an increase of the atomic number of the neutral atom. It is important to stress that the transition frequency between two Rydberg non-hydrogenic nl- and $n'l'$-states with the given magnitudes of both the principal and orbital quantum numbers

$$\omega_{n'l',nl} \approx \frac{2Ry|\delta_l - \delta_{l'} + n' - n|}{\hbar n^3} \qquad (24)$$

may be considerably smaller than between two neighboring hydrogen-like levels n and $n \pm 1$ due to mutual compensation of the quantum defect values.

    The high-precision measurements of the spectral-line wavelengths by the methods of tunable-dye-laser spectroscopy gave the opportunity to obtain the accurate values of the energy level positions and quantum defects for all alkali-metal atoms in a wide range of the principal quantum number.

In accordance with the quantum defect theory [5] the quantum defect magnitudes δ in (23) can be represented with high precision by the modified Ritz formula

$$\delta\left(n\right) = \sum_{k=0}^{N} p_{2k} \left(\frac{1}{n_*^2}\right)^k$$

$$= A + \frac{B}{(n-A)^2} + \frac{C}{(n-A)^4} + \frac{D}{(n-A)^6} \cdots . \qquad (25)$$

The coefficients p_{2k} $(k = 0, 1, 2, ..., N)$ or A, B, C, ... as well as the value of series limit T_{∞} are usually obtained by fitting the energies defined by formulae (23) and (25) to the energies determined from the experimental data. They can also be calculated if the form of the interaction potential between the Rydberg electron and the parent core is known. It is important to stress that the value of A corresponds to the asymptotic quantum defect $\delta\left(\infty\right)$ in the limit of large $n \to \infty$. Table 3 shows the expansion coefficients of quantum defects for the first few $n^2S_{1/2}$, $n^2P_{1/2}$, $n^2D_{3/2}$, $n^2F_{5/2}$, $n^2G_{7/2}$ states of atomic cesium. Fig. 3. clearly demonstrates the n-dependencies of the quantum defects of these levels in a wide range of the principal quantum number.

Table 3. Quantum defect expansion coefficients (25) for Cs I series fitted to experimental data of Ref. [52].

n^2L_J	$n^2S_{1/2}$	$n^2P_{1/2}$	$n^2D_{3/2}$	$n^2F_{5/2}$	$n^2G_{7/2}$
n	6 – 30	6 – 80	5 – 36	4 – 65	5 – 50
A	4.04935665	3.59158950	2.46631524	0.03341424	0.00703865
B	0.2377037	0.360926	0.013577	−0.198674	−0.049252
C	0.255401	0.41905	−0.37457	0.28953	0.01291
D	0.00378	0.64388	−2.1867	−0.2601	
E	0.25486	−0.198674	0.28953	−0.2601	
F			−56.6739		

If the orbital quantum number l of the Rydberg atom is not too small so that the external electron of the Rydberg atom is non-penetrating there is a simple theoretical way for representing a series of energy levels E_{nl}. In this case the deviation of the energy levels from their hydrogen-like positions is due almost entirely to the polarization of the ion parent core in the field of the external electron.

Thus the energy of the highly excited nl-level is given by the following expression [44]

$$E_{nl} = -\frac{Ry}{n^2}\left[1 + \frac{\alpha^2}{n^2}\left(\frac{n}{l+1/2} - \frac{3}{4}\right)\right]$$

$$-Rya_0\left[\alpha_d'\left\langle\frac{1}{r^4}\right\rangle_{nl} + \alpha_q'\left\langle\frac{1}{r^6}\right\rangle_{nl}\right], \tag{26}$$

where $\alpha = e^2/\hbar c$ is the fine-structure constant, α_d' [in $a_0{}^3$] and α_q' [in a_0^5] are the effective dipole and quadrupole polarizabilities of core, and the expectation values $\left\langle r^{-4}\right\rangle_{nl}$ and $\left\langle r^{-6}\right\rangle_{nl}$ are to be determined for the appropriate hydrogen-like state. In spite of the fact that the polarization formula (26) does not include the fine-structure splitting it can be applied to the centres of gravity of the doublet terms of an alkali-metal atom.

At present there is extensive material on the energy level positions and quantum defect values for different series of alkali-metal atoms. Detailed experimental data for Li , Na , K , Rb, and Cs were obtained at the end of the seventies and in the eighties (see [45–52] and references therein). Some recent data can be found in the papers [53] (Li), [54, 55] (Na), [56] (Na, K, Rb, Cs), and [57] (Cs).

Most of the experimental measurements of the rate coefficients of elementary processes involving Rydberg atoms were carried out with alkali-metal elements in selectively excited nl-states with the definite magnitudes of both the principal n and orbital l quantum numbers. When the orbital quantum number is not too large ($l \ll n$) the radiation lifetime τ_{nl} turns out to be approximately proportional to n_*^3.

$$\tau_{nl} = \tau_l n_*^\beta, \qquad n_* = n - \delta_l. \tag{27}$$

Here n_* is the effective principal quantum number of the nl-state, τ_l is the n-independent constant whose value is determined by the orbital quantum number l, and the value of the β index is close to 3. The magnitudes of the τ_l (ns) and β coefficients for the first few Rydberg s-, p-, d-, f-, and g-series of alkali-metal atoms (fitted to available experimental data in Ref. [43]) are presented in Table 4.

To demonstrate the typical values of the radiation lifetime for the Rydberg states with the given magnitudes of the principal and orbital quantum numbers we present in Fig. 5 the results of calculations by Eq. (27) for the Na(ns) atom. The parameters β and $\tau_{l=0}$ were taken from Table 4. It can be seen that $\tau_{ns} \sim 1$ μs, 10 ms, and 1 s at $n_* \approx 10$, 100, and 1000, respectively.

Table 4. The τ_l (ns) and β coefficients in Eq. (27) for the radiation lifetimes of the first few Rydberg nl-states of alkali-metal atoms obtained in Ref. [43].

Atom	Constants	s	p	d	f	g
Li	τ_l			0.74	1.22	2.00
	β			2.96	2.94	2.95
Na	τ_l	1.80		0.94	1.20	2.05
	β	2.89		3.06	2.93	2.94
K	τ_l		3.47		1.48	2.09
	β		3.06		2.70	2.92
Rb	τ_l	1.44	2.57		0.97	2.15
	β	2.95	3.03		2.86	2.90
Cs	τ_l	3.11	2.47		1.01	2.33
	β	2.6	3.13		2.76	2.85

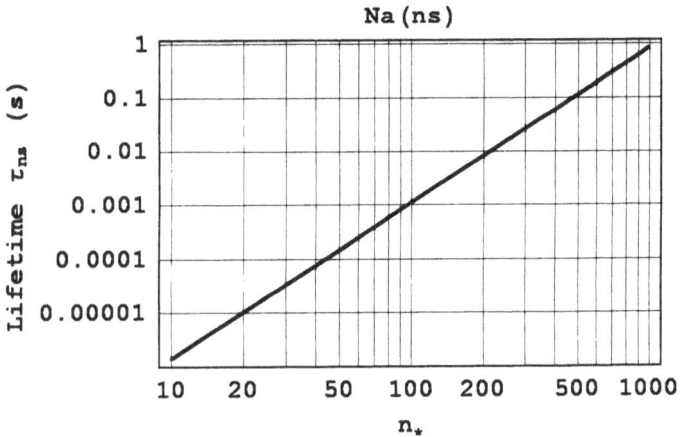

Figure 5. The radiation lifetime τ_{ns} (s) of Rydberg Na(ns) atom as a function of the principal quantum number.

2.3 Complex Rydberg Atom

A number of basic features in the behavior of the quantum defect values on the atomic number of element, on the magnitude of the orbital angular momentum and others remain the same for complex Rydberg atoms. However, the structure of highly excited levels and spectra of complex Rydberg atoms have a number of peculiarities such as the presence of a few level series with different ionization limits and autoionizing states, electron–electron correlation effects etc. In particular, an important feature of Rydberg alkali-earth atoms (Be, Mg, Ca, Sr, and Ba) is due to strong interseries interaction in the region, where the moduli of their quantum defects become close to each other. The detailed discussion of identified effects is beyond the scope of the present review. Structure and spectra of Rydberg alkali-earth atoms were the subject of intensive spectroscopical studies and theoretical calculations on the basis of the multichannel quantum defect theory [5] during the last decade. The available results are reflected in the books by *Gallagher* [2] and *Connerade* [4] as well as in the review article [7]. A number of additional data and theoretical results concerning the autoionizing states and the channel mixing of Rydberg series of alkali-earth atoms can be found in [58] for Mg, in [59, 60] for Ca, in [61, 62] for Ba, and in [64, 65] for Sr. The oscillator strengths of Ba I and Sr I Rydberg transitions are presented in [63].

Here we briefly discuss some features associated with the energy spectra of Rydberg rare-gas atoms since there is a large number of experimental and theoretical works on elementary processes involving the Rydberg He, Ne, Ar and Xe atoms (see [1, 3] and references therein). The ground state of the He atom is $1s^2\,{}^1S_0$. The excitation of one of the s-electrons leads to two series of Rydberg states, i.e. the singlet - $1snl\,{}^1L_J$ and the triplet - $1snl\,{}^3L_J$ series. The quantum defects δ for some Rydberg series of He are given in Table 6. It can be seen that as in the case of Rydberg alkali-metal atoms the quantum defect magnitudes strongly decrease with an increase of the orbital quantum number l of an excited nl-electron. Some recent data on Rydberg states of helium are presented in Ref. [66].

Table 6. The quantum defects δ of the Rydberg $1snl\,{}^{2S+1}L$ levels of He atom [67, 68].

Series	$1sns\,{}^1S$	$1sns\,{}^3S$	$1snp\,{}^1P$	$1snp\,{}^3P$	$1snd\,{}^{1,3}D$	$1snf\,{}^{1,3}F$
δ	0.14	0.30	0.01	0.07	0.003	0.0003

The configuration of the outer electronic p-shell of all other rare gas atoms in the ground state is $n_0 p^6\ {}^1S_0$, where $n_0 = 2$, 3, 4, and 5 for Ne, Ar, Kr, and Xe, respectively. The Rydberg series appear, when one of the $n_0 p$-electrons is excited into the nl-state with a high principal quantum number. The binding energy of the nl-electron is much less than the binding energy of the $n_0 p$-electrons of the ion-core $X^+ \left(n_0 p^5\right)$. The spin–orbit interaction of these $n_0 p^5$-electrons is greater than their electrostatic interaction with an excited nl-electron. The Rydberg states of rare-gas $X\left[n_0 p^5 \left({}^2P_j\right) nl\,[K]_J\right]$ atoms are characterized by quantum numbers n, l, K and J. Here $J = K \pm 1/2$ and $j = 3/2$, $1/2$ are the quantum numbers of total angular momenta of Rydberg atom X and its ion-core X^+, respectively. The quantum number $K = |j - l|$, ..., $j+l-1$, $j+l$ corresponds to the angular momentum $\mathbf{K} = \mathbf{j}+\mathbf{l}$ (see [69]). For the heavy Ar, Kr, Xe atoms the Rydberg $nl\,[K]_J$ sublevels with different K and J values are considerably split within the given nl-level at $l = 0$, 1, and 2. These $nl\,[K]_J$ sublevels have large quantum defects δ, while the Rydberg $nl\,[K]_J$ sublevels with $l \geq 3$ are practically hydrogen-like. The typical energy spectra of Rydberg Ar and Xe atoms are shown in Fig. 6, in which we present the structure of $nl\,[K]_J$ states converging to one ionization limit $n_0 p^5 \left({}^2P_j\right)$ with $j = 3/2$.

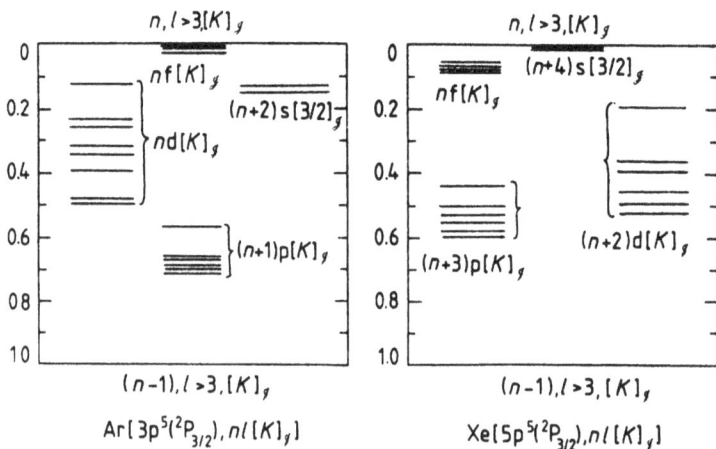

Figure 6. The structure of Rydberg $nl\,[K]_J$ states of Ar and Xe converging to one ionization $n_0 p^5 \left({}^2P_j\right)$ limit with $j = 3/2$.

At present there are several new experimental works aimed towards the detailed studies of different Rydberg series of the rare-gas atoms including the autoionizing states. For example, *Klar et al.* [70] have performed the high-resolution measurement and quantum-defect analysis for the Rydberg nd $J = 1, 2$, and 3 series of neon. The two-photon optogalvanic Rydberg spectra of Ne and Ar have been obtained in Refs. [71, 72] and for Kr in Ref. [73]. The inner-shell doubly excited states of $Ar(2p^5 3p^5 nln'l')$ and $Ne(1s2p^5 nln'l')$ were recently observed in [74]. The Rydberg series of krypton excited from the $5p[3/2]_{1,2}$ states were studied in [75], while quantitative investigations of the decay of Kr I $3d^9 np$ resonances are presented in [76]. Double-resonant excitation of the $5p^5 nf'$ $(J = 2)$ autoionizing states of Xe^* has been obtained in Ref. [77] by using synchronized radiation pulses.

2.4   Features of Rydberg Atom–Neutral Collisions

Before going into a consistent description of the general physical approaches in collision theory involving Rydberg atoms we briefly discuss the main characteristic features of elementary collisional processes. In the case of Rydberg-atom–neutral-atom collisions the characteristic radii of interaction potentials of the perturber with both the highly excited electron and the ionic core are, usually, small in comparison with the orbital radius. Therefore, the collisional processes of Rydberg atoms with atomic projectiles (as well as with non-polar and polar molecules) may often be regarded as two independent processes caused by the scattering of the incident particle on the highly excited (valence) electron and by the scattering on the ionic core. The relative role of the scattering mechanisms is determined by the type of colliding particles and their relative velocity, by the energy defect of transition and by the range of the principal quantum number.

   The cross sections of the Rydberg-atom–neutral-atom collisions can be of the order of the geometrical cross section of highly excited atom provided the principal quantum number is not too large. It is possible only for elastic and quasielastic processes (i.e. for processes without change of the highly excited electron energy or with a very small energy defect). The values of the cross sections of the inelastic bound–bound and bound–free transitions reveal the strong damping as the energy transferred to the Rydberg atom is increased. Usually, both the excitation and ionization processes with large values of the energy transferred to the highly excited electron result from the perturber–core scattering in contrast to the quasielastic and inelastic transitions with small and intermediate energy defects, for which the traditional mechanism of scattering of the quasifree electron by a neutral atom is predominant. On the other hand, the large cross sections of the inelastic n-changing transitions and ionization in collisions with molecules are the result of quasiresonant energy transfer from the internal rotational degrees of freedom of the projectile to the Rydberg electron (or the ion-pair formation in the final channel in collisions with electron-attaching molecules).

2.5 Lippman–Schwinger Formalism

The general formulation of a scattering problem may be given within the framework of a stationary Schrödinger equation with an appropriate outgoing scattering condition. For collisions involving a Rydberg atom (ion) A and incident structureless particle B the stationary wave function $\Psi(r, R)$ of the total system A+B is the solution of the following equation

$$H\Psi = E\Psi, \qquad H = H_0 + V, \qquad H_0 = -\frac{\hbar^2}{2\mu}\Delta_R + H_A. \qquad (28)$$

Here H and E are the total Hamiltonian and energy of system, respectively; H_0 is the unperturbed Hamiltonian, which consists of the highly excited atom Hamiltonian H_A and the relative kinetic energy operator $\left(-\hbar^2\Delta_R/2\mu\right)$ of the colliding particles A+B ($\mu = M_A M_B/(M_A + M_B)$ is being their reduced mass). Here we have removed the Hamiltonian H_B of projectile because we consider it as a structureless particle. In the zero-order approximation over the projectile–atom interaction V the solution of the Schrödinger equation

$$H_0\Psi^{(0)} = E\Psi^{(0)}, \qquad \Psi_{\alpha q}^{(0)}(r, R) = e^{iq \cdot R}\psi_\alpha(r) \qquad (29)$$

is a product of the plane wave $\phi_q(R) = \exp(iq \cdot R)$ describing the relative motion of the colliding particles with the wave-vector q, and eigenfunction $\psi_\alpha(r)$ of the Rydberg atom (ion) Hamiltonian

$$H_A\psi_\alpha = E_\alpha\psi_\alpha , \qquad H_A = -\frac{\hbar^2}{2\mu_{eA^+}}\Delta_r + U , \qquad (30)$$

whereas α and E_α denote the set of quantum numbers character-izing its eigenstate and energy, respectively. Here U is the inter-action potential between the Rydberg electron and ion core (the Coulomb-type at large separations $U \to -e^2/r$ at $r \to \infty$), and $\mu_{eA+} = mM_{A+}/(m + M_{A+})$ is the reduced mass of the (e, A^+)-pair, i.e. $\mu_{eA+} \approx m$. The unperturbed wave function of colliding particles A+B is normalized by the condition

$$\left\langle \Psi^{(0)}_{\alpha'q'} \left| \Psi^{(0)}_{\alpha q} \right\rangle = \int \int \phi^*_{q'}(\mathbf{R}) \psi^*_{\alpha'}(\mathbf{r}) \psi_\alpha(\mathbf{r}) \phi_q(\mathbf{R}) \, d\mathbf{r} \, d\mathbf{R} \right.$$
$$= (2\pi)^3 \, \delta \, (\mathbf{q}' - \mathbf{q}) \, \delta_{\alpha'\alpha} \, . \tag{31}$$

The radius vector $\mathbf{R}$ of the incident particle B in (29) is taken relative to the centre of mass of the Rydberg atom A (i.e. to the centre of mass of the $(e-A^+)$-pair), while $\mathbf{r}$ is the vector of the valence-electron–ion-core $(e - A^+)$ separation.

The exact wave function $\Psi^+_{\alpha q}(\mathbf{r}, \mathbf{R})$ of the system A+B is a so-lution of the Schrödinger equation (28) for the total Hamiltonian H=H$_0$+V including the projectile–atom interaction V, which leads to the inelastic transitions between the Rydberg atomic states $|\alpha\rangle \to |\alpha'\rangle$. This solution satisfies the following boundary condition

$$\Psi^+_{\alpha q}(\mathbf{r}, \mathbf{R}) \xrightarrow[R \to \infty]{}$$
$$\Psi^{(0)}_{\alpha q}(\mathbf{r}, \mathbf{R}) + \sum_{\alpha'} f_{\alpha'\alpha} (\mathbf{q}', \mathbf{q}) \, \frac{\exp(iq'R)}{R} \, \psi_{\alpha'}(\mathbf{r}) \, , \tag{32}$$

whereas the wave numbers $q \equiv q_\alpha$ and $q' \equiv q_{\alpha'}$ and the atomic energies before and after collision are given by the energy conservation law

$$\mathsf{E} = E_\alpha + \frac{\hbar^2 q^2}{2\mu} = E_{\alpha'} + \frac{\hbar^2 (q')^2}{2\mu}, \tag{33}$$

where E is the total energy of system A+B. As is evident from (32), at large distances the wave function of this system involves the unperturbed wave function (29) and the linear combination of the spherical divergent waves $R^{-1} \exp(iq'R) \psi_{\alpha'}(\mathbf{r})$ corresponding to all possible eigen states of the Rydberg atom. The coefficient $f_{\alpha'\alpha}(\mathbf{q}', \mathbf{q})$ is the scattering amplitude for the inelastic transition between highly excited states $|\alpha\rangle \to |\alpha'\rangle$. (It should be noted that for the ionization

process the final function $\psi_{\alpha'}$ is a function of continuous spectrum of
Rydberg atom A). As a result, the differential cross section of scat-
tering for the $|\alpha\rangle \rightarrow |\alpha'\rangle$ transition into the interval $d\Omega = \sin\theta d\theta d\varphi$
of solid angles is equal to

$$d\sigma_{\alpha'\alpha}(\mathbf{q}',\mathbf{q}) = \frac{q'}{q}\left|f_{\alpha'\alpha}(\mathbf{q}',\mathbf{q})\right|^2 d\Omega . \tag{34}$$

Thus, the Schrödinger equation (28) with the boundary condition
(32) and formula (34) give a consistent formulation of the scattering
problem. However, a direct solution of this equation is failed by both
the analytical and numerical methods. The next sections will be
devoted to various approximated approaches which are widely used
in the collision theory involving Rydberg atoms and ions. Here we
also present one more general formulation of scattering problem based
on the Lippman–Schwinger equation which turns out to be useful for
the understanding and analysis of a number of approximate methods
(e.g. the quantum impulse approximation).

By adopting the Lippman–Schwinger formalism, the exact solution of the Schrödinger equation (28) with the outgoing scattering
condition (32) can be rewritten in the two equivalent operator forms
(see, for example, [78, 79])

$$\Psi_{\alpha\mathbf{q}}^+ = \Psi_{\alpha\mathbf{q}}^{(0)} + \mathsf{G}^+(\mathsf{E})\,\mathsf{V}\Psi_{\alpha\mathbf{q}}^{(0)} ,$$

$$\mathsf{G}^+(\mathsf{E}) = (\mathsf{E} - \mathsf{H} + i0)^{-1} , \tag{35}$$

$$\Psi_{\alpha\mathbf{q}}^+ = \Psi_{\alpha\mathbf{q}}^{(0)} + \mathsf{G}_0^+(\mathsf{E})\,\mathsf{V}\Psi_{\alpha\mathbf{q}}^+ ,$$

$$\mathsf{G}_0^+(\mathsf{E}) = (\mathsf{E} - \mathsf{H}_0 + i0)^{-1} . \tag{36}$$

Here the operators $\mathsf{G}(\mathsf{E}+i0)$ and $\mathsf{G}_0(\mathsf{E}+i0)$ considered as functions of
complex variable are the Green resolvents for the total H and for the
unperturbed H_0 Hamiltonians, respectively. To clarify an explicit
integral form of the Lippman–Schwinger equation for the exact wave
function $\Psi_{\alpha\mathbf{q}}^+(\mathbf{r},\mathbf{R})$ in the coordinate representation it is convenient
to rewrite (36) in terms of the unperturbed Green function of the
system

$$G_0^+ (\mathbf{R}, \mathbf{r}; \mathbf{R}', \mathbf{r}'; E)$$

$$= -\frac{\mu}{2\pi\hbar^2} \sum_\alpha \frac{\exp\left(iq_\alpha |\mathbf{R} - \mathbf{R}'|\right)}{|\mathbf{R} - \mathbf{R}'|} \psi_\alpha^* (\mathbf{r}') \psi_\alpha (\mathbf{r}), \qquad (37)$$

$$q_\alpha = (2\mu/\hbar) (E - E_\alpha)^{1/2},$$

which satisfies the following equation

$$\left(E + \frac{\hbar^2}{2\mu} \Delta_\mathbf{R} - H_A\right) G_0^+ (\mathbf{R}, \mathbf{r}; \mathbf{R}', \mathbf{r}'; E)$$

$$= \delta (\mathbf{R} - \mathbf{R}') \delta (\mathbf{r} - \mathbf{r}') G_0^+ (\mathbf{R}, \mathbf{r}; \mathbf{R}', \mathbf{r}'; E). \qquad (38)$$

For the local interaction potential V, the final result takes the form

$$\Psi_{\alpha\mathbf{q}}^+ (\mathbf{r}, \mathbf{R}) = \Psi_{\alpha\mathbf{q}}^{(0)} (\mathbf{r}, \mathbf{R})$$

$$+ \iint G_0^+ (\mathbf{r}, \mathbf{R}; \mathbf{r}', \mathbf{R}'; E) V (\mathbf{r}', \mathbf{R}') \Psi_{\alpha\mathbf{q}}^+ (\mathbf{r}', \mathbf{R}') d\mathbf{r}' d\mathbf{R}'. \qquad (39)$$

Using the well known expression for the Green's function at large separations $R \to \infty$

$$G_0^+ (\mathbf{R}, \mathbf{r}; \mathbf{R}', \mathbf{r}'; E)$$

$$= -\frac{\mu}{2\pi\hbar^2} \sum_\alpha \frac{\exp(iq_\alpha R)}{R} \psi_\alpha^* (\mathbf{r}') \exp\left(-iq_\alpha (\mathbf{n}_\mathbf{R} \cdot \mathbf{R}')\right), \qquad (40)$$

the basic formula (39) for the exact wave function of colliding particles can be reduced to its asymptotic form (32) in terms of the amplitude $f_{\alpha'\alpha} (\mathbf{q}', \mathbf{q})$ of inelastic scattering, where $\mathbf{n}_\mathbf{R} = \mathbf{R}/R$ is the unit vector along the radius vector $\mathbf{R}$. Then, we derive the final expression for scattering amplitude in the coordinate representation

$$f_{\alpha'\alpha} (\mathbf{q}', \mathbf{q}) = -\frac{\mu}{2\pi\hbar^2} \left\langle \Psi_{\alpha'\mathbf{q}'}^{(0)} |V| \Psi_{\alpha\mathbf{q}}^+ \right\rangle$$

$$= -\frac{\mu}{2\pi\hbar^2} \iint d\mathbf{r} \, d\mathbf{R} \exp\left(-i\mathbf{q}' \cdot \mathbf{R}\right) \psi_{\alpha'}^* (\mathbf{r}) V (\mathbf{r}, \mathbf{R}) \Psi_{\alpha q}^+ (\mathbf{r}, \mathbf{R}). \qquad (41)$$

A number of general results of stationary scattering theory can be also formulated in terms of the so-called scattering T-operator (or

T-matrix), which transforms the unperturbed wave function $\Psi_{\alpha\mathbf{q}}^{(0)}$ of the total system to the exact solution $\Psi_{\alpha\mathbf{q}}^{+}$ of scattering problem. In accordance with the basic equations (35) and (36) this operator has the form

$$T\,(E)\,\Psi_{\alpha\mathbf{q}}^{(0)} = V\Psi_{\alpha\mathbf{q}}^{+}\,, \qquad T\,(E) = V + VG^{+}\,(E)\,V\,. \qquad (42)$$

The Lippman–Schwinger equation for the $T(E)$-operator and for the Green $G(E)$-resolvent

$$T\,(E) = V + VG_0^{+}\,(E)\,T\,(E)\,,$$
$$G^{+}\,(E) = G_0^{+}\,(E) + G_0^{+}\,(E)\,VG^{+}\,(E) \qquad (43)$$

proceeds directly from (35)–(42) if we take into account the following operator equality $G\,(E)V = G_0\,(E)T(E)$. As is apparent from (41) and (42), the scattering amplitude (33) for the inelastic $\alpha \to \alpha'$ transition can be expressed in terms of the T-matrix elements on the energy shell (see (34)) taken over the unperturbed wave functions (29) of the system. This relationship is given by

$$f_{\alpha'\alpha}\,(\mathbf{q}',\mathbf{q}) = -\tfrac{\mu}{2\pi\hbar^2}T_{\alpha'\alpha}\,(\mathbf{q}',\mathbf{q})$$
$$\equiv -\tfrac{\mu}{2\pi\hbar^2}\left\langle \Psi_{\alpha'\mathbf{q}'}^{(0)} \,|T\,(E)|\, \Psi_{\alpha\mathbf{q}}^{(0)} \right\rangle\,, \qquad (44)$$

if the wave function $\Psi_{\alpha\mathbf{q}}^{(0)}$ is normalized by the condition (31). By introducing Dirac's standard designations for the eigenvectors of a Rydberg atom and for a plane wave of scattering

$$|\alpha\rangle \equiv |\psi_\alpha(\mathbf{r})\,\rangle\,, \qquad |\mathbf{q}\rangle \equiv |\psi_\mathbf{q}(\mathbf{R})\rangle\rangle = |\exp\,(i\mathbf{q}\cdot\mathbf{R})\rangle\,, \qquad (45)$$

we can represent the T-matrix element in the following form

$$T_{\alpha'\alpha}\,(\mathbf{q}',\mathbf{q}) = \langle \mathbf{q}',\alpha'\,|T\,(E)|\,\mathbf{q},\alpha\rangle = \left\langle \mathbf{q}',\alpha'\,|V|\,\Psi_{\alpha\mathbf{q}}^{+}\right\rangle\,, \qquad (46)$$

which is especially convenient for further analysis. The second equation in (46) is in full agreement with (41) and (42).

Using the well known expansion of the Green resolvent over the eigenfunctions of Rydberg atom $|\alpha\rangle = \psi_\alpha(\mathbf{r})$ and over the plane waves of scattering $|\mathbf{q}\rangle = \psi_\mathbf{q}(\mathbf{R})$

$$G_0^+(E) = \sum_\alpha \int \frac{d\mathbf{q}}{(2\pi)^3} \frac{|\mathbf{q},\alpha\rangle\langle\mathbf{q},\alpha|}{E - E_\alpha - \hbar^2 q^2/2\mu + i0} , \tag{47}$$

the Lippman–Schwinger operator equation (43) can be rewritten in the following integral form for the transition matrix elements

$$\langle \mathbf{q}', f | \mathsf{T}(E) | \mathbf{q}, i \rangle = \langle \mathbf{q}', f | \mathsf{V} | \mathbf{q}, i \rangle$$

$$+ \sum_\alpha \int \frac{d\mathbf{q}''}{(2\pi)^3} \frac{\langle \mathbf{q}',f|\mathsf{V}|\mathbf{q}'',\alpha\rangle\langle\mathbf{q}'',\alpha|\mathsf{T}(E)|\mathbf{q},i\rangle}{E - E_\alpha - \hbar^2(q'')^2/2\mu + i0} . \tag{48}$$

The square of the matrix element of the scattering T-operator directly yields the probability or the cross section of the collision process. For example, the differential cross section of the $|\,i, \mathbf{q}\rangle \rightarrow |\,f, \mathbf{q}'\rangle$ transition into the interval of the wave vectors $d\mathbf{q}'$ is expressed in terms of the T-matrix on the energy shell (34) in accord with the following equation

$$d\sigma_{fi}(\mathbf{q}', \mathbf{q})$$

$$= \frac{2\pi}{\hbar} \left(\frac{\mu}{\hbar q}\right) |\mathsf{T}_{fi}(\mathbf{q}', \mathbf{q})|^2 \, \delta\left(E_f + \mathcal{E}' - E_i - \mathcal{E}\right) \frac{d\mathbf{q}'}{(2\pi)^3} , \tag{49}$$

$$q = \mu V/\hbar , \qquad \mathcal{E} = \hbar^2 q^2/2\mu .$$

Here $\mathcal{E}$ and $\mathcal{E}'$ are the kinetic energies and V and V' are the relative velocities of the incident particle with respect to the Rydberg atom before and after collision. Integration of (49) over all possible values of the final wave numbers q' and over the solid scattering angles $d\Omega_{\mathbf{q}'\mathbf{q}} = \sin\theta \, d\theta \, d\varphi$ leads to the basic formula

$$\sigma_{fi}(q) = \left(\frac{\mu}{2\pi\hbar^2}\right)^2 \frac{q'}{q} \int |\mathsf{T}_{fi}(\mathbf{q}', \mathbf{q})|^2 \, d\Omega_{\mathbf{q}'\mathbf{q}} \tag{50}$$

for the integral cross section σ_{fi} of the bound–bound $i \rightarrow f$ transition between Rydberg states. For the bound–free $i \rightarrow f$ transition the σ_{fi} quantity on the left hand-side of (50) should be replaced, as usual, by the differential ionization cross section $d\sigma_{fi}/dE_f$ per unit energy interval of the ejected electron provided its final wave function ψ_f of continuous spectrum is normalized on the δ-function of energy.

The form of equation (48) is such that we may easily establish the relationship between the imaginary part of the transition matrix element for elastic scattering in the forward direction $\theta_{\mathbf{q}'\mathbf{q}}$ and the

total cross section $\sigma_i^{tot} = \sum\limits_f \sigma_{fi}$ of all inelastic and elastic processes.
Indeed, using (46) and the well known relation for the integration path, and assuming $(f = i)$ and $(\mathbf{q}' = \mathbf{q})$ in (48), we have

$$\mathrm{Im}\left\{\mathsf{T}_{ii}\left(\mathbf{q},\mathbf{q}\right)\right\} = -\tfrac{\hbar^2 q}{2\mu}\sigma_i^{tot}\left(q\right),$$

$$\mathrm{Im}\left\{\mathsf{f}_{\alpha'\alpha}\left(q,\theta_{\mathbf{q}'\mathbf{q}}\right)\right\} = \tfrac{q}{4\pi}\sigma_i^{tot}\left(q\right). \tag{51}$$

The second relation in (51) was rewritten in terms of the usual scattering amplitude (44) in accord with (44). This is the so-called optical theorem, which reflects the basic analytical properties of the scattering amplitude.

2.6   Interaction of a Rydberg Atom with a Neutral Particle

The most important phenomena in collisions involving a Rydberg atom (ion) and structureless neutral or charged projectile may be described in terms of the local interaction potential $V=V_{AB}$, which is the sum of both the interaction between the incident particle B and highly excited electron V_{eB} and the ion core interaction V_{A+B}. In other words, we shall assume the total potential interaction to be of the form

$$V = V_{eB}\left(\mathbf{r}_{eB}\right) + V_{A+B}\left(\mathbf{R}_{A+B}\right) \tag{52}$$

neglecting, as usual, the additional small polarization interaction between the ion core and projectile which results from the induced electric field caused by the Rydberg electron. This effect is fully absent in the case of an electron projectile.

The radius vectors $\mathbf{r}_{eB}$ and $\mathbf{R}_{A+B}$ of a projectile B relative to the Rydberg electron e and to the ion core A^+ can be expressed in terms of its position vector $\mathbf{R}$ relative to the $\left(e, A^+\right)$ centre of mass, and the vector r of the electron–core separation, which are involved in the general equations of scattering theory

$$\mathbf{r}_{eB} = \mathbf{R} - \left(M_{A+}/M_A\right)\mathbf{r}, \qquad \mathbf{R}_{A+B} = \mathbf{R} + \left(m/M_A\right)\mathbf{r}, \tag{53}$$

where $M_A = m + M_{A+}$ is the total mass of the Rydberg atom. Since the electron mass is small compared to the mass of the Rydberg atom

$(m \ll M_A)$ it is natural to expand the core-projectile interaction in power series of the ratio $(m\mathbf{r}/M_A)$. As a result, we have

$$\mathsf{V}_{A+B}\left(\mathbf{R}_{A+B}\right) = \mathsf{V}_{A+B}\left(\mathbf{R}\right) + (m/M_A)\,\mathbf{r}\cdot\nabla_{\mathbf{R}}\mathsf{V}_{A+B}\left(\mathbf{R}\right) + ... \quad , \quad (54)$$

and, hence, the matrix element of the core–perturber interaction over the Rydberg atom wave functions takes the form

$$\langle\alpha'|\,\mathsf{V}_{A+B}\,|\alpha\rangle = \mathsf{V}_{A+B}\left(\mathbf{R}\right)\delta_{\alpha'\alpha}$$
$$+ (m/M_A)\,\langle\psi_{\alpha'}\left(\mathbf{r}\right)|\,\mathbf{r}\,|\psi_{\alpha}\left(\mathbf{r}\right)\rangle\cdot\nabla_{\mathbf{R}}\mathsf{V}_{A+B}\left(\mathbf{R}\right) + ... \quad . \tag{55}$$

The second term in (54) corresponds to the inertial force $\mathbf{F}_e = m d^2\mathbf{R}_{A+}/dt^2$ acting on the Rydberg electron in the noninertial frame of reference moving with the ion core A^+. Since this term contains a factor $(m/M_A) \ll 1$ its contribution to the total transition amplitude is sufficiently small that it can usually be neglected. Therefore, for most collision problems having practical importance, we can neglect the so-called non-inertial effects in collisions of Rydberg atoms with neutral or charged particles. Situations, in which these effects become important, will be discussed in Sect. 7.

Thus, neglecting the weak noninertial effects we can count off the projectile position vector $\mathbf{R}$ from the centre of mass of the ion core A^+. In this approximation the total interaction of Rydberg atom with projectile can be written as

$$\mathsf{V} \approx \mathsf{V}_{eB}\left(\mathbf{R} - \mathbf{r}\right) + \mathsf{V}_{A+B}\left(\mathbf{R}\right), \tag{56}$$

whereas the potential V_{eB} of the electron–projectile interaction depends only on the difference of $\mathbf{r}_{eB} = \mathbf{R} - \mathbf{r}$, while the core–projectile interaction does not contain the dependence on the Rydberg electron coordinates. Hence, the matrix elements $\langle\psi_{\alpha'}\left(\mathbf{r}\right)|\mathsf{V}_{A+B}\left(\mathbf{R}\right)|\psi_{\alpha}\left(\mathbf{r}\right)\rangle$ of the core-projectile interaction are rigorously equal to zero (if $\alpha' \neq \alpha$) due to the orthogonality of the Rydberg atom wave functions. As a result, in collisions of a Rydberg atom with a structureless particles the cross section of the inelastic transition $(\alpha \neq \alpha')$ is mainly determined by the interaction V_{eB} between the Rydberg electron and the projectile. The core–perturber interaction V_{A+B} is certainly important in the process of elastic scattering $(\alpha' = \alpha)$ and electron Rydberg ion collision since the potential V_{A+B} changes the trajectory of the incident particle.

2.7   Basic Concept of Quasifree Electron Model

Most of the available methods for the description of collisions involving a Rydberg atom A^* and the atoms or molecules in the ground or low excited states are based on the key assumption that processes caused by scattering of the neutral projectile B on the highly excited electron and on the parent core A^+ may be treated independently. Thus, the resultant three-body problem is rather simplified because the perturbing atom or molecule does not interact simultaneously with both the outer electron and ion core. This is true for a wide range of the principal quantum numbers n due to the orbital radius $r_n \sim n^2 a_0$ of the Rydberg atom being large as compared to the characteristic r_{eB} and R_{A+B} dimensions of both the electron-perturber and core-perturber interaction regions. Then, neglecting the interference effects and multiple scattering, the resultant cross section of the Rydberg-atom–neutral collision is determined by additive contribution of the electron-perturber and core-perturber scattering mechanisms. In many situations the contribution of the scattering of a highly excited electron on the perturbing particle turns out to be predominant. Then, the theoretical analysis of collisional processes involving a highly excited atom and neutral particle is based on the quasifree-electron model first proposed by *Fermi* [80] in order to explain the spectral line shift behavior of the Rydberg atomic series in a buffer gas. This model has been formulated in its final form as a result of intensive investigations of elastic, quasielastic, inelastic and ionizing collisions between the Rydberg atoms and neutral projectiles (see [3, 36] and review articles by *Hickman* et al. [81], *Matsuzawa* [82], and *Flannery* [83]).

The main idea of the quasifree-electron model is that the collision of the Rydberg atom A^* with an atom or molecule B is treated as an elastic or inelastic binary encounter between the valence (slow and free) electron and neutral projectile. The parent core A^+ is only a spectator in this encounter, so that its interaction with the neutral perturber is entirely neglected. However, it is responsible for the generation of the electron momentum distribution function in the given Rydberg atomic state. The momentum distribution function is determined by the interaction $U(r)$ of the electron with the parent core A^+ (i.e. primarily by their Coulomb interaction at large distances r).

Only sufficiently small electron energies $\epsilon = \hbar^2 k^2 / 2m \ll Ry$ and the wave numbers $k a_0 \ll 1$ are responsible for thermal collisions between the Rydberg atoms and neutral particles. The characteristic range of k, which makes the main contribution to one or another process, depends essentially on both the principal quantum number n and the energy ΔE transferred to the highly excited electron in the binary $e - B$ encounter. In pure elastic and quasielastic collisions (when $\Delta E = 0$) the characteristic values of k are about the mean orbital wave number $k_n \sim 1/na_0$ of the Rydberg electron. This means that $ka_0 \sim k_n a_0 \approx 0.2$, 0.1, and 0.03 and, hence, the averaged Rydberg electron energies are equal to $\epsilon_n \approx Ry/n^2 \sim 0.5$, 0.1, and 0.01 eV for $n = 5$, 10, and 30, respectively. At $n \sim 100$ (when $k_n a_0 \sim 10^{-2}$) the milli- and submilli-electron-volt energy diapason is becoming important, while for very high values of $n \sim 1000$ the averaged kinetic energy of Rydberg electron corresponds to the micro-electron-volt energy region. It is important to stress, that the inelastic n-changing excitation (deexcitation) and ionization processes with large energy ΔE transferred to the translational motion of colliding atoms are determined by much greater values of $k \sim |\Delta E|/2\hbar V$ than the mean orbital momentum $1/(na_0)$. Nevertheless, the characteristic momenta of k in such transitions are also satisfied by the condition $kr_0 \ll 1$ if the principal quantum number is not too small. Here r_0 is the effective radius of the short-range interaction of electron with target atom or molecule B, whose value is usually a few atomic units. As follows from the analysis of different elementary processes involving Rydberg atoms and neutral targets, commonly of practical use are only the energies $\epsilon = \hbar^2 k^2/2m$, which do not exceed $0.1 - 1$ eV $(ka_0 < 0.1 - 0.3)$.

2.8   Low-Energy Electron Scattering by Neutral Targets

Electron scattering by atomic [79, 84] and molecular [85, 86] targets has been the subject of intensive theoretical and experimental research for many years (see also [87, 88] and the recent review articles [36, 89, 90]). Since the de Broglie wavelength $1/k$ of an ultra-low-energy electron is large compared to the effective range r_0 of the electron–atom or electron–molecule short-range interaction, one of the most adequate methods for describing the electron scattering in the sub-thermal energy region is based on the effective range theory. This theory gives an appropriate expansion of the scattering amplitude in the vicinity of zero energy $(0 \leq k \ll 1/r_0)$. It is mainly

determined by the type of long-range interaction between an electron
and target atom or molecule (i.e. by the polarization, quadrupole or
dipole potentials). The short-range interaction ($r < r_0$) is usually described by means of several parameters such as the scattering length
L and the effective radius r_0. It is well known that the dominant
terms of the elastic electron–atom scattering amplitude and cross
section at $k \to 0$ are determined by the contribution of the s-wave
alone and can be expressed in terms of one parameter ($f_{eB} = -L$
and $\sigma_{eB}^{el} = 4\pi L^2$, where L is the scattering length).

Until recently this simplest description of electron scattering at
ultra-low energies was used in most of the theoretical works on collisions involving Rydberg atoms and neutral atomic targets. However,
detailed studies showed that the scattering length approximation fails
to describe reliably even thermal collisions of highly excited atoms
with the ground-state heavy-rare gas atoms, particularly so, for inelastic transitions with large energy transfer. It also does not hold
for collisions with a neon atom which has an anomalously low value
of the scattering length. Moreover, it certainly becomes inapplicable
for collisions of Rydberg atoms with strongly polarizable targets (e.g.
alkali-metal atoms). The effective range of the long-range (polarization) part of electron interaction with these target systems is characterized by the Weisskopf radius $r_W = (\pi \alpha / 4 k a_0)^{1/3}$. Therefore, at
large values of atomic polarizability α it may become comparable to
the electron wavelength $1/k$ even for submilli-electron-volt energies.
A similar situation occurs for electron scattering by polar molecules
which have a substantial permanent dipole moment, for which the
effective range of the long range interaction turns out to be particularly large. This leads to the appearance of anomalously large values
of elastic electron–atom scattering as well as rotationally elastic and
inelastic processes in electron–molecule scattering at very low energies. They are primarily the result of the presence of real discrete
or virtual levels of very weakly bound negative atomic or molecular
ions or the shape resonances on the quasidiscrete (quasistationary)
levels. These phenomena and their revelation in collisions involving
Rydberg atoms were the subject of intensive research in recent years
(see [36, 89, 90] and references therein).

In this section we shall briefly discuss the behavior of the scattering amplitude and the differential cross section of ultra-slow electron
scattering by the rare gas and alkali-metal atoms as well as by the
polar and non-polar molecules. We shall present below only some

final formulae, which are necessary for the next analysis of the Rydberg atom–neutral collisions. Besides, in addition to the discussion of the most important physical effects we shall provide the text with several figures and tables containing the basic parameters of ultra-low electron scattering by the rare gas and alkali-metal atoms

2.8.1 Effects of Electron–Atom Scattering

Basic expressions for the elastic scattering amplitude and for the differential cross section averaged over the two possible $S_+ = s_B + 1/2$ and $S_- =| s_B - 1/2 |$ values of the total spin S of the B + e system are given by

$$f_{eB}^{(S)}(k, \theta) = \sum_{\ell} (2\ell + 1) f_{\ell}^{(S)}(k) P_{\ell}(\cos \theta),$$

$$f_{\ell}^{(S)} = \left[k \cot \left(\eta_{\ell}^{(S)} \right) - ik \right]^{-1}, \tag{57}$$

$$\frac{d\sigma_{eB}}{d\Omega} = |f_{eB}(k, \theta)|^2 = \sum_{S=S_+, S_-} C(S) \left| f_{eB}^{(S)}(k, \theta) \right|^2. \tag{58}$$

Here $f_{\ell}^{(S)}(k)$ and $\eta_{\ell}^{(S)}(k)$ are the amplitude and the phase shift of the partial wave with the orbital momentum ℓ; θ is the scattering angle; and

$$C(S) = \frac{(2S + 1)}{2(2s_B + 1)} \tag{59}$$

is the spin factor, s_B is the spin of the target B. Note that $S_+ = S_- = 1/2$ and, hence, $C(S_+) = C(S_-) = 1/2$ for the ground-state rare gas atoms, for which $s_B = 0$. At the same time, there are the triplet $(S_+ = 1, C(S_+) = 3/4)$ and singlet $(S_- = 0, C(S_-) = 1/4)$ waves for electron scattering by the ground-state alkali-metal atoms.

Potential Scattering by Rare Gas Atoms.   In the case of potential scattering of slow electron by the atom B with sufficiently small polarizability α (e.g. the rare gas atoms) the modified effective range theory has been formulated in a series of papers by *Spruch* et al. [91] and *O'Malley* [92]. For the rare-gas atoms the basic parameters of this theory (such as scattering length L and atomic polarizability α) are presented in Table 7.

Table 7. Electron scattering lengths L (in a_0) and polarizabilities α (in a_0^3) of the ground-state rare gas atoms (Z is the nuclear charge)

Atom	Z	α (RD)	L (ERT)	L (SRL)
He	2	1.383 [a1)] 1.384 [a2)]	1.19 [d)] 1.17 [e)]	1.11 $\pm$ 0.2 [f)]
Ne	10	2.67 [b)] 2.70 [a2)] 2.92 [a3)]	0.214 [d)] 0.204 [e)] 0.2065 [b)]	0.192 $\pm$ 0.003 [f)] 0.24 $\pm$ 0.01 [g)]
Ar	18	11.08 [a1)] 11.075 [a2)] 14.0 [a4)]	-1.70 [d)] -1.55 [e)] -1.593 [c)]	-1.4 ± 0.1 [f)]
Kr	36	16.74 [a1)] 16.77 [a2)] 16.60 [a4)]	-3.70 [d)] -3.5 [e)] -3.478 [c)]	-3.05 ± 0.06 [f)] -4.0 ± 0.3 [g)]
Xe	54	27.29 [c)] 27.30 [a2)] 27.4 [a3)]	-6.50 [d)] -6.5 [e)] -6.527 [c)]	-6.1 ± 0.3 [g)]

RD – reference data for dipole polarizability: [a1)] from [99]; [a2)] experiment, [a3)] quantum theory, [a4)] statistical theory – from [100].
ERT – extrapolation of experimental data on electron scattering to zero energy using effective range theory: [b)] Gulley et al. [95], [c)] Weyhreter et al. [94], [d)] O'Malley and Crompton [92], [e)] Golovanivskii and Kabilan [93].
SRL – measurements of pressure shift of high-Rydberg levels in rare gases: [f)] Heber et al [97], [g)] Thompson et al. [98].

Within the framework of the theory [91, 92] the scattering partial phase shifts $\eta_\ell(k)$ are given by the expressions

$$\frac{\tan[\eta_0(k)]}{k} = -L - \frac{\pi\alpha}{3a_0}k - \frac{4\alpha L}{3a_0}k^2\ln(ka_0) + D_0k^2$$

$$+F_0k^3 + O(k^4), \tag{60}$$

$$\frac{\tan[\eta_\ell(k)]}{k} = \frac{\pi\alpha k}{a_0(2\ell+3)(2\ell+1)(2\ell-1)} + D_\ell k^2 + O(k^3), \qquad \ell \geq 1.$$

One can see that the values of η_ℓ fall strongly with an increase of the orbital momentum ℓ for electron–target-atom relative motion. The low-energy expansions for the amplitude and differential cross section of electron–atom scattering are as follows:

$$\mathrm{Re}\left\{f_{\mathrm{eB}}\left(k,\theta\right)\right\} = -L - \tfrac{\pi\alpha}{2a_0} \, k \, \sin\left(\tfrac{\theta}{2}\right) - \tfrac{4\alpha L}{3a_0} \, k^2 \, \ln\left(ka_0\right)$$

$$-bk^2 + O\left(k^3\right), \tag{61}$$

$$\left|f_{\mathrm{eB}}\left(k,\theta\right)\right|^2 = L^2 + \tfrac{\pi\alpha L}{a_0} \, k \, \sin\left(\tfrac{\theta}{2}\right) + \tfrac{8\alpha L^2}{3a_0} \, k^2 \, \ln\left(ka_0\right)$$

$$+Bk^2 + O\left(k^3\right). \tag{62}$$

Here L is the scattering length of the target atom B; D_ℓ, F_0, b and B are the constant coefficients ($D_\ell > 0$ for $\ell = 0$, $D_\ell < 0$ for $\ell = 1$ and all D_ℓ values with $\ell \geq 2$ are usually taken to be $D_\ell = 0$). The energy range in which these expressions are valid is limited by the condition $\alpha k^2/a_0 \ll 1$. The specific feature of the expansion (61), (62) is the presence of the second linear k term and the third logarithmic term, which appear due to the long-range polarization interaction $V^{\mathrm{l.r.}} = -\alpha e^2/2r^4$. The second term, which is proportional to the momentum transfer $Q = |k' - k| = 2k \sin\left(\theta/2\right)$ in the electron–atom collision, vanishes in the forward direction $\theta = 0$.

The theory [91, 92] provides a successful explanation of all phenomena which occur in electron–rare-gas-atom scattering at small energies. For He, Ne, Ar, Kr and Xe a detailed analysis of the phase shifts and the cross sections was made by *Golovanivskii and Kabilan* [93], *Weyhreter* et al. [94], and *Gulley* et al. [95, 96]. The results for the momentum transfer cross sections, obtained [93] by numerical simulation of the Electron-Cyclotron-Resonance experimental curves, are presented in Fig. 7 by full curves. The dashed curves represent the scattering length approximation $\sigma^{\mathrm{tr}}\left(\epsilon\right) = 4\pi L^2$.

It can be seen that the scattering length approximation (in which the elastic scattering cross section is independent of the electron energy $\sigma_{\mathrm{eB}}^{\mathrm{el}} = 4\pi L^2 = const$ and is determined by the contribution of partial s-wave) may be reliably used in a wide energy region $\epsilon < 1$ eV for the helium atom alone. For the heavy rare gas atoms Ar, Kr and Xe there are deep Ramsauer–Townsend minima, which appear in the range of $\epsilon \approx 0.1 - 1$ eV due to negative values of the scattering lengths. For a neon atom, which has an anomalously low value of the scattering length, the momentum transfer cross section $\sigma_{\mathrm{eB}}^{\mathrm{tr}}$ grows with an increase of the electron momenta and considerably exceeds the value of $4\pi L^2$ in the range above 0.01 eV. It should be noted that the magnitudes of the scattering lengths found by extrapolation of electron–atom cross sections to zero energy using the modified effective range theory [91, 92] are in quite reasonable agreement (see Table 7) with the data obtained in the experiments [97, 98] on the pressure shift of the high-Rydberg atomic series by rare gas atoms.

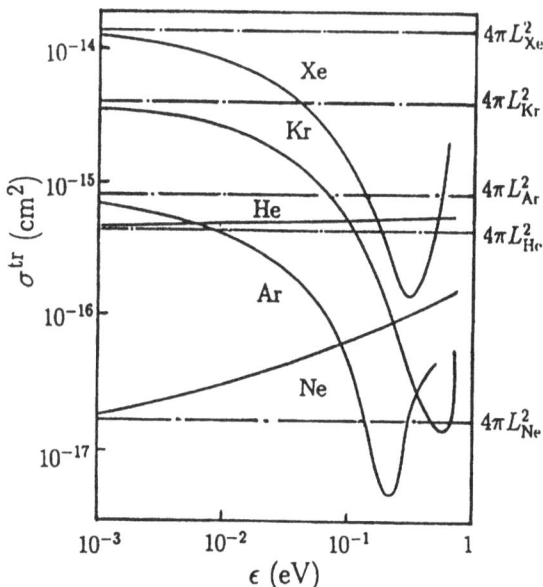

Figure 7. The results of Ref. [93] for the momentum transfer cross sections $\sigma^{tr}(\epsilon)$ for elastic electron scattering by rare gas atoms.

Resonance Scattering by Alkali-Metal Atoms. For highly polarizable alkali-metal atoms, the theory [91, 92] fails to describe the scattering amplitude in the energy region above a few meV. An important feature of electron scattering by the ground state alkali atom B is the presence of the ^3P-shape resonance on the quasidiscrete triplet level ($E_r > 0$) of the corresponding negative ion B^-. This leads to enormously high values of the cross sections (see Fig. 8) in the range of about 10–100 meV depending on the specific value of the polarizability of an alkali-metal atom (Table 8).

The appearance of the shape resonance for the p-scattering wave is the result of the potential barrier in the effective electron–atom interaction, which includes both the polarization and centrifugal terms

$$V_{eff}(r) = -\alpha e^2/2r^4 + \hbar^2\ell(\ell+1)/2mr^2, \quad (\ell = 1).$$

At high polarizability $\alpha \sim 160 - 400\ a_0^3$ this barrier occurs at large distances $r_{max} = (\alpha/a_0)^{1/2}$ and has a small height $V_{eff}^{max} = Ry(a_0^3/\alpha)$. Thus, possible quasidiscrete level can exist only in the ultra-low-energy region $E_r < V_{eff}^{max}$ (below about $30-80$ meV for alkali atoms).

Table 8. Polarizabilities α (in a_0^3) of the ground-state alkali-metal atoms.

Atom	Z	ED	QT	ST	RD	RP
Li	2	164	169	164	162	165
Na	11	162.5	159		162	166
K	19	293	290		286.8	303
Rb	37	319		250.3	310	328
Cs	55	402	310	316.9	385	402

ED, QT, and ST: experimental data, quantum theory, and statistical theory results from [100, 101], respectively. RD: reference data from [99]. RP: data used in calculations [104, 103] of the ^3P-resonance parameters.

The very presence of the ^3P-shape resonances proceeds from the close-coupling [84, 108] and variational [104] calculations of the partial phase shifts for electron scattering by Li, Na, K, and Rb as well as from the effective range theory [102, 103]. Recent relativistic R-matrix calculations [107] taking into account the electron correlation effects showed that there is also the quasidiscrete ^3P-level of negative Cs$^-$-ion (It is split by the spin-orbit interaction into three fine-structure components with total angular momentum values $J = 0, 1$ and 2, for which the shape resonance positions are $E_r = 1.78$, 5.56, and 12.76 meV, respectively). There is also a direct experiment [109] on measuring the elastic electron scattering cross sections, which clearly demonstrates that the ^3P-resonances occur for Na and Rb at low-energy range. A resonance feature of electron scattering by K, Rb, and Cs atoms is also confirmed by comparison of calculated cross sections for impact broadening (see Refs. [105, 106] and [103, 110]) and quenching [105, 111] of Rydberg atomic levels in alkali vapors with corresponding experimental data [112, 113] and [114].

Consider the influence of the ^3P-resonance on the behavior of the differential cross section for elastic scattering. It is convenient to separate the contributions of the resonance f_{eB}^r and potential f_{eB}^P scattering in the basic expression for the total amplitude $f(\epsilon, \theta)$ of triplet $(S_r = S_+ = 1)$ and singlet $(S_- = 0)$ scattering

$$f_+(\epsilon, \theta) = f_+^P(\epsilon, \theta) + (2\ell_r + 1)f_+^r(\epsilon)P_{\ell_r}(\cos\theta), \quad (63)$$
$$f_-(\epsilon, \theta) \equiv f_-^P(\epsilon, \theta), \quad (64)$$

where ℓ_r is the orbital moment of the resonance scattering partial wave. With the quasidiscrete level $(E_r > 0)$ being in the vicinity of zero energy the specific form of the resonance scattering amplitude is given by [37]

$$f_+^r(\epsilon) = -\frac{\hbar}{\sqrt{2m}} \frac{\gamma\epsilon^{\ell_r}}{\left(\epsilon - \epsilon_0 + i\gamma\epsilon^{\ell_r + 1/2}\right)} . \quad (65)$$

Here ϵ_0 and γ are the parameters $(\epsilon_0, \gamma > 0)$ defining the position of the maximum E_r in electron–atom scattering $\sigma_{max}^r = \sigma_{el}^r(E_r) \approx 9\pi/k_r^2$ and the width Γ_r of the resonance $(E_r \sim \epsilon_0, \Gamma_r/2 \sim \gamma\epsilon_0^{\ell_r + 1/2})$.

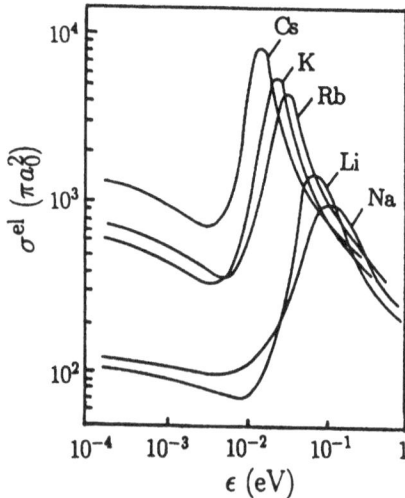

Figure 8. Total cross sections $\sigma^{el}(\epsilon)$ for elastic electron scattering by alkali-metal atoms obtained using the effective range theory for K, Rb, and Cs [103] and by a variational method for Li and Na [104].

In the case of a ^3P-resonance ($\ell_r = 1$, and spin factor $C\left(S_+\right) = 3/4$) its contribution to the total differential scattering cross section averaged over spins (58) can be written as [105]

$$\frac{d\sigma_{eB}^r}{d\Omega} = \frac{27}{4} \left| f_+^r\left(\epsilon\right) \right|^2 \cos^2 \theta, \tag{66}$$

$$\left| f_+^r\left(\epsilon\right) \right|^2 = \frac{\hbar^2}{2m} \frac{\gamma^2 \epsilon^2}{\left(\epsilon - \epsilon_0\right)^2 + \gamma^2 \epsilon^3} . \tag{67}$$

Then, the real and imaginary parts of the scattering amplitude

$$f_{eB}^r(\epsilon, \theta) = \left(2\ell_r + 1\right) f_+^r(\epsilon) \cos \theta \tag{68}$$

are given by the following expressions

$$\mathrm{Re}\left\{ f_+^r\left(\epsilon\right) \right\} = \frac{\sqrt{2m}\left(\epsilon_0 - \epsilon\right)}{\hbar \gamma \epsilon} \left| f_+^r\left(\epsilon\right) \right|^2, \tag{69}$$

$$\mathrm{Im}\left\{ f_+^r\left(\epsilon\right) \right\} = \frac{\sqrt{2m\epsilon}}{\hbar} \left| f_+^r\left(\epsilon\right) \right|^2 . \tag{70}$$

Comparison of the simple analytic expression (66) with the results [103, 104] of numerical calculations shows that it provides quite a reasonable quantitative description of the ^3P-resonance curve in the energy region $\epsilon \sim E_r$ for all alkali atoms.

As is evident from Fig. 8, the ^3P-resonance gives a major contribution to the electron–alkali-atom scattering in a wide energy region $\left|\epsilon - E_r\right| \sim \Gamma_r$ ($E_r \sim \Gamma_r \sim 10 - 100$ meV). However, in the vicinity of zero energy $\epsilon \ll E_r$ the main role is played by the potential scattering. It can be described at $\alpha k^2/a_0 \ll 1$ by the standard expression (62) of the effective range theory [91, 92] averaged over spins:

$$\frac{d\sigma_{eB}^P}{d\Omega} = L_1^2 + \frac{\pi\alpha}{a_0} L_2\, k \sin\left(\frac{\theta}{2}\right) + \frac{8\alpha}{3a_0} L_1^2\, k^2 \ln\left(ka_0\right)$$
$$+ Bk^2 + O\left(k^3\right). \tag{71}$$

Here L_1 and L_2 are the effective scattering lengths:

$$L_1^2 = \frac{3}{4}L_+^2 + \frac{1}{4}L_-^2, \qquad L_2 = \frac{3}{4}L_+ + \frac{1}{4}L_- . \tag{72}$$

L_+ and L_- are the triplet and singlet scattering lengths, respectively. Since the value of L_2 in (72) is negative (see Table 9), the

Ramsauer–Townsend minimum is observed for any alkali atom in the region between 1–20 meV. It is important to stress that available experimental data of *Heinke* et al. [112] for the averaged over spins scattering lengths of the heavy alkali-metal atoms $L_2^{Rb} = -13.7\ a_0$ and $L_2^{Cs} = -18.3\ a_0$, obtained from measurements of the pressure shift of high-Rydberg levels by Rb and Cs, are in good agreement with the corresponding results ($L_2^{Rb} = -12.2\ a_0$ and $L_2^{Cs} = -17.6\ a_0$) found from the effective range theory [103].

Table 9. Positions E_r (in meV) and widths Γ_r (in meV) of the ^3P-resonances, singlet L_- and triplet L_+ electron scattering lengths (in a_0) of the ground-state alkali-metal atoms.

Atom	L_-	L_+	E_r	Γ_r
Li	3.65 [e]	−5.66 [e]	59 [a]	77 [a]
			60 [b]	57 [b]
Na	4.23 [e]	−5.91 [e]	83 [a]	188 [a]
			83 [b]	85 [b]
K	0.57 [a]	−15.4 [a]	19 [a]	16 [a]
	0.55 [e]	−15.0 [e]	20 [c]	21 [c]
Rb	2.03 [a]	−16.9 [a]	23 [a]	25 [a]
			28 [c]	31 [c]
Cs	−2.4 [a]	−22.7 [a]	9.5 − 12.6 [a]	5.7 − 9.1 [a]
	−4.04 [e]	−25.3 [e]	9.14 [d]	6.04 [d]

[a] effective-range theory [102, 103, 110]; [b] variational calculations [104] for Li and Na; [c] results [105, 106] for K and Rb obtained from experiments with Rydberg atoms; [d] relativistic R-matrix calculations [107] for Cs (J-averaged values); [e] two-state close coupling results [108] for L_- and L_+.

Real and Virtual States of Negative Ions. The behavior of the amplitude for elastic electron scattering by a neutral target B is also appreciably changed compared to the simplest law $f_{eB} = -L = const$ if there is a discrete or virtual level of corresponding negative ion B^- with energy close to zero. If this level corresponds to the electron momentum ℓ, then the low-energy expansion for the contribution of partial ℓ-wave to the scattering amplitude can be written as

$$f_{eB}^{(\ell)}(\epsilon, \theta) \approx (2\ell + 1) \frac{(-1)^{\ell+1} |E_0|^\ell}{\beta (\epsilon + |E_0|)} P_\ell (\cos \theta), \qquad \epsilon \sim |E_0|, \quad (73)$$

in accord with the general theory [37]. Here $|E_0| = \hbar^2 \alpha^2 / 2m$ is the energy of real ($\alpha > 0$) or virtual ($\alpha < 0$) bound state ($E_0 < 0$),and β is the parameter of low energy expansion. In the case of s-level ($\ell = 0$) the scattering amplitude is an isotropic function and the corresponding total cross section is given by

$$\sigma_{eB} = \frac{4\pi}{k^2 + \alpha^2} \qquad (74)$$

at $\epsilon \sim |E_0|$. Therefore, the presence of bound or virtual level of weakly bound negative ion (e.g. supported by the long-range interaction) leads to the scattering cross section, which turns out to be significantly greater than the value of $4\pi r_0^2$ determined by the characteristic radius r_0 of the short-range interaction.

During the last decade considerable experimental and theoretical efforts have been devoted to the study of very weakly bound energy ^2P-levels of negative atomic ions such as Ca$^-$ [116]. Sr$^-$, and Ba$^-$ [115] (see [89]) for more details and references on recent papers in Sects. 5 and 6). For example, the binding energy of Ca$^-$ turns out to be equal to 19.73 meV and 24.55 meV for the $^2P_{3/2}$ and $^2P_{1/2}$ states, respectively [116]. At present there is a series of experimental and theoretical works devoted to studies of the role of such weakly-bound states of negative ions in thermal collisions involving Rydberg atoms and ground-state atoms having small electron affinities. The theory of such processes will be described in Sect. 5.7 and 6.4.

2.8.2   Effects of Electron–Molecule Scattering

The main features of the electron–molecule scattering are due to the long-range dipole or quadrupole interaction and to the presence of rotational degrees of freedom. For instance, the rotational excitation and deexcitation of a molecule play a very important role at small energies and give a substantial contribution to the total and momentum transfer cross sections.

Non-polar Molecules.   Consider first the scattering of low-energy electrons ($k \ll 1/r_0$) by a non-polar diatomic molecule B in the $^1\Sigma$-state, for which a long-range potential is given by the expression

$$V^{l.r.}(r, \Theta) = -\left(\frac{eQ}{r^3} + \frac{\alpha_2 e^2}{2r^4}\right) P_2(\cos\Theta) - \frac{\alpha_0 e^2}{2r^4} + O(1/r^5) . \qquad (75)$$

Here Θ is the angle of the radius vector of the incident electron r relative to the vector of internuclear N-axis; Q is the quadrupole moment. The symmetrical α_0 and orientation-dependent α_2 polarizabilities are determined by the main magnitudes $\alpha_\parallel$ and $\alpha_\perp$ of the polarizability tensor, i.e. $\alpha_0 = \left(\alpha_\parallel + 2\alpha_\perp\right)/3$ and $\alpha_2 = 2\left(\alpha_\parallel - \alpha_\perp\right)/3$. For the potentials with the r^{-n} asymptotic dependence, the contribution to the scattering amplitude f_{eB}, which arises from the large distances (when $|V| \ll \hbar^2/mr^2$), can be calculated in the perturbation theory, if $n > 2$ [37]. However, the validity condition of perturbation theory for the short-range interaction may fail to hold. The low-energy expansion of the scattering amplitude has been found by *O'Malley* [92] in the case when the electron–point quadrupole interaction is the leading term at large distances. The final expressions for the differential cross sections $d\sigma_{eB}^{j'j}/d\Omega = (k'/k)\left|f^{j'j}\left(\mathbf{k}',\mathbf{k}\right)\right|^2$ of pure elastic $j \to j$ scattering and for the rotational $j \to j \pm 2$ transitions can be written as

$$\frac{d\sigma_{eB}^{jj}}{d\Omega} = L^2 + \frac{\pi\alpha_0}{2a_0}QL$$

$$+\frac{j(j+1)}{(2j+3)(2j-1)}\left[\frac{4a_0^2}{45}\left(\frac{Q}{ea_0^2}\right)^2 + \frac{\pi\alpha_2 Q}{60}\left(\frac{Q}{ea_0^2}\right)\right], \tag{76}$$

$$\frac{d\sigma_{eB}^{j'j}}{d\Omega} = \frac{j_>(j_> - 1)}{(j+j_>+1)(j+j_>-1)}\frac{k'}{k}\left[\frac{2a_0^2}{15}\left(\frac{Q}{ea_0^2}\right)^2 + \frac{\pi\alpha_2 Q}{40}\left(\frac{Q}{ea_0^2}\right)\right]. \tag{77}$$

Here $Q = |k' - k|$ is the momentum transfer,

$$(k')^2 = k^2 - 2m(E_{j'} - E_j)/\hbar^2 ,$$

j and $j' = j \pm 2$ are the rotational quantum numbers of the molecule before and after collision (E_j and $E_{j'}$ are its internal rotational energies), $j_> = \max(j,j')$, and

$$L = -\lim_{k \to 0}\left(k^{-1}\tan\left[\eta_0(k)\right]\right)$$

is standard scattering length. As is apparent from (76), in the presence of the quadrupole interaction the contribution of elastic scattering at $k \to 0$ to the cross sections gives not only an S-wave, but also all the highest partial scattering waves.

Polar Molecules. In this case the asymptotic form of the interaction potential is given by

$$V^{l.r.}(r, \Theta) = -\frac{e\mathcal{D}}{r^2} \cos\Theta - \left(\frac{e\mathcal{Q}}{r^3} + \frac{\alpha_2 e^2}{2r^4}\right) P_2(\cos\Theta)$$

$$-\frac{\alpha_0 e^2}{2r^4} + O(1/r^5) .$$

(78)

The main features of electron–polar-molecule scattering are due to the presence of a leading dipole term. It is predominant for rotational excitation (deexcitation) processes, provided that the dipole moment $\mathcal{D}$ is not too small. The quadrupole and polarization interactions in (78) give a significant contribution to the scattering amplitude if the $\mathcal{Q}k/ea_0$ and $\alpha_0 k^2/a_0$ parameters become of the same order of magnitude as $\mathcal{D}/ea_0$. Because of a sufficiently slow fall (as r^{-2}) of dipole potential at asymptotic region of r, the distant collisions between the low-energy electron and polar molecule are, on the whole, more important in rotational $j \to j \pm 1$ transitions than the close collisions. Thus, a simple description of such processes can be made on the basis of the first order perturbation theory with the use of point-dipole interaction. As has been found by *Massey* [117], in the Born approximation the differential $d\sigma_{eB}^{j'j}/d\Omega$ cross section of the rotational excitation (deexcitation) of the polar molecule is given by

$$\frac{d\sigma_{eB}^{j'j}}{d\Omega} = \frac{k'}{k} \left|f_{eB}^{j'j}(\mathbf{k}', \mathbf{k})\right|^2 = \frac{4}{3} \frac{j_>}{2j+1} \frac{k'}{k} \left(\frac{\mathcal{D}}{ea_0}\right)^2 \frac{1}{Q^2} ,$$

(79)

where $j_> = \max\{j, j'\}$ and

$$Q^2 = k^2 + k'^2 - 2kk' \cos\theta .$$

In the adiabatic approximation, when the energy of the incident electron is not very low $\epsilon \gg |\Delta E_{j'j}| \sim j\mathcal{B}_0$ (where $\mathcal{B}_0$ is the rotational constant of molecule), one may neglect the difference between the k and k' values in (79). Then, the simple expressions of *Altshuler* [118] for the differential cross section, averaged over the orientations of the molecular axis, directly follow from (79)

$$\frac{d\sigma_{eB}}{d\Omega} = \sum_{j'=j\pm 1} \frac{d\sigma_{eB}^{j'j}}{d\Omega} = \left(\frac{\mathcal{D}}{ea_0}\right)^2 \frac{2}{3k^2(1-\cos\theta)} .$$

(80)

These expressions (79) and (80) for the first-order Born approximation describe the contribution of the dipole interaction alone. It is
worthwhile to point out that the divergence of the total cross section for electron–point dipole scattering at small values of electron
momentum $(k \to 0)$ is due to the long-range potential behavior $1/r^2$
but not to the Born-approximation [37]. This divergence is due to
the use of the non-rotating polar molecule approximation, with the
rotational constant taken to be $\mathcal{B}_0 \to 0$.

The first-order Born approximation with the point-dipole is certainly satisfactory for rotational excitation (deexcitation) $j \to j \pm 1$
processes at relatively small dipole moment $\mathcal{D} < (0.5 - 1)ea_0$ in the
whole low-energy range of practical importance [85, 86]. As follows
from the close coupling calculations, it yields reasonable cross sections even for $\mathcal{D} \sim ea_0$. On the other hand, at low energies the
cross section for elastic electron–polar molecule scattering is mainly
determined by the lowest partial S- and P-waves. Hence, it is very
sensitive to the value of the short-range interaction and can not be
considered within the framework of perturbation theory. The differential cross section of inelastic $j \to j \pm 1$ transitions is also determined
by the intermediate or the short-range parts of interaction at large
scattering angles (i.e. at high momentum transfer Q), while at the
small scattering angles it is due to the long-range dipole potential.

At large values of dipole moment $\mathcal{D} > (0.5 - 1)\,ea_0$ there are some
interesting effects in the behavior of elastic scattering and rotational
deexcitation of a molecule at ultra-low electron energies. They are the
results of the shape resonances on quasidiscrete levels or the presence
of weakly bound or virtual dipole supported states of negative ions of
polar molecules (see [89]). For example, detailed studies of electron
scattering by the strongly polar HF, CH_3Cl, and $C_6H_5NO_2$ molecules
at sub-milli-electron-volt energies (down to a few microelectronvolts)
were recently performed in [119, 120, 121] using the experimental data
with the Rydberg atoms at very high principal quantum numbers (up
to 1100). It was suggested that at ultra-low energy region the electron
scattering by these molecules are strongly influenced by the presence
of dipole-supported virtual states.

In Fig. 9 (panel a) we present the results [121] for the dependence
of the rate constant $v\sigma_j(v)$ of rotational deexcitation of $HF(j)$ with
different values of $j = 0, ..., 7$ on the incident electron velocity $v = \hbar k/m$. Analysis of rotationally inelastic $j \to j'$ transitions was based
on the close-coupling calculations with the cut-off dipole, quadrupole

and polarization interaction. The presence of virtual state $(j = 0)$ with the energy 1.36 meV is assumed.

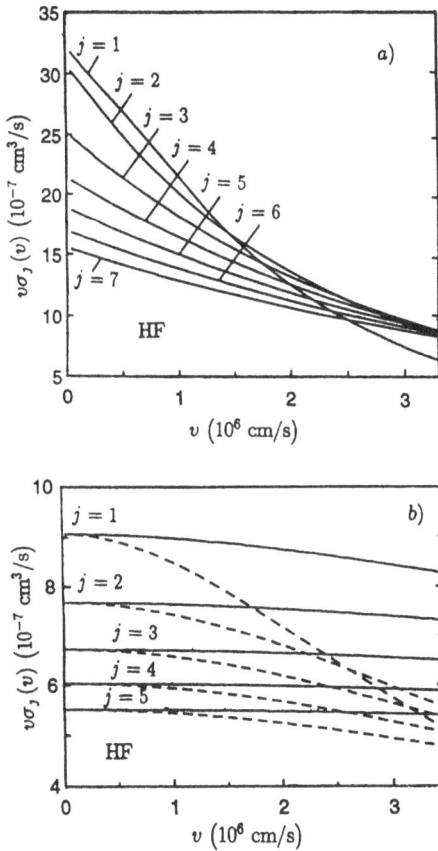

Figure 9. The velocity dependence of the rate constants $v\sigma_j(v)$ for rotational deexcitation of $HF(j)$ by electron impact. Full curves on panel a are the close-coupling calculations [121]. Full curves on panel b are calculations [121] in the Born approximation.

It was shown that for the incident partial s-electron wave the dipole-allowed transitions $j \to j - 1$ are predominant in the deexcitation of molecule (the contribution of the $j \to j - 2$ transitions is about 4 %). The deexcitation rate constants increase with decreasing velocity, whereas at low v they turn out to be larger for rotational states with small j than those for high-j values. At higher v the reverse is true. As is evident from Fig. 9 (panel b) similar behavior follows from the Born approximation. However, the close coupling results are substantially higher than the Born rate constants and have a more pronounced electron velocity dependence, which is the result of the presence of a virtual state with small energy.

It should be noted that a series of interesting works in recent years was devoted to detailed experimental studies of electron scattering from polar and non-polar molecules at very low electron energies. In particular, several authors have carried out the absolute cross section measurements for ultra-slow electron scattering from CO [122]; N_2, P_2, As_2, Sb_2 [124]; CH_4 [123]; and O_3 [125, 126].

3   Impact-Parameter Approach

Due to the peculiar properties of highly excited states the theoretical description of collisional processes involving Rydberg atoms differs significantly from the case of the atoms in the low excited or ground states. Here we present an efficient time-dependent approach to collisions of a Rydberg atom with a neutral atomic projectile based on the impact-parameter method for the heavy particle relative motion and semiclassical approximation for the Rydberg atom wave functions. The goal of this section is to formulate a general semiclassical technique which will be the basis for describing collisional processes of Rydberg atoms with neutral particles in the following sections. We give a brief account of the time-dependent close coupling method, first order perturbation theory and its unitarized version, and binary-encounter theory in the coordinate representation. We consider here the projectile as a structureless neutral particle and use the Fermi pseudopotential model of zero-range for the interaction potential between the highly excited electron and perturbing atom. This makes the main ideas more transparent. Simple analytic formulae for probabilities, cross sections, and rate coefficients presented in this section allow us to demonstrate their dependencies on the main physical parameters such as principal quantum number, relative velocity of colliding particles and transition energy defect.

3.1 General Overview

The general quantum problem of collision between a highly excited atom and projectile is significantly simplified if their relative motion can be considered as classical motion along a given trajectory. Such a classical description would be valid provided the de Broglie wavelength $1/q$ of the projectile is much smaller than the characteristic radius R_{int} of the interaction potential $(qR_{\text{int}} \gg 1)$ and the energy $\Delta \mathcal{E} = \hbar^2 \left((q')^2 - q^2\right)/2\mu$ transferred to the translational motion of colliding particles is small as compared to the kinetic energy $\mathcal{E} = \hbar^2 q^2/2\mu$ of the projectile $(\Delta \mathcal{E} \ll \mathcal{E})$. The first condition shows that the action function $S \sim \hbar q R_{\text{int}}$ is much greater than $\hbar$ and hence clearly demonstrates the classical character of motion in accord with the basic correspondence principle [37]. The second one means that the classical description of projectile motion along a given trajectory is certainly inapplicable for large values of transition energy defects and, in particular, near the threshold of the process $\mathcal{E} \sim \Delta \mathcal{E}$, where the quantum effects in scattering becomes important. For practical estimates it is convenient to rewrite the condition $\Delta \mathcal{E} \ll \mathcal{E}$ in terms of the projectile-Rydberg atom relative velocity V, orbital electron velocity $v_n \sim v_0/n$ and its binding energy $|E_n| = Z^2 Ry/n^2$

$$V \gg v_n \left(\frac{m}{\mu} \frac{\Delta \mathcal{E}}{|E_n|}\right)^{1/2}. \tag{81}$$

It is important to stress that the range of applicability of the first condition $qR_{\text{int}} \gg 1$ depends on the principal quantum number, relative velocity of colliding particles, transition energy defect etc. For a collision with a neutral projectile (atom or molecule) its motion relative to the ion core of the Rydberg atom may be considered as classical if $(\mu V/\hbar) R_{\text{int}} \gg 1$, where μ is the reduced mass of atom and projectile. Since the characteristic radius R_{int} of the core–projectile interaction is usually about or greater than a few atomic radii a_0 it means that a classical description of projectile motion would certainly be valid in the range of about or greater than thermal energies. The quantum character of projectile motion is realized only at small relative velocities (subthermal energies), particularly so, for collisions of atoms with a small reduced mass. A more detailed discussion concerning the applicability of the classical and quasiclassical descriptions of atom–atom and atom–molecule collisions is given in [37].

The semiclassical impact-parameter method combined with the Fermi pseudopotential model was originally developed by *Gersten* [127], *Omont* [128], *Derouard* and *Lombardi* [129], and *de Prunelé* and *Pascale* [130] in the studies of quasielastic l-mixing and impact broadening of highly-excited atomic states by neutral atoms (see [81] for more details). Later *Kaulakys* [131] applied this approach to derive simple expressions for the broadening and shift of isolated Rydberg ns-states induced by elastic collisions with the rare gas atoms. The results obtained in [131] generalize the asymptotic (high n) expressions for the elastic scattering contribution to the shift [80] and broadening [128] of Rydberg ns-states in the range of intermediate and sufficiently low values of n. Further development of the Fermi pseudopotential model was carried out by *Lebedev* and *Marchenko* [132], who presented a semiclassical theory for the inelastic $nl \rightarrow n'$ and $n \rightarrow n'$ transitions with changing orbital and principal quantum numbers within the framework of perturbation theory. Analytic expressions for the cross sections and rate constants obtained in this paper were applied to get a quantitative description of experimental data on the inelastic quenching of Rydberg atomic states by rare gas atoms and to demonstrate the sharp dependence of the n, l-changing quenching processes on the value of the quantum defect δ_l. Later the results [132] for the inelastic $nl \rightarrow n'$ transitions were used [105] in order to evaluate the contribution of the electron-perturber potential scattering to the broadening of high-Rydberg nl-levels of alkali-metal atoms perturbed by the ground-state parent atoms Li, Na, K, and Rb.

Using simple semiclassical expressions for the inelastic [132] and elastic [131] Rydberg-atom–neutral collisions, *Sun* and *West* [133] calculated the impact broadening cross sections of the Rb(ns) by rare gas atoms. In another paper, *Sun* et al. [134] applied the multichannel-quantum defect method in order to generalize the above mentioned semiclassical results on the Rydberg atom having more than one valence electron, and to explain the irregular broadening rates of the Sr($5sns\,^1S_0$) series induced by collisions with Xe atoms. Simultaneously, *Sirko* and *Rosinski* [135] carried out numerical calculations of the J-mixing ($n^2D_{3/2} \rightarrow n^2D_{5/2}$) cross sections for the Rydberg states of Cs* by rare gas atoms on the basis of the semiclassical Fermi pseudopotential model in the weak coupling approximation ($n > 10 - 15$). An analytic semiclassical description of the $nlJ \to nlJ'$ transitions between the fine-structure components of the Rydberg atom and of pure elastic $nlJ \to nlJ$ scattering has been performed by *Lebedev* [136, 137] who applied the normalized version of perturbation theory and the JWKB-approximation. The results obtained are valid for weak and close coupling, i.e. at high, intermediate and sufficiently low n. Recently, an efficient unitarized semiclassical approach was developed [138, 139, 140], for the simultaneous description of both the quasielastic l-mixing and the inelastic n, l-changing processes. It provides a successful quantitative explanation of major phenomena in these processes in the dependence on the energy ΔE transferred to the Rydberg atom, the relative velocity V, and the parameter of electron–perturber interaction in a wide range of the principal quantum number n.

3.2   Time-Dependent Close Coupling Method

Given the classical character of projectile–Rydberg-atom relative motion along a given trajectory $\mathbf{R}(t)$, one of the most efficient theoretical methods for calculating transition probabilities and cross sections is based on the time-dependent impact-parameter approach. According to this approach the projectile B affects the Rydberg atom A by means of the perturbation interaction $\mathsf{V}(\mathbf{r}, \mathbf{R}(t))$, which contains an implicit dependence on the time t through its position vector $\mathbf{R}(t)$ relative to the center of mass of the ion core A^+ and highly excited electron e. (For most problems the radius vector of projectile is taken with respect to the nuclei of A^+). This allows us to remove the kinetic energy operator from the basic time-dependent Schrödinger equation for the total quantum wave function of the system. As a result, within the framework of the impact-parameter time-dependent approach the wave function $\psi(\mathbf{r}, \mathbf{R}(t))$ of colliding particles is determined from the following equation

$$i\hbar \frac{\partial}{\partial t} \psi(\mathbf{r}, \mathbf{R}(t)) = [H_A + \mathsf{V}(\mathbf{r}, \mathbf{R}(t))]\, \psi(\mathbf{r}, \mathbf{R}(t)). \qquad (82)$$

Since the main contribution to the cross sections of Rydberg atom-neutral (charged) particle collisions is determined by large separations R between the ionic core A^+ and projectile B (i.e. by large values of the impact parameter ρ), their relative motion is usually taken to be rectilinear, so that

$$\mathbf{R}(t) = \boldsymbol{\rho} + \mathbf{V}t \ . \tag{83}$$

The solution of the Schrödinger equation (82) can be obtained by Dirac's method of variation of constants according to which the time-dependent wave function $\psi\,[\mathbf{r}, \mathbf{R}(t)]$ of colliding particles is expanded in the eigenfunctions $\psi_\alpha\,(\mathbf{r})$ of the unperturbed Rydberg atom Hamiltonian H_A

$$\psi\,[\mathbf{r}, \mathbf{R}\,(t)] = \sum_\alpha b_\alpha\,(t)\,\psi_\alpha\,(\mathbf{r}) \tag{84}$$

Substitution of this expression into the time dependent Schrödinger equation (82) leads to a system of close coupled differential equations for transition amplitudes $b_\alpha\,(t)$

$$i\hbar\frac{db_\alpha}{dt} = E_\alpha b_\alpha + \sum_{\alpha'} \mathsf{V}_{\alpha\alpha'}\,[\mathbf{R}(t)]\,b_{\alpha'} \ . \tag{85}$$

Here E_α is the eigenenergy of the Rydberg atom Hamiltonian H_A, and

$$\mathsf{V}_{\alpha'\alpha}\,[\mathbf{R}(t)] = \int \psi^*_{\alpha'}\,(\mathbf{r})\,\mathsf{V}\,(\mathbf{r}, \mathbf{R}(t))\,\psi_\alpha\,(\mathbf{r})\,d\mathbf{r} \tag{86}$$

is the matrix element of the interaction potential, which depends on time only through the position vector of the projectile $\mathbf{R}(t)$. The standard replacement of variables

$$b_\alpha\,(t) = a_\alpha(t)\exp(-iE_\alpha t/\hbar), \tag{87}$$

transforms this system to the form

$$i\hbar\frac{da_\alpha}{dt} = \sum_{\alpha'} \mathsf{V}_{\alpha\alpha'}\,[\mathbf{R}(t)]\,\exp(i\omega_{\alpha'\alpha}t)\,a_{\alpha'} \ , \tag{88}$$

$$\omega_{\alpha'\alpha} = (E_{\alpha'} - E_\alpha)/\hbar \ , \tag{89}$$

which is called the system of the close coupled equations in the interaction representation.

A solution of (88) satisfying the initial conditions

$$\lim_{t\to-\infty} a_\alpha^{(i)}(\rho, V; t) = \delta_{i\alpha} \ ,$$

directly yields the result for the probability W_{fi} of the $i \to f$ transition and for the corresponding cross section after integration over the impact parameters

$$W_{fi}(\rho, V) \;=\; \lim_{t \to \infty} \left| a_f^{(i)}(\rho, V; t) \right|^2, \tag{90}$$

$$\sigma_{fi}(V) \;=\; 2\pi \int_0^\infty W_{fi}(\rho, V)\, \rho\, d\rho. \tag{91}$$

This formula can be rewritten in terms of the S-matrix of scattering in the impact-parameter representation. The solution set of differential equations (88) forms its matrix elements

$$\mathsf{S}_{fi}(\rho, V) \;=\; a_f^{(i)}(\rho, V; t = \infty), \tag{92}$$

$$\sigma_{fi}(V) \;=\; 2\pi \int_0^\infty |\mathsf{S}_{fi}(\rho, V)|^2\, \rho\, d\rho. \tag{93}$$

The S-matrix is unitary $\mathsf{SS}^\dagger = \mathsf{I}$ (where I is the unitary matrix) so that its elements satisfies the relation $\sum_f |\mathsf{S}_{fi}(\rho, V)|^2 = 1$. Hence the total cross section σ_i^{tot} of elastic σ_i^{el} and inelastic σ_i^{in} scattering

$$\sigma_i^{\text{tot}} = \sigma_i^{\text{el}} + \sigma_i^{\text{in}}, \qquad \sigma_i^{\text{in}} = \sum_{f \neq i} \sigma_{fi} \tag{94}$$

can be written as

$$\sigma_i^{\text{tot}}(V) \;=\; 4\pi \int_0^\infty \left(1 - \mathrm{Re}\left\{\mathsf{S}_{ii}(\rho, V)\right\}\right) \rho\, d\rho, \tag{95}$$

$$\sigma_i^{\text{el}}(V) \;=\; 2\pi \int_0^\infty |1 - \mathsf{S}_{ii}(\rho, V)|^2\, \rho\, d\rho, \tag{96}$$

$$\sigma_i^{\text{in}}(V) \;=\; 2\pi \int_0^\infty \left(1 - |\mathsf{S}_{ii}(\rho, V)|^2\right) \rho\, d\rho. \tag{97}$$

Note also that the S-matrix of scattering is expressed through the T-matrix, introduced previously in Sect. 2.4, by the following relation $\mathsf{S} = \mathsf{I} + \mathsf{T}$.

It is important to stress that the system (88) (or (85)) consists of an infinite number of differential equations. Usually, they are cut off and are considered in some finite dimensional space of states. The choice of the Rydberg levels with different magnitudes of the principal quantum number, involved in the description of collision dynamics, depends on the specific problem, accuracy, on the range of n and transition frequencies. However, an important feature of any process involving Rydberg atom is a great number of closely spaced quasidegenerate sublevels.

3.3   Unitarized Perturbation Theory

In the range of weak coupling, when the transition probability is sufficiently small $W_{fi} \ll 1$, it can be calculated using the basic formula of first-order perturbation theory. This formula directly follows from the system of the close coupling equations (88) by substitution of the relation $a_{\alpha}^{(i)}(t = -\infty) = \delta_{i\alpha}$ into its right-hand side

$$
W_{fi}(\rho, V) = \left| a_f^{(i)}(\rho, V; t = +\infty) \right|^2 ,
$$

$$
a_f^{(i)}(\rho, V; t = \infty) = -\tfrac{i}{\hbar} \int\limits_{-\infty}^{\infty} \mathsf{V}_{fi}\left[\mathbf{R}(t) \right] \exp\left(i\omega_{fi}t \right) dt,
$$

(98)

where $a_f^{(i)}(\rho, V; t = \infty)$ is the transition amplitude in the impact-parameter representation. These results are also called Born's expressions for the transition amplitude and probability in the impact-parameter representation.

If the interaction potential is not small, the transition probability calculated according to (98) may become greater than unity and, hence, the standard version of perturbation theory breaks down. In order to restore the unitarity property for the transition probability and to reasonably describe the behavior of the cross section in the range of close coupling states *Seaton* [141] suggested a normalized version of perturbation theory. Here we present only the simplest way for normalization of the transition probability in the range of close coupling which turns out to be effective in practical calculations of processes involving Rydberg atoms. It consists of separation of the whole range of impact parameters into two regions ($0 \leq \rho \leq \rho_0$) and ($\rho_0 < \rho$) with qualitatively different behavior of the transition probability

$$W_{fi}(\rho, V) = \begin{cases} W_{fi}^{\mathrm{B}}(\rho, V), & \rho \geq \rho_0, \\ c, & \rho < \rho_0. \end{cases} \tag{99}$$

At large impact parameters $\rho > \rho_0$ the coupling between Rydberg states is weak, and the probability can be calculated using formula (98) of the Born approximation, i.e. $W_{fi} = W_{fi}^{\mathrm{B}}$. However, at small $\rho < \rho_0$, due to strong coupling between Rydberg states, first-order perturbation theory leads to overestimated values of the transition probability and it is normalized to a constant c of the order of unity. The magnitude of the impact parameter ρ_0, which separates the region of weak coupling from that of close coupling is to be found from the relation $W_{fi}^{\mathrm{B}}(\rho_0) = c$. As a result, the cross section of the $i \rightarrow f$ transition is given by

$$\sigma_{fi}(V) = c\pi\rho_0^2 + 2\pi \int\limits_{\rho_0}^{\infty} W_{fi}^{\mathrm{B}}(\rho, V)\, \rho\, d\rho. \tag{100}$$

The choice of normalizing constant c contains some ambiguities and is determined by the specific problem.

3.4 Semiclassical Description of Collision Processes

3.4.1 Zero-Range Pseudopotential Model

Below we present the basic expressions of first-order perturbation theory (98) for the transition probabilities between the $nlJ \rightarrow n'l'J'$ and $nl \rightarrow n'l'$ states of the Rydberg atom with one valence electron over and above the closed shell of the parent core. Within the framework of the Fermi pseudopotential model for the short-range part of the electron–atom interaction

$$\mathsf{V}(\mathbf{r} - \mathbf{R}) = \frac{2\pi\hbar^2 L}{m} \delta(\mathbf{r} - \mathbf{R}), \tag{101}$$

the transition matrix elements

$$\mathsf{V}_{fi}[\mathbf{R}(t)] = \int \psi_f^*(\mathbf{r})\, \mathsf{V}[\mathbf{r} - \mathbf{R}(\mathbf{t})]\, \psi_i(\mathbf{r})\, d\mathbf{r} \tag{102}$$

over the wave functions of Rydberg atom A^* are given by

$$\mathsf{V}_{fi}[\mathbf{R}(t)] = \frac{2\pi\hbar^2 L}{m} \psi_f^*[\mathbf{R}(t)]\, \psi_i[\mathbf{R}(t)]. \tag{103}$$

The use of this expression and the impact-parameter approach allows us to derive simple analytic formulae for the cross sections of elastic, quasielastic and inelastic Rydberg-atom–neutral collisions. For slightly polarizable perturbing atoms (for example, He) these formulae provide an appropriate qualitative description of such collisions and yield reliable quantitative results.

In the case of transitions between $nlJ \rightarrow n'l'J'$ states with the given magnitudes of the principal and orbital quantum numbers and total angular momenta the matrix elements (103) should be calculated over the wave functions in the $nlsJM$-representation, i.e.

$$|nlJM\rangle = \mathcal{R}_{n_*l}(r)\, Y_{JM}^{ls}(\theta_{\mathbf{r}}, \varphi_{\mathbf{r}}) , \tag{104}$$

$$Y_{JM}^{ls}(\mathbf{n}_r) = \sum_{m\sigma} C_{lms\sigma}^{JM} Y_{lm}(\theta_{\mathbf{r}}, \varphi_{\mathbf{r}}) \chi_{s\sigma} . \tag{105}$$

Here $\mathcal{R}_{n_*l}(r)$ is the radial wave function of the $|nlJM\rangle$-state with the effective quantum number $n_* = n - \delta_{lJ}$, δ_{lJ} being its quantum defect; $Y_{JM}^{ls}(\theta_{\mathbf{r}}, \varphi_{\mathbf{r}})$ is the spherical spinor with the total angular momentum of J $(J = |l - 1/2|$ or $l + 1/2)$ and its z-projection of M; $\chi_{s\sigma}$ is the spin function of the electron $(s = 1/2)$ with the z-projection of $\sigma \equiv s_z = \pm 1/2$; and $C_{lms\sigma}^{JM}$ is the Clebsch–Gordan coefficient. The radial part of the coordinate wave function is normalized by the relation

$$\int\limits_0^\infty \mathcal{R}_{n'l}\mathcal{R}_{nl}r^2 dr = \delta_{nn'} \tag{106}$$

Further, we use an expansion of the zero-range pseudopotential (101) over the spherical harmonics and the technique of non-reduced tensor operators [142] for calculation of the matrix elements $\langle n'l'J'M'|\mathsf{V}|nlJM\rangle$. Then, the use of the basic formula (98) of the first-order perturbation theory for the probability

$$W_{nlJ}^{n'l'J'} = \frac{1}{2J+1} \sum_{MM'} W_{nlJM}^{n'l'J'M'}$$

of the $nlJ \rightarrow n'l'J'$ transition and the summation of the matrix element squares over all possible M and M' magnitudes, leads to the following semiclassical result [36]

$$W_{nlJ}^{n'l'J'}(\rho, V) = \frac{\hbar^2 L^2}{4m^2} \sum_{\text{œ}=|l'-l|}^{l'+l} A_{l'J',lJ}^{(\text{œ})} \int_{-\infty}^{\infty} dt \int_{-\infty}^{\infty} dt'$$

$$\times \mathcal{R}_{n'_*l'}\left[R(t)\right] \mathcal{R}_{n_*l}\left[R(t)\right] \mathcal{R}_{n'_*l'}\left[R(t')\right] \mathcal{R}_{n_*l}\left[R(t')\right] \tag{107}$$

$$\times \exp\left[-i\omega_{fi}(t - t')\right] P_{\text{œ}}\left(\cos\Theta_{\mathbf{R'R}}\right),$$

where $\omega_{f_i} = |E_{n'l'J'} - E_{nlJ}|/\hbar$, and the angular $A_{l'J',lJ}^{(\text{œ})}$ coefficients are expressed in terms of $3j$- and $6j$-symbols

$$A_{l'J',lJ}^{(\text{œ})} = (2l + 1)(2l' + 1)(2J' + 1)(2\text{œ} + 1)$$

$$\times \left\{ \begin{array}{ccc} l' & J' & 1/2 \\ J & l & \text{œ} \end{array} \right\}^2 \left(\begin{array}{ccc} l' & \text{œ} & l \\ 0 & 0 & 0 \end{array} \right)^2. \tag{108}$$

Upon use of a straight-line trajectory, the relation between the internuclear $R \equiv R(t)$ (or $R' \equiv R(t')$) distances of heavy A$^+$ and B particles and time moment t (or t') are reduced to

$$R(t) = \sqrt{\rho^2 + V^2 t^2}, \qquad dt = \frac{R dR}{V\sqrt{R^2 - \rho^2}}. \tag{109}$$

It is convenient to choose the coordinate system so that the ion core A$^+$ is placed at its origin O, the Z-axis is perpendicular to the collision plane of A$^+$ and B particles and the X and Y axes are directed in parallel to the impact parameter vector $\boldsymbol{\rho}$ and relative velocity $\mathbf{V}$, respectively. Then, the Rydberg electron $\theta_{\mathbf{r}}$ and $\varphi_{\mathbf{r}}$ angles at the points $\mathbf{r} = \mathbf{R}$ and $\mathbf{r}' = \mathbf{R}'$ and the included angle $\Theta_{\mathbf{R'R}} = \varphi_{\mathbf{R'}} - \varphi_{\mathbf{R}}$ of the $\mathbf{R}$ and $\mathbf{R}'$ vectors in (107) are given by

$$\theta_{\mathbf{R}} = \frac{\pi}{2}, \quad \varphi_{\mathbf{R}} = \arctan\left(\frac{Vt}{\rho}\right), \quad \cos\Theta_{\mathbf{R'R}} = \frac{\rho^2 + V^2 tt'}{RR'}. \tag{110}$$

A similar semiclassical expression for the probability

$$W_{n'l',nl} = \frac{1}{2l+1} \sum_{mm'} W_{nlm}^{n'l'm'}$$

of the $nl \to n'l'$- transition directly follows from (107) if we neglect the energy $|\Delta E_{J'J}| = 2Ry\,|\delta_{J'} - \delta_J|/n_*^3$ of the fine-structure splitting as compared to the energy defect $|\Delta E_{n'l',nl}| = 2Ry \, |n'_* - n_*| / n_*^3$ of the initial nl and final $n'l'$-levels ($|\Delta \delta_{J'J}| \ll 1$). The final result is given by [36, 134]

$$
W_{n'l',nl}(\rho, V) = \frac{(2l'+1)\hbar^2 L^2}{4m^2} \int\limits_{-\infty}^{\infty} dt \int\limits_{-\infty}^{\infty} dt'
$$

$$
\times \exp\left[-i\omega_{fi}(t-t')\right] P_{l'}(\cos\Theta_{\mathbf{R'R}}) \, P_l(\cos\Theta_{\mathbf{R'R}}) \tag{111}
$$

$$
\times \mathcal{R}_{n'_*l'}[R(t)] \, \mathcal{R}_{n_*l}[R(t)] \, \mathcal{R}_{n'_*l'}[R(t')] \, \mathcal{R}_{n_*l}[R(t')] \,,
$$

where $n_* = n - \delta_l$, and $n'_* = n' - \delta_{l'}$.

3.4.2 Radial Wave Function: JWKB-Approximation

Theoretical calculations of the probabilities and cross sections for the various types of the bound–bound transitions between the highly excited states are significantly simplified if we use the quasiclassical description of the Rydberg atom. In the JWKB-approximation the radial wave function is given by

$$
\mathcal{R}_{nl}(r) = \left(\frac{2}{\pi n^3 a_0^2}\right)^{1/2} \frac{\cos\Phi_r}{r k_r^{1/2}}, \quad \Phi_r = \int\limits_{r_1}^{r} k_r \, dr - \pi/4, \tag{112}
$$

$$
k_r^2 = \frac{2m}{\hbar^2}\left[E + \frac{e^2}{r} - \frac{\hbar^2}{2m}\frac{(l+1/2)^2}{r^2}\right], \tag{113}
$$

where the phase Φ_r of the quasiclassical wave function can be expressed in analytical form for the Coulomb field

$$
\Phi_r = k_r r - n \arcsin\left[\left(1 - r/n^2 a_0\right)/\epsilon\right]
$$

$$
-(l+1/2)\arcsin\left[\left(1 - (l+1/2)^2 a_0/r\right)/\epsilon\right] \tag{114}
$$

$$
+\left[n - (l+1/2)\right]\frac{\pi}{2} - \frac{\pi}{4} \,,
$$

and the left and right turning points are given by

$$
r_{1,2} = n^2 a_0 (1 \pm \epsilon), \qquad \epsilon = \left[1 - (l+1/2)^2/n^2\right]^{1/2}. \tag{115}
$$

In (114) the quantum factor $l(l+1)$ in the centrifugal potential is replaced by $(l+1/2)^2$, which corresponds to the Langer correction in the JWKB-approximation (see [37, 143] for more details).

In the case of $\Delta n \ll n$ and $\Delta l \ll l, l' \ll n$ the first term of expansion of the phase difference $\Delta \Phi_r = \Phi_r(n + \Delta n, l + \Delta l) - \Phi_r(n, l)$ over $\Delta n = n' - n$ and $\Delta l = l' - l$ takes the form

$$\Delta \Phi_r = \frac{E_{n'} - E_n}{2Ry} \int\limits_{r_1}^{r} \frac{dr}{a_0^2 k_r} - (l+1/2)\,\Delta l \int\limits_{r_1}^{r} \frac{dr}{r^2 k_r}$$

$$= -\Delta n \left\{ \arcsin\left[\left(1 - r/n^2 a_0\right)/\epsilon\right] + k_r r - \pi/2 \right\} \tag{116}$$

$$-\Delta l \left\{ \arcsin\left[\left(1 - (l+1/2)^2 a_0/r\right)/\epsilon\right] + \pi/2 \right\}.$$

This expression is particularly important in calculations of the transition matrix elements between closely spaced Rydberg states.

3.4.3   Transitions Between Hydrogen-like Sublevels

The basic semiclassical formula (111) becomes particularly simple in the case of pure elastic scattering ($l' = l$, $n' = n$) of the Rydberg atom by a neutral particle and for quasielastic $nl \to nl'$ transitions wich change the orbital angular momentum alone $l' \neq l$, $n' = n$

$$A^*(nl) + B \to A^*(nl') + B. \tag{117}$$

In the simplest case of elastic scattering of a perturbing atom by the Rydberg atom in the ns-state, when $\omega = 0$ and $l' = l = 0$, it can be rewritten as

$$W_{ns}^{\text{el}}(\rho, V) = \frac{\hbar^2 L^2}{4m^2} \left| \int\limits_{-\infty}^{\infty} \mathcal{R}_{ns}^2 \left[R(t)\right] dt \right|^2$$

$$= \frac{\hbar^2 L^2}{m^2 V^2} \left| \int\limits_{\rho}^{\infty} \mathcal{R}_{ns}^2(R) \frac{R\,dR}{(R^2 - \rho^2)^{1/2}} \right|^2. \tag{118}$$

Then, using the JWKB-approximation for the radial wave function $R_{ns}(r)$ of the Rydberg atom (112) and calculating the integral over the classically allowed range of r $(r = R, 0 \leq R \leq 2n_*^2 a_0)$, on the assumption that $\langle \cos^2 \Phi_R \rangle = 1/2$, we obtain

$$W_{ns}^{el}(\rho, V) = \frac{L^2}{\pi^2 n_*^6 \rho a_0} \left(\frac{v_0}{V}\right)^2 K^2[k(\rho)] , \qquad (119)$$

$$K(k) = \int_0^{\pi/2} \left(1 - k^2 \sin^2 \theta\right)^{-1/2} d\theta, \qquad (120)$$

$$k(\rho) = 2^{-1/2} \left(1 - \frac{\rho}{2n_*^2 a_0}\right)^{1/2} . \qquad (121)$$

Here $K(k)$ is the complete elliptic integral of the first kind [144], the value of which is slowly varying $(\pi/2 \leq K[k(\rho)] \leq 1.8)$ in the range of $0 \leq \rho \leq 2n_*^2 a_0$. Due to the exponential decrease of the wave function $\mathcal{R}_{nl}(r)$ in the classically forbidden range, the probability falls strongly in the range of $\rho > 2n_*^2 a_0$. Therefore, its contribution to the cross section $\sigma_{ns}^{el}(V)$ can be neglected. Expression (119) is valid in the range of weak coupling, in which the probability obtained in first-order perturbation theory is small $W_{ns}^{el}(\rho, V) \ll 1$. Hence, it becomes inapplicable at small values of $\rho \leq \rho_0$ $(\approx a_0 \left(v_0 L/2a_0 V n_*^3\right)^2)$, where ρ_0 is defined by the condition $W_{ns}^{el}(\rho_0, V) \approx 1$. At large values of $n_* \gg (v_0 |L|/4Va_0)^{1/4}$ the range of impact-parameters $0 \leq \rho \leq \rho_0$ corresponding to the close coupling is small $(\rho_0 \ll 2n_*^2 a_0)$ and, hence, its contribution can also be neglected.

In the general case, the contribution of the perturber–quasifree-electron scattering mechanism to the cross section

$$\sigma_{nl',nl} = \frac{1}{2l+1} \sum_{mm'} \sigma_{nlm}^{nl'm'} \qquad (122)$$

of quasielastic $(\omega_{nl',nl} = 0)$ $nl \rightarrow nl'$ transition may be described at $l, l' \ll n$ by simple analytic formula [145, 129]

$$\sigma_{nl',nl} = \frac{2\pi C_{l'l} L^2 v_0^2}{V^2 n_*^4} , \qquad n_* \gg \left(\frac{v_0 |L|}{V a_0}\right)^{1/4} , \qquad (123)$$

which is applicable in the range of weak coupling. The magnitudes of the $C_{l'l}$ coefficients in the case of pure elastic scattering $(nl \rightarrow nl$ transition) are as follows $C_{ss} = 0.58$; $C_{pp} = 1.03$; $C_{dd} = 1.28$; and $C_{ff} = 1.46$.

3.5   Binary Encounter Theory: Fermi Interaction

3.5.1   Transitions from Selectively Excited Level

We shall now present the main results of semiclassical theory [132] for the probabilities $W_{n',nl} = \sum_{l'} W_{n'l',nl}$, cross sections $\sigma_{n',nl} = \sum_{l'} \sigma_{n'l',nl}$ and rate constants $K_{n',nl} = \sum_{l'} K_{n'l',nl}$ of the inelastic $nl \rightarrow n'$ transitions from a given nl-state to all degenerate $n'l'$-sublevels of another hydrogen-like n'-level

$$A^*(nl) + B \rightarrow A^*(n') + B.$$

These quantities are needed, for one, to evaluate quenching and broadening cross sections in inelastic collisions between an atomic projectile B and a selectively excited Rydberg atom $A^*(nl)$. An analytic semiclassical description of such $nl \rightarrow n'$ transitions can be made using the JWKB-approximation for the initial and final radial wave functions of Rydberg atom. Now we substitute the expression (112) into (111) and use only the first term $\Phi_{R'} - \Phi_R = (d\Phi_R/dR)(R' - R)$ of the series expansion for the semiclassical phase difference

$$\Phi_{R'} - \Phi_R = k_R(R' - R), \quad k_R = \sqrt{\mathfrak{E}_R^2 - \frac{(l+1/2)^2}{R^2}}, \quad (124)$$

$$\mathfrak{E}_R = \sqrt{\frac{2}{Ra_0} - \frac{1}{n_*^2 a_0^2}} \;. \quad (125)$$

Then, neglecting the rapidly-oscillating terms of $\cos(\Phi_{R'} + \Phi_R)$ and using (124), the probability of the $nl \rightarrow n'$ transition can be written as

$$W_{n',nl} = \frac{(v_0/a_0)^2 L^2}{2\pi^2 n_*^3 (n')^3} \int\limits_{-\infty}^{\infty} dt \int\limits_{-\infty}^{\infty} dt'$$

$$\times \exp\left[i\omega_{n',nl}(t - t')\right] P_l\left(\cos\Theta_{\mathbf{R'R}}\right) \frac{\cos\left[k_R(R' - R)\right]}{R^2 k_R} \quad (126)$$

$$\times \sum_{l'} (l' + 1/2) P_{l'}\left(\cos\Theta_{\mathbf{R'R}}\right) \frac{\cos\left[k_R'(R' - R)\right]}{R^2 k_R'} \;.$$

The major contribution to the sum $W_{n',nl} = \sum_{l'} W_{n'l',nl}$ over all orbital momenta in the final state n' is determined by large values

$l' \gg 1$. Thus, we may use an approximate expression for the Legendre polynomial

$$P_{l'}\left(\cos\Theta_{\mathbf{R'R}}\right) \approx J_0\left[(l' + 1/2)\,\Theta_{\mathbf{R'R}}\right] \quad (\Theta_{\mathbf{R'R}} \ll 1)\,. \tag{127}$$

This is equivalent to the application of the JWKB-approximation for the angular wave function $Y_{l'm'}$ in the final $n'l'm'$-state. Moreover, summation over all possible l' values may be replaced in (126) by integration over dl'. As a result, we have

$$\sum_{l'} (l' + 1/2)\,P_{l'}\left(\cos\Theta_{\mathbf{R'R}}\right) \frac{\cos[k'_R(R'-R)]}{R^2 k_R}$$

$$= \frac{\sin\left(\infty'_R\left[(R'-R)^2 + R^2\Theta_{\mathbf{R'R}}^2\right]^{1/2}\right)}{\left[(R'-R)^2 + R^2\Theta_{\mathbf{R'R}}^2\right]^{1/2}}\,. \tag{128}$$

The use of a series expansion for the $R' - R$ and $\Theta_{\mathbf{R'R}} = \varphi_{\mathbf{R'}} - \varphi_{\mathbf{R}}$ values (see (110)) in the first order of accuracy, yields

$$R' - R = V\tau\sin\varphi_{\mathbf{R}}, \qquad \sin\varphi_{\mathbf{R}} = Vt/R, \tag{129}$$

$$R\Theta_{\mathbf{R'R}} = V\tau\cos\varphi_{\mathbf{R}}, \qquad \cos\varphi_{\mathbf{R}} = \rho/R, \tag{130}$$

where $\tau = t' - t$ is the time interval.

Henceforth, we shall consider small orbital momenta $l \ll n$ of the initial Rydberg nl-states, because they are of greatest importance for experimental applications (i.e. ns, np, nd and nf-states). In this case the main contribution to the transition probability $W_{n',nl}$ is due to small enough time intervals $\tau \underset{\sim}{<} na_0/V$. Therefore, the characteristic values of the polar angles difference $\Theta_{\mathbf{R'R}} = V\tau\rho/R^2 \underset{\sim}{<} 1/n \ll 1$, and we may assume $P_l\left(\cos\Theta_{\mathbf{R'R}}\right) = 1$ in (127), if the initial orbital momentum $l \ll n$. Thus, with the aid of (110)–(130), the inelastic transition probability $W_{n',nl}$ in the first-order perturbation theory can be described by the expression

$$W_{n',nl}(\rho, V) = \frac{2(v_0/a_0)^2 L^2}{\pi^2 n_*^3 (n')^3 V^2} \int\limits_{R_{\min}}^{R_{\max}} \frac{dR}{Rk_R(R^2-\rho^2)^{1/2}}$$

$$\times \int\limits_0^\infty \frac{d\tau}{\tau}\,\sin(\infty'_R V\tau)\,\cos(\omega_{n',nl}\tau)\,\cos\left(\frac{V\tau k_R(R^2-\rho^2)^{1/2}}{R}\right). \tag{131}$$

Here $R_{\max} = r_2$ and $R_{\min} = \max\{\rho, r_1\}$; while r_1 and r_2 are the classical turning points (see (114), so that $r_2 \approx 2n_*^2 a_0$ and $r_1 \approx (l + 1/2)^2 a_0/2 \ll r_2$ at $l \ll n$ (i.e. we may assume $r_1 \approx 0$). Besides, it is evident that $k_R \approx \mathrm{æ}_R$ at $l \ll n$.

The cross section $\sigma_{n',nl}(V)$ of the $nl \to n'$ transition is determined from (131) by integrating the probability over the classically allowed range of the impact parameter. The result of the calculation for small orbital momentum $l \ll n$ of the initial Rydberg nl-state, is given by the simple analytic formula [132]

$$\sigma_{n',nl}(V) = \frac{2\pi L^2 v_0^2}{V^2 (n')^3} f_{n',nl}(\lambda), \qquad \lambda = \frac{n_* a_0 \omega_{n',nl}}{V}, \qquad (132)$$

$$f_{n',nl}(\lambda) = \frac{2}{\pi}\left[\arctan\left(\frac{2}{\lambda}\right) - \left(\frac{\lambda}{2}\right)\ln\left(1 + \left(\frac{2}{\lambda}\right)^2\right)\right]. \qquad (133)$$

Here $\omega_{n',nl} = |\Delta E_{n',nl}|/\hbar$ and $\Delta E_{n',nl} = E_{n'} - E_{nl}$ are the transition frequency and energy defect between the initial nl-state with quantum defect δ_l and final degenerate $n'l'$-states of hydrogen-like n'-level, respectively (i.e. $E_{nl} = -Ry/(n - \delta_l)^2$ and $E_{n'} = -Ry/n'^2$). The dimensionless parameter λ characterizes the inelasticity of the collisional $nl \to n'$ transition. This expression can be also derived [146] in the impulse approximation with a constant electron–perturber scattering amplitude $f_{eB} = -L$ using the simple formula of the binary-encounter theory [147] for atomic form factor $F_{n',nl}(Q)$ of the $nl \to n'$ transition.

The asymptotic behavior of the $f_{n',nl}(\lambda)$ function at small and large inelasticity parameters, is given by

$$\begin{aligned} f_{n',nl}(\lambda) &\to 1 , & (\lambda \to 0), \\ f_{n',nl}(\lambda) &\approx (8/3\pi)\lambda^{-3}, & (\lambda \gg 1). \end{aligned} \qquad (134)$$

Therefore, for high $n \gg (|\delta_l + \Delta n| v_0/V)^{1/2}$ (when $\lambda \ll 1$) the cross sections for inelastic $nl \to n'$ transitions ($\Delta n = n' - n$) tend to the quasielastic limit $\omega_{n',nl} \to 0$ of weakly coupled states

$$\sigma_{n',nl}(V) = \frac{2\pi L^2 v_0^2}{V^2 (n')^3} \qquad (\lambda \to 0) \qquad (135)$$

because of the reduction in the energy defect $\Delta E_{n',nl} \propto 1/n^3$ with increasing n. Thus, the analytic formula (132) contains *Omont's*

result [128] for the sum $\sigma_{n',nl} = \sum\limits_{l'} \sigma_{n'l',nl}$ of pure quasielastic $nl \rightarrow$
nl' transitions with a change in the orbital angular momentum alone
$(n' = n, \omega_{nl',nl} = 0)$ as a special case. However, in the range of low
$n \ll (|\delta_l + \Delta n| v_0/V)^{1/2}$, it leads to the rapid power fall $(\sigma_{n',nl} \propto n^3)$
in the cross section for an inelastic $nl \rightarrow n'$ transition

$$\sigma_{n',nl}(V) \approx \frac{16L^2 V n^3}{3v_0 |\delta_l + n' - n|^3} \qquad (\lambda \gg 1) \qquad (136)$$

with a decrease in the principal quantum number (i.e. with increas-
ing in the energy defect $\Delta E_{n',nl} = 2Ry |\delta_l + \Delta n|/n^3$). For a given
relative velocity V, the cross section $\sigma_{n',nl}(V)$ reaches its maximum

$$\sigma_{n',nl}^{\max}(V) \approx \frac{0.1\sigma_{\mathrm{eB}}^{\mathrm{el}}}{|\delta_l + n' - n|^{3/2} (V/v_0)^{1/2}} \qquad (\lambda \approx 1.2) \qquad (137)$$

at $n_{\max} \approx 0.9 \left(|\delta_l + \Delta n| v_0/V\right)^{1/2}$. Here $\sigma_{\mathrm{eB}}^{\mathrm{el}} = 4\pi L^2$ is the total
cross section of elastic electron-atom scattering. One can see, that
the particular magnitudes of $n_{\max}$ and $\sigma_{n',nl}^{\max}$ depend significantly
on the value of the electron–perturber scattering length L, on the
relative velocity V of colliding atoms, and on the quantum defect δ_l
of Rydberg state.

It is important to stress that for non-hydrogenic atoms the cross
section (132) depends on the orbital momentum l only through the
frequency of the $nl \rightarrow n'$ transition. This dependence disappears
in the quasielastic limit $\omega_{n',nl} \rightarrow 0$ or at very high velocity in full
agreement with (135).

The cross section $\langle \sigma_{n',nl} \rangle_T$ of an inelastic transition averaged over
the Maxwellian distribution of relative velocities of colliding A* and
B particles is most interesting for the experimental applications at
thermal energies. For the deexcitation process $nl \rightarrow n'$ (when $E_{nl} >$
$E_{n'}$), the result of the calculation [132] can be written as

$$\langle \sigma_{n',nl} \rangle_T = \frac{\langle V\sigma_{n',nl} \rangle_T}{\langle V \rangle_T} = \frac{2\pi L^2 v_0^2}{V^2(n')^3} \varphi_{n',nl}(\lambda_T),$$

$$\varphi_{n',nl}(\lambda_T) = \exp\left(\frac{\lambda_T^2}{4}\right) \mathrm{erfc}\left(\frac{\lambda_T}{2}\right) \qquad (138)$$

$$-\frac{\lambda_T}{\pi} \int\limits_0^\infty \ln\left(1 + \frac{4}{\lambda_T^2} u\right) \exp\left(-u\right) \frac{du}{u^{1/2}}.$$

Here $V_T = (2kT/\mu)^{1/2}$ is the velocity corresponding to the kinetic energy $\mathcal{E} = \mu V^2/2 = kT$ of the relative motion of colliding atoms for a given gas temperature T, $\langle V \rangle_T = (8kT/\pi\mu)^{1/2}$ is the mean thermal velocity, and erfc (z) is the additional probability integral.

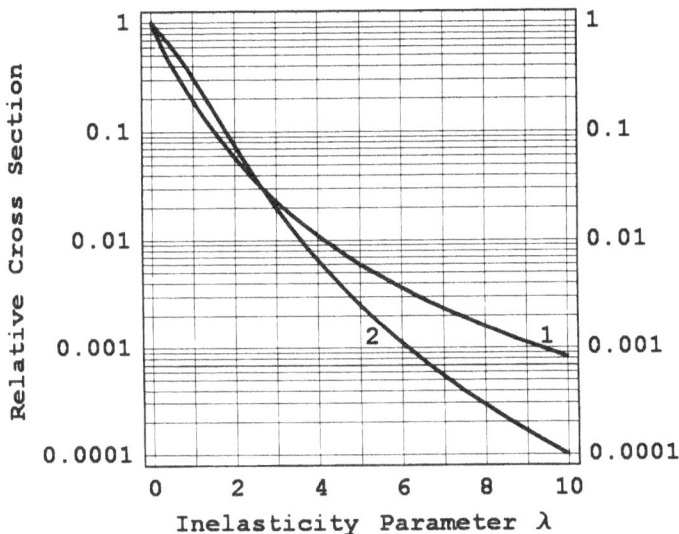

Figure 10. The dependencies of the relative cross sections $f_{n',nl}(\lambda)$ (curve 1) and $F_{n'n}(\lambda)$ (curve 2) of the $nl \to n'$ and $n \to n'$ transitions on the value of inelasticity parameter $\lambda = na_0\omega/V$.

Figure 10 shows the plot of the $f_{n',nl}(\lambda)$ function (curve 1). The dimensionless $f_{n',nl}$ and $\varphi_{n',nl}$ quantities characterize the behavior of the relative cross section $\sigma_{n',nl}(\lambda)/\sigma_{n',nl}(\lambda = 0)$ and its Maxwell-averaged value as a function of the inelasticity parameter λ. This fact directly follows from the comparison of expressions (133), (135). We also present in Fig. 10 the behavior of the relative cross section $F_{n'n}(\lambda) = \sigma_{n'n}(\lambda)/\sigma_{n'n}(\lambda = 0)$ (curve 2) corresponding to the inelastic $n \to n'$ transitions between the hydrogen-like levels. One can see, that the relative cross sections reveal a strong drop with increase of the λ value. The results of calculations of the Maxwell-averaged relative cross sections $\varphi_{n',nl}(\lambda_T)$ and $\phi_{n'n}(\lambda_T)$ are given in Ref. [132].

3.5.2   Transitions Between Hydrogen-like Levels

Let us now consider inelastic transitions $(\omega_{n'n} \neq 0)$ between the Rydberg hydrogen-like levels $n \to n'$

$$A^*(n) + B \to A^*(n') + B,$$

when different degenerate lm-sublevels corresponding to a given n are equally populated in accordance with their statistical weights $g_{nl} = 2l + 1$. This situation occurs even for relatively low buffer-gas densities because of high cross section values for the quasielastic l-mixing process $(\omega_{nl',nl} = 0)$. The most interesting quantities are, then, the total probability $W_{n'n}$, the cross section $\sigma_{n'n}$ and rate constant $K_{n'n}$ of inelastic transition from level n to n' summed over all final l' and averaged over the initial l degenerate states of the Rydberg atom. The first numerical calculations of the rate constants for inelastic $n \to n'$ transitions with a change in the principal quantum number were carried out by *Bates* and *Khare* [148] and *Flannery* [149] within the framework of the semiclassical theory in the momentum representation which was described in the review [83].

Simple analytic formulae for the probability $W_{n'n}$, the cross section $\sigma_{n'n}$, and the rate constant $K_{n'n}$ of inelastic $n \to n'$ transition, which are valid in a wide range of n and energy defects $\Delta E_{n'n}$, were obtained [132] using the semiclassical impact-parameter method presented above. The final result for the cross section of inelastic $n \to n'$ transitions can be written as

$$\sigma_{n'n}(V) = \frac{2\pi L^2 v_0^2}{V^2 (n')^3} F_{n'n}(\lambda), \quad \lambda = \frac{n a_0 \omega_{n'n}}{V},$$

$$F_{n'n}(\lambda) = \frac{2}{\pi} \left[\arctan\left(\frac{2}{\lambda}\right) - \frac{2\lambda(3\lambda^2 + 20)}{3(4 + \lambda^2)^2} \right]. \tag{139}$$

The same expression was simultaneously derived [150, 146] in the impulse approximation using the binary-encounter approximation for the form factor of the $n \to n'$ transition and the scattering length approximation (when $f_{eB} = -L$ and $\sigma_{eB}^{el} = 4\pi L^2$). The plot of the $F_{n'n}$ function against the inelasticity parameter λ is shown in Fig. 10.

Integration of (139) over the Maxwellian distribution of the colliding atoms relative velocity, leads to the following expressions for the rate constants of excitation $K_{nn'}$ ($n' \to n$ transition) and deexcitation $K_{n'n} = \langle V\sigma_{n'n} \rangle_T$ ($n \to n'$ transition) of the Rydberg A* atom by a neutral atom B ($n > n'$) [132]

$$K_{n'n}(T) = \frac{v_0^2 \sigma_{eB}^{el}}{\pi^{1/2} V_T n^3} \phi_{n'n}(\lambda_T), \qquad \phi_{n'n}(\lambda_T) = \exp\left(\frac{\lambda_T^2}{8}\right)$$

$$\times \left[\exp\left(\frac{\lambda_T^2}{8}\right) \operatorname{erfc}\left(\frac{\lambda_T}{2}\right) - \frac{\lambda_T^2}{(2\pi)^{1/2}} D_{-3}\left(\frac{\lambda_T}{2^{1/2}}\right) \right. \tag{140}$$

$$\left. - \frac{5\lambda_T}{\pi^{1/2}} D_{-5}\left(\frac{\lambda_T}{2^{1/2}}\right) \right].$$

$$K_{nn'} = \frac{n^2}{(n')^2} K_{n'n} \exp\left(-\frac{|\Delta E_{n'n}|}{kT}\right), \tag{141}$$

where $D_{-\nu}(z) \equiv U(\nu - 1/2, z)$ is the Weber parabolic cylinder function (see [144]), $\lambda_T = na_0 \omega_{n'n}/V_T$ and $V_T = (2kT/\mu)^{1/2}$. The asymptotic behavior of the $\phi_{n'n}(\lambda_T)$ function is given by

$$\phi_{n'n}(\lambda_T) \approx 1 - 8\lambda_T/3\sqrt{\pi}, \qquad (\lambda_T \ll 1),$$
$$\phi_{n'n}(\lambda_T) \approx 2^6/\lambda_T^5\sqrt{\pi}, \qquad (\lambda_T \gg 1). \tag{142}$$

Thus, in the asymptotic region of very high $n \gg (|\Delta n| v_0/V)^{1/2}$ the inelastic $n \to n'$ transition rate constant (140) reduces to the simple expression

$$K_{n'n}(T) = \left(\frac{\mu Ry}{\pi mkT}\right)^{1/2} \frac{v_0 \sigma_{eB}^{el}}{n^3}, \qquad (\lambda_T \to 0), \tag{143}$$

which corresponds to the quasielastic limit ($\Delta E_{n'n} \to 0$) considered first by *Flannery* [149]. However, in the region of $n \ll (|\Delta n| v_0/V)^{1/2}$ (where $\lambda_T \gg 1$), the general formula (140) yields [132]

$$K_{n'n}(T) = \frac{2^6 n^7}{\pi |\Delta n|^5} \left(\frac{2kT}{\mu v_0^2}\right)^2 v_0 \sigma_{eB}^{el}, \qquad (\lambda_T \gg 1), \tag{144}$$

i.e. the rate constant falls rapidly $K_{n'n} \propto n^7$ with n decreasing. The $K_{n'n}$ value reaches its maximum for $\lambda_T \approx 1$.

3.6   Diffusion of Rydberg Electron in Energy Space

It is interesting to note that formula (139) yields the well known result
of *Pitaevskii* [151] for the diffusion coefficient $D\left(|E_n|\right)$ in the energy
space of a weakly bound electron (moving in the Coulomb field of
the ion core with the energy $E_n < 0$) in the process of three-body
electron-ion recombination in a buffer gas. In accord with [39] it is
determined by the averaged value of the energy transfer square ΔE^2
to the highly excited electron in its collisions with neutral atoms

$$D\left(|E_n|\right) = \frac{1}{2}\frac{\partial}{\partial t}\overline{\Delta E^2} = \frac{1}{2}\left\langle \sum_{n'} \Delta E_{n'n}^2 NV\sigma_{n'n}\left(V\right)\right\rangle_T. \qquad (145)$$

Here the straight line denotes the averaging over a great number of
collisions during the time t, which is equivalent to the averaging over
the Maxwellian velocity distribution for a given gas temperature T. In
the second equality we have rewritten (145) in terms of the transition
cross section between the Rydberg levels $n \to n'$ and the gas density
N. Let us substitute (139) into (145) and replace the summation over
all final levels n' by integration over the energy transfer $\Delta E_{n'n} = 2Ry\Delta n/n^3$ (or over the inelasticity parameter λ). Then, the diffusion
coefficient takes the form

$$D\left(|E_n|\right) = \left(\frac{m^2 e^6}{\hbar^3}\right)\frac{N\sigma_{\mathrm{eB}}^{\mathrm{tr}}}{n^3}\left\langle V^2\int\limits_0^\infty \lambda^2 \mathsf{F}_{n'n}(\lambda)d\lambda\right\rangle_T, \qquad (146)$$

where $\sigma_{\mathrm{eB}}^{\mathrm{tr}}$ is the momentum transfer cross section in an electron–
atom collision, which is equal to $\sigma_{\mathrm{eB}}^{\mathrm{tr}} = \sigma_{\mathrm{eB}}^{\mathrm{el}} = 4\pi L^2$ in the ultra-low
energy limit $\epsilon \to 0$. Calculating the integral and using the relations
$\left\langle V^2\right\rangle_T = 3kT/\mu$ and $|E_n| = me^4/2\hbar^2 n^2$, we finally obtain the result

$$D\left(|E_n|\right) = \frac{32\sqrt{2}NkT\sigma_{\mathrm{eB}}^{\mathrm{tr}}}{3\pi\mu}|E_n|^{3/2} \qquad (147)$$

derived in [151] with the use of classical mechanics for calculation
of the energy transfer and the microcanonical distribution function
for the electron energy in the Coulomb field. This expression was
applied in [151] in order to evaluate the coefficient of the three-body
recombination of an electron with an atomic ion in the three-body
collision with a neutral atom.

4   Impulse Approximation Approach

In the present section we shall present general techniques for calculation of the cross sections for bound–bound and bound–free transitions induced by scattering of quasifree Rydberg electron by a neutral target within the framework of the quantal impulse approximation and binary-encounter approaches. These approaches will allow us to express the cross section of the Rydberg atom-neutral collisions in terms of the amplitude $f_{eB}(\mathbf{k}', \mathbf{k})$, differential $d\sigma_{eB}(k, \theta)/d\Omega$ or total $\sigma_{eB}(k)$ cross sections for scattering of the free electron by a perturbing atom or molecule. It will provide the general treatment of Rydberg-atom–neutral collisions within the framework of the quasifree electron model and will be valid for any arbitrary form of the electron-perturber scattering amplitude.

4.1   Historical Sketch

The impulse approximation was originally proposed by *Fermi* [152] and *Chew* and coworkers [153] in nuclear physics. In particular, it was widely used for describing high-energy neutron scattering by complex nuclei (see [78] for more details). The quantal impulse approximation and its semiclassical version (the so-called binary-encounter approach in the momentum representation) found wide employment in collision theory involving Rydberg atoms [36, 82, 83]. The purely classical impulse-approximation approach to description of the energy transfer processes in collisions of a Rydberg atom with neutral and charged particles is also well known. According to [78] there are two classes of interactions of the composite system, for which the impulse approximation would be valid, they are the weak binding and quasiclassical binding. The first one corresponds to the case when the kinetic energy $\mathcal{E} = \mu V^2/2$ of the incident particle B relative to the Rydberg atom appreciably exceeds the binding energy $|E_n| = Ry/n_*^2$ of the outer electron e with its parent core A^+. This is a standard condition for the applicability of the impulse approximation. However, actually the impulse approximation approach may be also used for the cross section calculations if the weak binding condition is broken down but the collision time of the projectile with the Rydberg electron is small compared to the characteristic period of the electron orbital motion. This second case corresponds to the so-called quasiclassical binding condition, when the interaction potential between the Rydberg electron and its ionic core assumes a very small variation ΔU during the scattering of the incident particle on the highly excited electron. Therefore, the ionic core A^+ does not affect the process of the encounter and its role reduces only to the generation of the momentum space wave functions in the initial and final Rydberg states.

For collisions of Rydberg atoms with neutral particles, the impulse approximation [153] has been applied first by *Alekseev* and *Sobel'man* [154]. They obtained simple asymptotic formulae for the impact broadening and shift of highly excited levels in terms of the forward elastic scattering amplitude $f_{eB}(k, \theta = 0)$ of the quasifree electron e by a perturbing atom B and the momentum distribution function in the Rydberg nl-state. Further applications of the quantal impulse approximation to collisions involving the Rydberg atoms and atomic or molecular targets were stimulated by a series of papers by *Matsuzawa* [145, 155, 156] (see [82] for more details). He has derived some expressions for the ionization and excitation cross sections in terms of the squared amplitude $|f_{eB}(Q)|^2$ of electron-perturber scattering and the transition form factors $F_{f_1}(Q)$. They are valid given that the amplitude $f_{eB}^{\beta'\beta}$ of elastic $(\beta' = \beta)$ or inelastic $(\beta' \neq \beta)$ electron-perturber scattering is either constant $(f_{eB} = const)$ or a function of the momentum transfer Q alone, i.e. $f_{eB} = f_{eB}(Q)$. A simple expression for the ionization cross section of a highly excited atom by polar or quadrupolar molecules was independently derived by *Fowler* and *Preist* [157, 158].

Another efficient semiclassical impulse approach for describing the energy transfer $n \to n'$ and ionization of Rydberg atoms by neutral atoms and molecules has been developed by *Pitaevskii* [151], *Bates* and *Khare* [148] and especially by *Flannery* [149, 159]. It is based on classical mechanics for the calculation of the energy and momentum transfer in the binary encounter of the quasifree electron with a neutral perturber and on the quantal (or semiclassical) description of the Rydberg atom. The total ionization cross section in this approach [149, 159] can be expressed through the differential cross section $d\sigma_{eB}/d\Omega = |f_{eB}(\mathbf{k}', \mathbf{k})|^2$ of the electron–perturber scattering and the momentum distribution function of the Rydberg electron in the initial state by means of the fourfold multiple integral. Significant simplifications in the basic expression appear, when the scattering amplitude f_{eB} is a function only of the momentum transfer $Q = |\mathbf{k}' - \mathbf{k}|$ or it is independent of the scattering angle θ (see [83]).

Hickman [161, 162] and *Hugon* et al. [163, 164, 165] have used the
Born approximation combined with the Fermi pseudopotential much
for calculations of the l-mixing and n, l-changing processes with small
energy defects $\Delta E_{n'l',nl}$ at thermal collisions of high-Rydberg atoms
and the rare gas atoms. Such a perturbative approach is equivalent
to the quantal impulse approximation under the scattering length approximation $f_{eB} = -L = const$. The impulse-approximation calculations of the l-mixing and quasielastic state-changing cross sections
in the Rydberg-atom–rare-gas-atom collisions have been also carried
out by *Matsuzawa* and coworkers [145, 166, 167]. All these results
are applicable in the range of weak coupling (i.e. at high enough
n for quasielastic collisions). At low and intermediate n, the simple power behavior of the l-mixing cross sections ($\sigma_{nl',nl} \propto n^{-4}$ and
$\sigma_{nl}^{l-\text{mix}} = \sum_{l' \neq l} \sigma_{nl',nl} \propto n^{-3}$), predicted by the impulse approximation,
is broken down. As was shown by *Hahn* [168], the high-order non-
impulsive correction terms to the impulse approximation (due to the
interaction between the perturbing particle B and the parent core A^+
of Rydberg atom A^* and by the effects of multiple scattering ([78])
become important at $n \leq n_{\text{max}}$, where $n_{\text{max}} \sim 10 - 20$ for the l-
mixing thermal collisions with the rare gas perturbing atoms. Then,
the l-mixing cross section behaves like $\sigma_{nl}^{l-\text{mix}} \propto n^4$ at $n \leq n_{\text{max}}$).

Gounand and *Petitjean* [147] have extended the binary-encounter
approach (originally developed [156, 169] for the ionization and $n \to$
n' transitions with large values of Δn) to the inelastic $nl \to n'$ transitions from the initial nl-level to all degenerate sublevels of the final
hydrogen-like state. They have applied the impulse approximation
combined with a simple expression for the sum of squared form factors $\sum_{l'} F_{n'l',nl}(Q)$ for a theoretical analysis of quasielastic l-mixing,
inelastic n-changing and ionizing processes involving the Rydberg
atoms and neutral atoms or molecules [150, 170]. It is important
to stress that this impulse-approach combined with the scattering
length approximation $f_{eB} = -L$ and the binary-encounter approximation for atomic form factors was used by *Kaulakys* and *Petitjean*
and *Gounand* for an analytical description of the $n \to n'$ [146, 150]
and $nl \to n'$ [146] transitions between Rydberg states. The resultant
expressions are similar to the analytical formulae derived simultaneously in Ref. [132] by using the semiclassical impact-parameter
method.

    Further development of the collision theory of Rydberg atoms

with neutral particles on the basis of the impulse approximation was
made in a series of papers by *Lebedev* and *Marchenko* [105, 106] and
Lebedev [136, 137, 171], who have formulated a general approach for
describing various types of inelastic, quasielastic, elastic and ionizing collisions within the framework of the quasifree electron model in
the momentum representation. Quantal and semiclassical formulae
for the cross sections derived in these papers are applicable to the
electron–perturber scattering amplitude $f_{eB}^{\beta'\beta}(\mathbf{k}',\mathbf{k})$ of an arbitrary
form (i.e. when the amplitude $f_{eB}^{\beta'\beta}$ depends simultaneously on both
the electron momentum k and the scattering angle θ, in contrast to
the Born approximation or the standard version of the impulse approximation developed in [82]. A general semiclassical description of
the bound-bound $nl \to n'$ and $nl \to n'l'$ transitions in the momentum
representation was independently given by *Kaulakys* [172, 173] on
the basis of the free electron model. This version of the free-electron
model is somewhat similar to the semiclassical approach developed
in [148, 149]. Later *Lebedev* derived simple formulae for the cross sections and the rate constants of direct ionization [171] and developed
a general analytic approach [136, 137] for the description of the inelastic transitions between the fine-structure $nlJ \to nlJ'$ components
and of elastic Rydberg-atom–neutral scattering.

Some new applications of the impulse approximation approach
have been considered by *van Regemorter* and *Hoang-Bing* [174, 175]
for the broadening and shift of high member infrared atomic lines perturbed by collisions with the rare gas atoms and with neutral hydrogen. A number of recent theoretical works have also been devoted to
further development of purely classical and semiclassical methods for
atomic form factors of collisional $nl \to n'l'$ (*Vrinceanu* and *Flannery*
[176, 177], *Samengo* [178], *Fang* et al. [179]). As follows from these
results such methods are valid not only for large quantum numbers
but also for some other cases, which have been previously considered
only within the framework of quantum-mechanical formalism.

4.2   Quantum-Mechanical Technique

4.2.1   Momentum Representation

A general quantum treatment of the impulse approximation for description of collisional transitions can be given in terms of the scattering $\mathsf{T}(\mathsf{E})$-operator. Let us consider the Rydberg atom A as a weakly bound two-particle system consisting of the valence electron e and of its ionic core A^+ with a large orbital radius. In accordance with the basic idea of the impulse approximation our goal here is to represent the scattering amplitude from a complex system (Rydberg atom) as a superposition of scattering amplitudes from "free" independent scatterers (highly excited electron and ion core) which have the same momentum distribution as the initially bound particles. This means that we may neglect the Coulomb-type electron-core interaction U during the scattering of an incident particle on both the highly excited electron e and on the ion core A^+. At the same time the bound character of the Rydberg electron motion around its ion core appears in this approach due to the use of exact atomic wave functions in the initial $|i\,\rangle$ and final $|f\rangle$ states.

The basic treatment of the impulse approximation may be given using the wave functions $G_i(\mathbf{k})$ and $G_f(\mathbf{k}')$ in the momentum representation. The momentum space wave function is defined by the relation

$$G_\alpha(\mathbf{k}) = (2\pi)^{-3/2} \int \exp\left(-i\mathbf{k}\cdot\mathbf{r}\right)\psi_\alpha(\mathbf{r})\,d\mathbf{r}. \qquad (148)$$

Here $\mathbf{k} = \mathbf{p}/\hbar$ is the wave vector of an electron, while $\mathbf{p}$ is its momentum.

For the Rydberg $|nlm\rangle$ -state with the given magnitudes of the principal n, orbital l, and magnetic m quantum numbers the radial parts $g_{nl}(k)$ and $\mathcal{R}_{nl}(r)$ of the wave functions in the momentum and coordinate spaces

$$
\begin{aligned}
G_{nlm}(\mathbf{k}) &= g_{nl}(k)\,Y_{lm}(\theta_\mathbf{k},\varphi_\mathbf{k}), & (149)\\
\psi_{nlm}(\mathbf{r}) &= \mathcal{R}_{nl}(r)\,Y_{lm}(\theta_\mathbf{r},\varphi_\mathbf{r}) & (150)
\end{aligned}
$$

are connected by the relation

$$g_{nl}(k) = (-i)^l \sqrt{\frac{2}{\pi}} \int\limits_0^\infty \mathcal{R}_{nl}(r)j_l(kr)\,r^2\,dr, \qquad (151)$$

where $j_l(z) = (\pi/2z)^{1/2} J_{l+1/2}(z)$ is the spherical Bessel function. The radial wave function g_{nl} satisfies the normalization condition:

$$\int_0^\infty g_{n'l}^* g_{nl}(k)(k)\, k^2 dk = \delta_{nn'} .$$

To derive the basic formulae of quantum impulse approximation it is convenient to rewrite the total Hamiltonian of the system A+B in the form

$$H = K_{tot} + V + U, \qquad V = V_{eB} + V_{A+B} , \tag{152}$$

$$K_{tot} = \frac{\hbar^2 q^2}{2\mu} + \frac{\hbar^2 \kappa^2}{2\mu_{eA+}} , \qquad \mu \equiv M_B = \frac{M_B(m + M_{A+})}{(M_{A+} + m + M_B)} . \tag{153}$$

This allows us to distinguish explicitly the electron–ion-core interaction potential U, the total kinetic energy operator K_{tot} of three particles $\left[(e, A^+) - B\right]$, and the two-body operators V_{eB} and V_{A+B} for the electron–projectile and core–projectile interaction potentials. Here the first term corresponds to the kinetic energy of projectile (B) motion relative to the center of mass of the (e, A^+)-pair, while the second one is the kinetic energy operator of electron motion relative to the ion core. Further, we present the Green's resolvent $G^+(E) = (E - H + i0)^{-1}$ as an expansion in power series of U and retain only its first term

$$G^+(E) \approx (E - K_{tot} - V + i0)^{-1} , \tag{154}$$

which involves the two-body operators V_{eB} and V_{A+B} for the electron-projectile and core-projectile interaction potentials and the term K_{tot}.

The next step consists of separating the variables describing the projectile motion relative to the Rydberg electron and its motion relative to the ion core. Starting from (154) and the basic expression for the scattering T-operator (42) one can present it in the form

$$T \approx T^{imp} = T_{eB} + T_{A+B} , \tag{155}$$

following the well known work by *Chew* and *Goldberger* [153]. Thus, in the impulse approximation the scattering operator contains only the scattering $T_{eB}(E)$ and $T_{A+B}(E)$ operators

$$T_{eB} = V_{eB} + V_{eB}\left(\frac{1}{E - K_{tot} - V_{eB} + i0}\right) V_{eB} , \tag{156}$$

$$T_{A+B} \;=\; V_{A+B} + V_{A+B} \left(\frac{1}{E - K_{tot} - V_{A+B} + i0} \right) V_{A+B} \quad (157)$$

involving the two-body (electron–perturber) V_{eB} and (core–perturber) V_{A+B} interactions, respectively. The removed terms in (155) describe the non-impulsive corrections to the impulse approximation and the effects of the multiple scattering. As is evident from (155), in the impulse approximation the contributions of the perturber-electron $(B–e)$ and perturber-core $(B–A^+)$ scattering to the transition matrix (46) and cross section can be calculated independently

$$T_{fi}^{imp} (\mathbf{q}', \mathbf{q}) = \langle \mathbf{q}', f \, |T_{eB}(E)| \, \mathbf{q}, i \rangle + \langle \mathbf{q}', f \, |T_{A+B}(E)| \, \mathbf{q}, i \rangle , \quad (158)$$

$$\sigma_{fi}^{imp} = \sigma_{fi}(e - B) + \sigma_{fi}(A^+ - B) . \quad (159)$$

It is worthwhile to remember that these matrix elements are taken over the plane waves $|\mathbf{q}\rangle = |\exp(i\mathbf{q} \cdot \mathbf{R})\rangle$ describing the projectile motion relative to the center of mass of the Rydberg atom, i.e. we work in the total center of mass system of the three particles $[B\text{-}(e, A^+)]$.

4.2.2 Description of Electron–Projectile Scattering

Let us first consider the major contribution to the transition amplitude induced by the electron–perturber scattering mechanism. The core–perturber contribution will be analyzed in Sect. 7. The fundamental step in deriving the basic formula of the impulse approximation is to use the expansion of the Rydberg atom wave functions in the initial $|i\rangle$ and final $|f\rangle$ states over the plane waves $\exp(i\boldsymbol{\kappa} \cdot \mathbf{r})$ describing the relative electron-core motion with the wave vector $\boldsymbol{\kappa}$

$$|\alpha\rangle = (2\pi)^{-3/2} \int G_\alpha(\boldsymbol{\kappa}) |\boldsymbol{\kappa}\rangle \, d\boldsymbol{\kappa} , \qquad |\alpha\rangle = |\psi_\alpha(\mathbf{r})\rangle ,$$

$$|\boldsymbol{\kappa}\rangle = \exp(i\boldsymbol{\kappa} \cdot \mathbf{r}) , \qquad \boldsymbol{\kappa} = \mu_{eA^+}\mathbf{v}_{eA^+}/\hbar . \quad (160)$$

Here $\mathbf{v}_{eA^+}$ is the relative velocity, $\mu_{eA^+} = mM_{A^+}/(m + M_{A^+})$ being the reduced mass of the (e, A^+)-pair, and the expansion coefficient $G_\alpha(\boldsymbol{\kappa})$ corresponds to the momentum space wave function introduced by the relation (148). Then, the transition matrix element takes the form

$$\langle \mathbf{q}', f \, |T_{eB}(E)| \, \mathbf{q}, i \rangle$$

$$= (2\pi)^{-3} \int \int G_f^*(\boldsymbol{\kappa}') \, G_i(\boldsymbol{\kappa}) \, \langle \mathbf{q}' \, |\langle \boldsymbol{\kappa}'| \, T_{eB}(E) \, |\boldsymbol{\kappa}\rangle| \, \mathbf{q}\rangle \, d\boldsymbol{\kappa} \, d\boldsymbol{\kappa}' . \quad (161)$$

Further we note that the momentum $\hbar q_{A+}$ of the ion core A^+ and its kinetic energy $\left(\hbar^2 q_{A+}^2/2M_{A+}\right)$ in the (total) center of mass system is not changed during the electron–perturber scattering in accordance with the basic assumption of the impulse approximation (i.e. A^+ is only a spectator in the e–B encounter). Moreover, the total kinetic energy operator of the system (153) can be rewritten in the equivalent form

$$K_{tot} = \frac{\hbar^2 q_{A+}^2}{2M_{A+}} + \frac{\hbar^2 k^2}{2\mu_{eB}} , \qquad M_{A+} = \frac{(m + M_B)\, M_{A+}}{(M_{A+} + m + M_B)} .$$

Here $\hbar k = \mu_{eB} v_{eB}$ is the electron momentum relative to the projectile B, and $\mu_{eB} = mM_B/(m + M_B)$ being their reduced mass. Thus, the transition matrix element of the T_{eB}-operator (156), taken over the plane waves, involves the delta function

$$\langle q'|\langle \kappa'|\, T_{eB}(E)\,|\kappa\rangle|\, q\rangle$$
$$= (2\pi)^3\, \delta(q_{A+}' - q_{A+})\, \langle k'\,|t_{eB}(\epsilon = \hbar^2 k^2/2\mu_{eB})|\, k\rangle . \qquad (162)$$

Expression (162) corresponds to the reduction of the three-body matrix elements in terms of the two-body scattering operator for electron–perturber scattering.

The relation between the wave vectors (q, κ) and (q_{A+}, k)

$$k = \left(\frac{\mu_{eB}}{m}\right)\kappa - \left(\frac{\mu_{eB}}{m}\right)q, \qquad (163)$$

$$v_{eB} = \left(\frac{\mu_{eA+}}{m}\right)v_{eA+} - V, \qquad (164)$$

$$q_{A+} = -\kappa - \frac{M_{A+}}{m + M_{A+}}q \qquad (165)$$

allows us to rewrite the delta-function as

$$\delta(q_{A+}' - q_{A+}) = \delta\left(\kappa - \kappa' + \frac{M_{A+}}{m + M_{A+}}Q\right). \qquad (166)$$

Here $\hbar Q = \hbar\,(q - q')$ is the momentum transfer vector in the collision of projectile with the Rydberg atom in the total center of mass of the three particles $\left[B - (e, A^+)\right]$, and $V = \hbar q/\mu$ is the relative velocity of projectile with respect to the center of mass of the (e, A^+)-pair. Then, substituting the relations (162) and (166) into (161) and performing the integration over κ', we obtain the final formula of the impulse approximation for the contribution of the electron–perturber scattering to the transition matrix element between the Rydberg atomic states $i \rightarrow f$

$$
\begin{aligned}
T_{fi}^{\text{eB}}(\mathbf{q}', \mathbf{q}) &= \langle G_f(\kappa') | \, t_{\text{eB}}(\mathbf{k}', \mathbf{k}; \epsilon) \, | G_i(\kappa) \rangle \\
&= \int G_f^*(\kappa') t_{\text{eB}}(\mathbf{k}', \mathbf{k}; \epsilon) G_i(\kappa) d\kappa .
\end{aligned}
\tag{167}
$$

The matrix element $t_{\text{eB}}(\mathbf{k}', \mathbf{k}; \epsilon)$ of the two-body operator for scattering of the free electron by a perturbing particle B in the final formula (167) has the form

$$
\begin{aligned}
&t_{\text{eB}}(\mathbf{k}', \mathbf{k}; \epsilon) \\
&= \left\langle \mathbf{k}' \left| \mathsf{V}_{\text{eB}} + \mathsf{V}_{\text{eB}} \left(\frac{1}{\epsilon - \mathsf{K}_{\text{eB}} - \mathsf{V}_{\text{eB}} + i0} \right) \mathsf{V}_{\text{eB}} \right| \mathbf{k} \right\rangle .
\end{aligned}
\tag{168}
$$

It is taken over the plane waves $(|\mathbf{k}\rangle = |\exp(i\mathbf{kr}_{\text{eB}})\rangle)$ describing the electron–perturber relative motion in the system of their center of mass. The relative momentum transfers in (167) are given by the relations

$$
\mathbf{k}' - \mathbf{k} = (\mu_{\text{eB}}/\mu) \left[(M_{\text{A}+}/M_{\text{A}}) + (m/\mu) \right] \mathbf{Q} ,
\tag{169}
$$

$$
\kappa' = \kappa + \mathbf{K}, \qquad \mathbf{K} = (M_{\text{A}+}/M_{\text{A}})\mathbf{Q},
\tag{170}
$$

where $\hbar\mathbf{Q} = \hbar(\mathbf{q} - \mathbf{q}')$ is the momentum transfer vector for the collision of the projectile B with the Rydberg atom A^* $(M_{\text{A}} = m + M_{\text{A}+}$ is its mass). Note also that in contrast to (156) the two-body scattering t_{eB}-operator (168) in the final formula of the impulse approximation contains only the kinetic energy operator $\mathsf{K}_{\text{eB}} = \hbar^2 \mathbf{k}^2 / 2\mu_{\text{eB}}$ for the electron–perturber relative motion and the interaction potential V_{eB} of the (e, B)-pair.

Expression (167) for the contribution of the electron–perturber scattering may be rewritten in the following equivalent form [36, 82]

$$
\begin{aligned}
\mathsf{f}_{fi}^{\text{eB}}(\mathbf{q}', \mathbf{q}) &= \tfrac{\mu}{\mu_{\text{eB}}} \langle G_f(\kappa') | f_{\text{eB}}(\mathbf{k}', \mathbf{k}) | G_i(\kappa) \rangle \\
&= \tfrac{\mu}{\mu_{\text{eB}}} \int G_f^*(\kappa') f_{\text{eB}}(\mathbf{k}', \mathbf{k}) G_i(\kappa) \, d\kappa .
\end{aligned}
\tag{171}
$$

Here $f_{f_i}(\mathbf{q}', \mathbf{q})$ is the standard inelastic scattering amplitude (44) of the Rydberg atom A^* by the projectile B associated with the corresponding matrix element $T_{f_i}(\mathbf{q}', \mathbf{q})$ of the transition operator on the energy shell (see (44)); and $f_{eB}(\mathbf{k}', \mathbf{k})$ denotes the two-particle (electron–perturber) scattering amplitude defined by the relation

$$f_{eB}(\mathbf{k}', \mathbf{k}) = -\tfrac{\mu_{eB}}{2\pi\hbar^2}\, t_{eB}(\mathbf{k}', \mathbf{k}; \epsilon)$$

$$= -\tfrac{\mu_{eB}}{2\pi\hbar^2} \int d\mathbf{r}_{eB}\, \exp\left(-i\mathbf{k}' \cdot \mathbf{r}_{eB}\right) V_{eB}\, \psi_{\mathbf{k}}^+(\mathbf{r}_{eB}). \tag{172}$$

Note that in the impulse approximation, the electron–perturber scattering amplitude $f_{eB}(\mathbf{k}', \mathbf{k})$ in the basic equation (171) (or the matrix element $t_{eB}(\mathbf{k}', \mathbf{k}; \epsilon)$ of the scattering $t_{eB}(\epsilon)$-operator) should be taken, generally, both on $(k = k')$ and off the energy shell. The final expression of the quantal impulse approximation for the contribution of the perturber-quasifree electron scattering to the cross section $\sigma_{fi}(q)$ of the $i \to f$ transition can be presented as

$$\sigma_{f_i} = \left(\frac{\mu}{\mu_{eB}}\right)^2 \frac{q'}{q} \int \left|\langle G_f(\boldsymbol{\kappa}')|\, f_{eB}(\mathbf{k}', \mathbf{k})\, |G_i(\boldsymbol{\kappa})\rangle\right|^2 d\Omega_{\mathbf{q}'\mathbf{q}}, \tag{173}$$

where $\mathbf{q}' = \mathbf{q} - \mathbf{Q}$. It is worthwhile to point out that the shift of the Rydberg atom wave function in the momentum space $\boldsymbol{\kappa} \to \boldsymbol{\kappa}'$ is performed by the translational operator

$$G_\alpha(\boldsymbol{\kappa} + \mathbf{K}) = \exp\left(i\mathbf{K} \cdot \hat{\mathbf{r}}\right) G_\alpha(\boldsymbol{\kappa}), \tag{174}$$

where $\hat{\mathbf{r}} = i\partial/\partial\boldsymbol{\kappa}$ is the radius vector operator of the valence electron.

In the most interesting case of collisions between the Rydberg atom and the heavy projectile (atom, ion or molecule with mass $M_B \gg m$) one can take into account that $\mu_{eB} \approx m$, $\mu_{eA+} \approx m$ and $M_A = m + M_{A+} \approx M_{A+}$ so that the relationships for the relative momenta and velocities of three particles can be rewritten as

$$\mathbf{k} \approx \boldsymbol{\kappa} - \left(\frac{m}{\mu}\right)\mathbf{q}, \quad \mathbf{v}_{eB} \approx \mathbf{v}_{eA+} - \mathbf{V}, \quad \boldsymbol{\kappa}' - \boldsymbol{\kappa} \approx \mathbf{q} - \mathbf{q}', \tag{175}$$

i.e. $\mathbf{K} \approx \mathbf{Q}$.

For collisions of a Rydberg atom with an electron $\mu_{eB} = m/2$, and we have

$$\boldsymbol{\kappa}' - \boldsymbol{\kappa} \approx \mathbf{Q}, \quad \mathbf{v}_{eB} \approx \mathbf{v}_{eA^+} - \mathbf{V}, \quad \mathbf{k}' - \mathbf{k} = \mathbf{q} - \mathbf{q}' = \mathbf{Q}. \qquad (176)$$

It is important to stress that the error of the identified relations is of the order of the mass ratio (m/M_A) and certainly is less than that of collision theory.

All results presented above describe the case of collisions between the Rydberg atom and a neutral or charged particle, when the projectile state is not changed during the scattering process. Actually, however, the final formulae of this section are also applicable in the case when transitions between highly excited atomic states are accompanied by the inelastic excitation (deexcitation) of projectile. Then, the basic formula of the impulse approximation for the transition amplitude $f_{fi}^{\beta'\beta}(\mathbf{q}', \mathbf{q})$ of the process

$$A(\alpha) + B(\beta) \to A(\alpha') + B(\beta') \qquad (177)$$

can be rewritten as

$$f_{fi}^{\beta'\beta}(\mathbf{q}', \mathbf{q}) = \tfrac{\mu}{\mu_{eB}} \langle G_{\alpha'}(\boldsymbol{\kappa}')| f_{eB}^{\beta'\beta}(\mathbf{k}', \mathbf{k}) |G_\alpha(\boldsymbol{\kappa})\rangle \ ,$$

$$f_{eB}^{\beta'\beta}(\mathbf{k}', \mathbf{k}) = - \tfrac{\mu_{eB}}{2\pi\hbar^2} \langle \mathbf{k}', \beta'| t_{eB}(\epsilon) |\beta, \mathbf{k}\rangle \ . \qquad (178)$$

Here $f_{eB}^{\beta'\beta}(\mathbf{k}', \mathbf{k})$ is the amplitude for the scattering of the free electron by a perturber

$$e + B(\beta) \to e + B(\beta') \ ,$$

and β is the set of quantum numbers characterizing its internal state (e.g. the vibrational-rotational states of perturbing molecule $\beta = v, j, j_z$). As in the case of pure elastic quasifree electron–perturber scattering the two-body matrix elements $t_{eB}^{\beta'\beta}(\mathbf{k}', \mathbf{k};\epsilon)$ should be taken in (178) both on and off $\left(E_\beta + \hbar^2\mathbf{k}^2/2\mu_{eB} \neq E_{\beta'} + \hbar^2(\mathbf{k}')^2/2\mu_{eB}\right)$ the energy shell.

The next approximations are connected with separate consideration of fast and slow collisions that will be considered below.

4.3   Binary Encounter Theory: General Treatment

In the following we present general formulae of binary encounter theory for the cross sections of the bound–bound $nl \to n'$ [105] and bound–free $nl \to E$ [171] transitions summed over possible quantum numbers l' and m' in the final state, and averaged over the magnetic m-sublevels of the initial nl-level. Consider first the process of direct ionization

$$A(nl) + B(\beta) \to A^+ + B(\beta') + e .\tag{179}$$

We shall proceed from the basic equation (49) of scattering theory for the differential cross section $d\sigma^{\beta'\beta}_{\alpha'\alpha}(\mathbf{q}',\mathbf{q})$ of inelastic transition $|\alpha,\beta,\mathbf{q}\rangle \to |\alpha',\beta',\mathbf{q}'\rangle$ and quantum impulse approximation (178). Integration of this equation over the solid angles $d\Omega_{\mathbf{q}'\mathbf{q}}$ of scattering and its summation over all final states of the Rydberg atom in the discrete spectrum and integration in the continuous spectrum will lead to the following expression for the differential cross section $d\sigma_{nl}(q',q)/d\mathcal{E}'$ per unit energy interval $d\mathcal{E}' = \hbar^2 q' dq'/\mu$ of the colliding A and B particles

$$\frac{d\sigma^{\beta'\beta}_{\alpha'\alpha}(q',q)}{d\mathcal{E}'} = \frac{\mu^2 q'}{\mu^2_{eB} q(2l+1)} \sum_m \int d\Omega_{\mathbf{q}'\mathbf{q}} \sum_{\alpha'} \delta\left(E_{\alpha'} - E_\alpha + \Delta E\right)$$

$$\times \left\langle G_\alpha(\boldsymbol{\kappa}) \left| \left[f^{\beta'\beta}_{eB}(\mathbf{k}',\mathbf{k}) \right]^* \right| G_{\alpha'}(\boldsymbol{\kappa}') \right\rangle \tag{180}$$

$$\times \left\langle G_{\alpha'}(\boldsymbol{\kappa}') \left| f^{\beta'\beta}_{eB}(\mathbf{k}',\mathbf{k}) \right| G_\alpha(\boldsymbol{\kappa}) \right\rangle .$$

Here ΔE is given by the relation

$$\Delta E = E_{\beta'} - E_\beta + \mathcal{E}' - \mathcal{E}.$$

Now we use the Fourier transform for the δ-function

$$\delta\left(E_{\alpha'} - E_\alpha + \Delta E\right) = \frac{1}{2\pi\hbar} \int\limits_{-\infty}^{\infty} \exp\left[\frac{i}{\hbar}\left(E_{\alpha'} - E_\alpha + \Delta E\right)t\right] dt$$

and the following relations for the Heisenberg evolution operators

$$\exp\left(-\frac{i}{\hbar} H_A t\right) |G_\alpha(\boldsymbol{\kappa})\rangle = \exp\left(-\frac{i}{\hbar} E_\alpha t\right) |G_\alpha(\boldsymbol{\kappa})\rangle .\tag{181}$$

As a result, using the completeness property

$$\sum_{\alpha'} |G_{\alpha'}\rangle \langle G_{\alpha'}| = 1$$

for the Rydberg atom wave functions expression (180) can be rewritten as

$$\frac{d\sigma_{nl}(q',q)}{d\mathcal{E}'} = \frac{\mu^2 q'}{\mu_{eB}^2 q(2l+1)} \sum_m \int d\Omega_{\mathbf{q'q}} \, \langle G_{nlm}(\boldsymbol{\kappa})| \hat{\mathcal{O}} |G_{nlm}(\boldsymbol{\kappa})\rangle \, ,$$

$$\hat{\mathcal{O}} = \frac{1}{2\pi\hbar} \int\limits_{-\infty}^{\infty} dt \, \exp\left(\tfrac{i}{\hbar}\Delta Et\right) \left[f_{eB}^{\beta'\beta}(\mathbf{k'},\mathbf{k}) \right]^* \tag{182}$$

$$\times \exp\left(\tfrac{i}{\hbar}H_A' t\right) f_{eB}^{\beta'\beta}(\mathbf{k'},\mathbf{k}) \exp\left(-\tfrac{i}{\hbar}H_A t\right) .$$

Thus, in the impulse approximation the differential cross section $d\sigma_{nl}/d\mathcal{E}'$ of ionization can be expressed in terms of some operator $\hat{\mathcal{O}}$, averaged over the initial $|i\rangle \equiv |nlm\rangle$ state of the Rydberg atom. This fact is in agreement with the general results of the theory of quasifree scattering on a system of weakly bound particles (see [78]). Here $H_A = H_A(\boldsymbol{\kappa},\hat{\mathbf{r}})$ and $H_A' = H_A(\boldsymbol{\kappa}',\hat{\mathbf{r}})$ are the Hamiltonians of the Rydberg atom A^* in the momentum representation for the initial $\boldsymbol{\kappa}$ and final $\boldsymbol{\kappa}' = \boldsymbol{\kappa} + \mathbf{K}$ electron momenta

$$H_A \equiv H_A(\boldsymbol{\kappa},\hat{\mathbf{r}}) = \hbar^2\kappa^2/2\mu_{eA+} + \hat{U}_{eA+} \, , \tag{183}$$

$\hat{U}_{eA+}(\hat{\mathbf{r}})$ is the potential energy operator of electron-core interaction, and $\hat{\mathbf{r}} = i\partial/\partial\boldsymbol{\kappa}$ is the radius vector operator of the valence electron.

Within the framework of the impulse approximation the evolution operator $\exp[-(i/\hbar)H_A t]$ in equation (182) commutes with the two-particle t_{eB} operator of the electron–perturber scattering. Then, using the Baker-Campbell-Hausdorf formula for expanding in series the product of evolution operators $\exp[(i/\hbar)H_A' t] \exp[-(i/\hbar)H_A t]$, we obtain

$$\exp[(i/\hbar)H_A' t] \exp[-(i/\hbar)H_A t]$$

$$= \exp\left\{ \tfrac{i}{\hbar}(H_A' - H_A)t + [H_A',H_A]\tfrac{t^2}{2\hbar^2} + \ldots \right\}$$

$$= \exp\left\{ \frac{i\hbar t((\kappa')^2 - \kappa^2)}{2\mu_{eA+}} + \frac{it^2}{2}\left(\frac{\mathbf{K}}{\mu_{eA+}}\right)\hat{\mathbf{F}}_{eA+} + \ldots \right\} . \tag{184}$$

82   V.S. LEBEDEV

Here $\widehat{\mathbf{F}}_{eA^+}$ is the operator of the force acting on the outer electron of
the Rydberg atom A^* by the ionic core A^+.

As follows from the comparison of the first and second terms in
(184), all terms (non-linear in time t), corresponding to the change
of the potential energy U_{eA^+} during the interaction of the colliding e and B particles, can be neglected if the following condition
$\hbar K_{eB} \gg \tau_{eB} F_{eA^+}$ is to be satisfied. Here $\hbar K_{eB}$ is the characteristic
momentum transferred to the outer electron e in its binary collision
with the perturbing particle B, and $(\tau_{eB} F_{eA^+})$ is the impulse of the
force $F_{eA^+} = e^2/r_{eA^+}^2$ acting on this electron by the ionic core A^+
during the collision time τ_{eB} of e and B particles. It should be noted
that this condition is one of the validity criteria of the impulse approximation and the quasifree electron model (see below Sect. 4.9).
Thus within the framework of the binary-encounter theory in the
impulse treatment, only the first term (linear in time t) of the series expansion should be retained in (184). This term is determined
by the change of the kinetic energy of the quasifree electron in its
collision with the perturber.

Then, substituting (184) into (182) and performing the integration over dt, we obtain the following general expression for a binary-
encounter theory for the differential cross section $d\sigma_{E,nl}(\mathcal{E})/dE$ of
ionization per unit energy interval of the ejected electron [171]

$$\frac{d\sigma_{E,nl}^{\beta'\beta}(\mathcal{E})}{dE} = \frac{\mu^2 q'}{\mu_{eB}^2 q(2l+1)} \sum_m \int d\Omega_{\mathbf{q'q}} \int d\boldsymbol{\kappa} \, |G_{nlm}(\boldsymbol{\kappa})|^2$$
$$\times \left| f_{eB}^{\beta'\beta}(\mathbf{k'},\mathbf{k}) \right|^2 \delta\left(\frac{\hbar^2[(\kappa')^2 - \kappa^2]}{2\mu_{eA^+}} + E_{nl} - E \right). \tag{185}$$

Here $E_{nl} = -Ry/(n-\delta_l)^2 < 0$ is the energy of the Rydberg electron
in the initial discrete state and $E > 0$ is the kinetic energy of the
ejected electron. The final kinetic energy $\mathcal{E}' = \hbar^2 q'^2/2\mu$ of colliding
A^* and B particles is determined from the law of energy conservation
for the direct ionization process

$$\mathcal{E} + E_\beta + E_{nl} = \mathcal{E}' + E_{\beta'} + E,$$
$$\mathcal{E} = \hbar^2 q^2/2\mu, \quad \mathcal{E}' = \hbar^2 q'^2/2\mu. \tag{186}$$

The total ionization cross section $\sigma_{nl}^{\beta'\beta}(\mathcal{E})$ is determined from (185) by integrating over all possible values of ejected electron energy

$$\sigma_{nl}^{\beta'\beta}(\mathcal{E}) = \int\limits_{0}^{E_{\max}} \left(\frac{d\sigma_{E,nl}(\mathcal{E})}{dE}\right) dE, \qquad (187)$$

Here the upper limit of integration is to be found from the relation

$$E_{\max} = \mathcal{E} - \left(|E_{nl}| - \Delta E_{\beta\beta'}\right), \quad \Delta E_{\beta\beta'} = E_{\beta} - E_{\beta'} \qquad (188)$$

and $\Delta E_{\beta\beta'}$ is the change of the internal energy of the perturbing particle B. As follows from (187), the direct ionization of the Rydberg atom is possible, when the kinetic energy $\mathcal{E} = \hbar^2 q^2/2\mu$ of colliding A^* and B particles relative motion satisfies the condition $\mathcal{E}_{\min} \leq \mathcal{E}$ (where $\mathcal{E}_{\min} = \max(0, |E_{nl}| - \Delta E_{\beta\beta'})$).

The general equation (185) of the binary-encounter theory relates the ionization cross section to the differential cross section for electron–perturber scattering

$$d\sigma_{eB}^{\beta'\beta}/d\Omega_{\mathbf{k'k}} = \left|f_{eB}^{\beta'\beta}(\mathbf{k'}, \mathbf{k})\right|^2$$

and with the momentum distribution function $|G_{nlm}(\boldsymbol{\kappa})|^2$ of the Rydberg electron in the initial atomic $|i\rangle \rightleftharpoons |nlm\rangle$ state. Thus, there is a definite analogy between this equation, obtained [171] in the impulse approximation directly from the quantum scattering theory, and the basic equation of semiclassical theory [83, 159]. However, an important feature of the semiclassical equation (185) for the ionization cross section is due to the presence of the delta function, which plays the role of a microcanonical distribution. This δ-function brings out only those momentum values from the entire momentum space which correspond to the classical energy transfer

$$\Delta E_{fi} = |E_{nl}| + E = \hbar^2(\boldsymbol{\kappa}'^2 - \boldsymbol{\kappa}^2)/2\mu_{eA^+}$$

for a given bound-free $nl \to E$ transition.

The general equation (185) of the binary-encounter theory may be also used in the case of the bound–bound $nl \to n'$ transition

$$A(nl) + B(\beta) \to A(n') + B(\beta'),$$

if we additionally introduce the quasicontinuum approximation for the hydrogen-like degenerate n' levels of the Rydberg atom [105,

106]. Then, in order to obtain the cross section $\sigma_{n',nl}^{\beta'\beta}(\mathcal{E})$ of inelastic (or quasielastic) $nl \to n'$ transition, the differential cross section $d\sigma_{E,nl}^{\beta'\beta}(\mathcal{E})/dE$ per unit energy interval of the quasicontinuous spectrum $E_{n'}$ should be multiplied by the factor $|dE_{n'}/dn'| = 2Ry/(n')^3$, while the final energy $E > 0$ on the right-hand side of (185) should be replaced by $E_{n'} < 0$, i.e.

$$\frac{d\sigma_{E,nl}^{\beta'\beta}(\mathcal{E})}{dE} \to \sigma_{n',nl}^{\beta'\beta} \left|\frac{dE_{n'}}{dn'}\right|^{-1} = \sigma_{n',nl}^{\beta'\beta} \frac{(n')^3}{2Ry}, \quad E \to E_{n'}. \quad (189)$$

The law of energy conservation for the identified process can be rewritten as

$$\frac{\hbar^2 q^2}{2\mu} + E_\beta - \frac{Ry}{(n-\delta_l)^2} = \frac{\hbar^2(q')^2}{2\mu} + E_{\beta'} - \frac{Ry}{(n')^2}. \quad (190)$$

It should be noted that a particular form of the general equation (185) is very convenient to perform some physical simplifications and for further analysis of the different special cases. We shall show later that it can be reduced to a rather simple form for the most interesting cases of slow $V \ll v_0/n$ and fast $V \gg v_0/n$ collisions between the Rydberg atom and neutral perturbing particle.

4.4 High-Energy Limit

Consider first the fast collisions ($V \gg v_0/n$) with the relative velocity V of neutral projectile B with respect to the center of mass of the Rydberg atom A being large as compared to the mean orbital electron velocity $v_n \sim v_0/n$. Then, the relative velocities and wave vectors of the Rydberg electron e and the perturber B can be approximately represented as [82, 83])

$$\mathbf{v}_{eB} \approx -\mathbf{V}, \quad \mathbf{k} \approx -(m/\mu)\mathbf{q}, \quad \mathbf{Q} \equiv \mathbf{q} - \mathbf{q}' \approx \mathbf{k}' - \mathbf{k} \quad (191)$$

since the first term in the right-hand side of (163) is small and can be neglected. This case corresponds to the weak binding condition in the impulse approximation (see [78]) owing to the kinetic energy of heavy particle relative motion $\mathcal{E} = \mu V^2/2$ is much greater than the Rydberg electron energy $|E_{nl}| = Ry/n_*^2$. Then, the asymptotic expression for the ionization cross section of the Rydberg atom by a neutral particle directly follows from the general equation (185) and can be written as

$$\sigma_{nl}^{\beta'\beta}(V) \underset{n\to\infty}{\longrightarrow} \int \left| f_{eB}^{\beta'\beta}(k,\theta) \right|^2 d\Omega = \sigma_{eB}^{\beta'\beta}(k \approx mV/\hbar). \tag{192}$$

Summation of (192) over all possible quantum numbers β' of the perturbing particle yields the well known result

$$\sigma_{nl,\beta}^{ion}(V) \underset{n\to\infty}{\longrightarrow} \sigma_{nl,\beta}^{tot}(V) = \sigma_{eB}^{tot}(k \approx mV/\hbar). \tag{193}$$

obtained by *Butler* and *May* [180] and *B.Smirnov* [181]. Hence, at very high $n \gg v_0/V$, the cross section $\sigma_{nl,\beta}^{ion}(V)$ of ionization

$$A(nl) + B \to A^+ + B + e$$

tends asymptotically $(n \to \infty)$ to the total cross section

$$\sigma_{nl,\beta}^{tot}(V) = \sum_{\alpha'\beta'} \sigma_{\alpha',nl}^{\beta'\beta}$$

of Rydberg-atom–neutral collisions. In this mechanism it is determined by the integral cross section $\sigma_{eB}(v_{eB} \approx V)$ of electron–perturber scattering. The contribution of all inelastic bound-bound $nl \to n'$ transitions to the total cross section $\sigma_{nl,\beta}^{tot}$ is very small as compared to the ionization and can be neglected. However, at thermal energies, the limit of fast collisions (191) between the Rydberg atom and a neutral particle is realized only for very high principal quantum numbers $n \gg v_0/V$, i.e. $n \gg 10^3 - 10^4$.

4.5   General Consideration of Slow Collisions

4.5.1   Transition Amplitude

Since the most experiments with Rydberg atoms and neutral species were performed at thermal energies $\mathcal{E} = \mu V^2/2$, the most important region of the principal quantum numbers $n \ll v_0/V$ corresponds to the case of slow collisions. Then, using the mass-disparity approximation $(M_B, M_A \gg m)$, the general relations (163)–(170) for the relative velocities and wave vectors of an electron e, ionic core A^+ and neutral perturber B can be rewritten as

$$\mathbf{v}_{eB} \approx \mathbf{v}_{eA^+}, \qquad \mathbf{v}_{A+B} \approx \mathbf{V}, \qquad \mathbf{k} \approx \boldsymbol{\kappa},$$

$$\mathbf{K} \equiv \boldsymbol{\kappa}' - \boldsymbol{\kappa} \approx \mathbf{k}' - \mathbf{k} \approx \mathbf{Q} \equiv \mathbf{q} - \mathbf{q}'. \tag{194}$$

Hence, for slow collisions the basic expression (171) of the impulse approximation for the scattering amplitude is given by

$$f_{f_\iota}^{\beta'\beta}(\mathbf{q} - \mathbf{Q}, \mathbf{q})$$

$$= \tfrac{\mu}{m} \left\langle G_{\alpha'}(\mathbf{k}+\mathbf{Q}) \left| f_{eB}^{\beta'\beta}(\mathbf{k}+\mathbf{Q},\mathbf{k}) \right| G_\alpha(\mathbf{k}) \right\rangle \qquad (195)$$

$$= \tfrac{\mu}{m} \int G_{\alpha'}^*(\mathbf{k}+\mathbf{Q}) f_{eB}^{\beta'\beta}(\mathbf{k}+\mathbf{Q},\mathbf{k}) G_\alpha(\mathbf{k}) \, d\mathbf{k} \, .$$

This approximation gives the opportunity to obtain the general formulae of the impulse approximation for the cross sections of various types of the bound-bound and bound-free transitions, which can be used in practical calculations.

4.5.2   Semiclassical Expressions for Cross Sections

Excitation and Deexcitation.   For slow collisions ($n \ll v_0/V$) the basic formula of the binary-encounter theory for the cross section of the inelastic (or quasielastic) $nl \to n'$ transitions (see (185) and (189)) may be reduced to a rather simple form [105]

$$\sigma_{n',nl}^{\beta'\beta} = \frac{\pi v_0^2}{V^2 (n')^3} \int\limits_{Q_{\min}}^{Q_{\max}} dQ \int\limits_{|k_0|}^{\infty} \frac{dk}{k} W_{nl}(k) \left| f_{eB}^{\beta'\beta}(k,Q) \right|^2 , \qquad (196)$$

$$k_0(Q) = \frac{\left[m\left(E_{n'} - E_{nl} \right)/\hbar^2 \right] - Q^2/2}{Q} , \qquad (197)$$

$$Q^2 = q'^2 + q^2 - 2qq' \cos\theta_{\mathbf{q'q}} , \qquad q = \mu V/\hbar, \qquad (198)$$

$$(q')^2 = q^2 + \left(2\mu/\hbar^2 \right) \left[(E_{n'} - E_{nl}) - (E_\beta - E_{\beta'}) \right] . \qquad (199)$$

Here $Q_{\min} = |q' - q|$ and $Q_{\max} = q' + q$ are the lower and upper limits of integration over the momentum $Q = |\mathbf{k'} - \mathbf{k}|$ transferred to the Rydberg atom in collision with a neutral particle, and $W_{nl}(k)$ is the momentum distribution function of a highly excited electron for a given atomic state nl, which is expressed in terms of the radial part $g_{nl}(k)$ of the momentum space wave function

$$W_{nl}(k) = k^2 |g_{nl}(k)|^2 . \tag{200}$$

This function is normalized by the condition

$$\int_0^\infty W_{nl}(k)\, dk = 1. \tag{201}$$

Formula (196) can be used in the general case, when the electron–perturber scattering amplitude $f_{eB}^{\beta'\beta}$ is a function not only of the momentum transfer Q, but also of the electron momentum k. Thus, this formula generalizes the well known result [156, 147] of the binary-encounter theory for the form factor

$$F_{n',nl}(Q) = \sum_{l'} F_{n'l',nl}(Q)$$

and for the corresponding cross section $\sigma_{n',nl}$ of the $nl \to n'$ transition.

The general expression (196) of the binary-encounter approximation becomes particularly simple in the case of slow collisions between the Rydberg atom and neutral atomic particle, when the internal energy of the perturber is not changed during the scattering, i.e. $\Delta E_{\beta'\beta} = 0$. In this case we can put $|k_0(Q)| \approx Q/2$ for all $Q \geq Q_{\min}$, since $|\Delta E_{n',nl}| \ll \hbar^2 Q_{\min}^2/2m$ for the most interesting range $n \ll (v_0/V)^{2/3}$ of the principal quantum numbers ($n \ll 100 - 500$ at thermal velocities). Then, using the substitution of the variables

$$k, Q \to k, \nu \qquad \left(\nu = \cos\theta = 1 - Q^2/2k^2\right) ,$$

the resulting expression for the cross section of inelastic $nl \to n'$ transition can be rewritten as [105])

$$\sigma_{n',nl}(V) = \frac{\pi v_0^2}{2^{1/2} V^2 (n')^3}$$

$$\times \int_{k_{\min}}^\infty dk\, W_{nl}(k) \int_{-1}^{\nu_{\max}} \frac{d(\cos\theta)}{\sqrt{1-\cos\theta}} |f_{eB}(k,\theta)|^2 , \tag{202}$$

where the limits of integration are to be found from the relations

$$\nu_{\max}(k) = 1 - 2k_{\min}^2/k^2, \qquad k_{\min} = Q_{\min}/2 \approx |\Delta E_{n',nl}|/2\hbar V .$$

It can be seen that the lower limit of integration over the quasi-free electron wave number dk (i.e. $|k_0(Q)| \approx Q/2$ or $k_{\min}$ in (196) and (202), respectively) corresponds to the case of the electron–perturber backward scattering $\theta = \pi$ and $\cos\theta = -1$ ($\mathbf{k}' = -\mathbf{k}$ and $Q = |\mathbf{k}' - \mathbf{k}| = 2k$). The analogous semiclassical expression for the cross section of the $nl \to n'$ transition was also derived in [172] within the framework of the quasifree electron model using the classical description of the energy and momentum transfer in the binary e-B encounter.

Ionization. The general expression of the binary-encounter theory for the total ionization cross section of a Rydberg atom in slow collision ($n \ll v_0/V$) with a neutral atom proceeds from the basic equation (185) and relations (194). After integration over the ejected electron energy E ($E_{\min} \leq E \leq E_{\max}$, where $E_{\min} = |E_{nl}| = Ry/n_*^2$ and $E_{\max} = \mathcal{E} - |E_{nl}|$), the final result is given by [171]

$$\sigma_{nl}^{\mathrm{ion}}(\mathcal{E}) = \frac{\pi a_0 v_0}{V} \int_{Q_1}^{Q_2} dQ \, \left(Q - Q_0 - \hbar Q^2/2\mu V\right)$$

$$\times \int_{Q/2}^{\infty} \frac{dk}{k} W_{nl}(k) \left|f_{\mathrm{eB}}(k, Q)\right|^2 ,$$

(203)

$$Q_{1,2} = \left(2\mu/\hbar^2\right)^{1/2} \left[\mathcal{E}^{1/2} \mp (\mathcal{E} - |E_{nl}|)^{1/2}\right],$$

$$Q_0 = |E_{nl}|/\hbar V.$$

An important feature of the ionization process by an atomic projectile is the presence of the threshold of ionization ($\mathcal{E}_{\min} = \hbar^2 q_{\min}^2/2\mu = Ry/n_*^2$) for all principal quantum numbers n. In contrast to the molecular projectiles (see below Sect. 6.5), the energy $\Delta E_{fi} = E + |E_{nl}|$ transferred to the Rydberg electron in the bound–free transition in this case is determined by the change $\Delta \mathcal{E}_{fi} = \mathcal{E} - \mathcal{E}'$ of the kinetic energy of the colliding atoms relative motion (i.e. $\Delta E_{fi} = \Delta \mathcal{E}_{if}$), since the internal energy of atomic projectile remains unchanged $\Delta E_{\beta\beta'} = 0$.

Momentum Distribution Functions. As follows from (196), (202), and (203) within the framework of the binary-encounter approach the cross sections of the $nl \to n'$ transitions and direct ionization are expressed in terms of the momentum distribution function of a Rydberg electron in the initial nl-state of the atom. In the case of small magnitudes of the orbital angular momentum $l \ll n$, the momentum distribution function is given by the following expression [182]

$$W_{nl}(k) = \left(\frac{4n_* a_0}{\pi}\right) \frac{1 - (-1)^l \cos[2n_*(\beta - \pi)]}{(x^2 + 1)^2}, \quad (204)$$

$$x = n_* k a_0, \quad \cos\beta = \frac{x^2 - 1}{x^2 + 1}, \quad (205)$$

where $n_* = n - \delta_l$ is the effective principal quantum number. Averaging over the fast oscillations reduces (204) to the well known result for the pure classical momentum distribution function at small orbital momentum $l \ll n$

$$W_{nl}(k) = |g_{nl}(k)|^2 k^2 = \frac{4n_* a_0}{\pi} \frac{1}{\left[1 + (nka_0)^2\right]^2}. \quad (206)$$

Explicit expressions for the momentum distribution functions of Rydberg nl-states with a quantum defect can be also obtained using recent results of [183] for nonhydrogenic wave functions in the momentum space.

For calculations of the cross sections averaged over all possible values of the orbital angular momentum for a given magnitude of n it is necessary to have the momentum distribution function $W_n(k)$ defined by the relation

$$W_n(k) = \frac{1}{n^2} \sum_{lm} \int |G_{nlm}(\mathbf{k})|^2 \, d\Omega_\mathbf{k}$$

$$= \frac{1}{n^2} \sum_{l=0}^{n-1} (2l + 1) |kg_{nl}(k)|^2, \quad (207)$$

and normalized to unity

$$\int_0^\infty W_n(k) \, dk = 1. \quad (208)$$

The resultant expression for (207) has been derived by *Fock* [184]

$$W_n(k) = \frac{32n^3 (ka_0)^2}{\pi \left[(nka_0)^2 + 1\right]^4}. \quad (209)$$

It is important to stress that the expression (209) is the same as purely classical expression for $W_n(k)$.

4.5.3   Quantal Expressions for Cross Sections

The solution of a number of problems in the theory of Rydberg-atom–neutral-particle collisions (for example, J-mixing processes and elastic scattering) need calculations of the transition cross sections between states with the given magnitudes of quantum numbers n, l, and total angular momentum J. In the range of $n \ll v_0/V$ corresponding to slow collisions, general formula of the impulse approximation for the $nlJ \to n'l'J'$ transitions was derived in [137] using the basic equation (195) for the scattering amplitude and expressions (148) and (151) for the momentum space wave functions (with spin $s = 1/2$)

$$G_{nlJM}(\mathbf{k}) = \sum_{m\sigma} C_{lms\sigma}^{JM} g_{nl}(k) Y_{lm}(\theta_{\mathbf{k}}, \varphi_{\mathbf{k}}) \chi_{s\sigma} \qquad (210)$$

of the initial $|\alpha\rangle = |nlJM\rangle$ and final $|\alpha'\rangle = |n'l'J'M'\rangle$ states. The method of calculation is based on the expansion (see [142]) of the spherical $j_{l'}(k'r')Y_{l'm'}(\theta_{\mathbf{k}'}, \varphi_{\mathbf{k}'})$ wave of the $(l' + 1/2)$-order over the bipolar harmonics of the l'-rank . The final general expression for the cross section can be written as

$$\sigma_{nlJ,\beta}^{n'l'J',\beta'}(V) = 2\pi \left(\frac{\hbar}{mV}\right)^2$$

$$\times \sum_{\text{œ}=|l'-l|}^{l'+l} A_{l'J',lJ}^{(\text{œ})} \int_{Q_{\min}}^{Q_{\max}} \left| \Phi_{n'l',nl}^{(\text{œ})}(Q) \right|^2 Q \, dQ , \qquad (211)$$

where the lower and upper limits of integration are given by

$$Q_{\min} = |q' - q|, \qquad Q_{\max} = q' + q , \qquad (212)$$

$$(q')^2 = q^2 + (2\mu/\hbar^2)[\Delta E_{\alpha'\alpha} + \Delta E_{\beta'\beta}] . \qquad (213)$$

Here $\Delta E_{\alpha'\alpha} = E_{n'l'J'} - E_{nlJ}$ and $\Delta E_{\beta'\beta} = E_{\beta'} - E_\beta$ are the energy changes of the Rydberg atom A^* and a neutral perturber B, respectively. The angular $A_{l'J',lJ}^{(\text{œ})}$ coefficients are defined by (108), while the radial integral $\Phi_{n'l',nl}^{(\text{œ})}(Q)$ is given by [137]

$$\Phi_{n'l',nl}^{(\infty)}(Q) = i^{l-l'} \left(\tfrac{2}{\pi}\right) \int\limits_0^\infty k^2 \, dk g_{nl}(k)$$

$$\times \int\limits_0^\infty (k')^2 \, dk' g_{n'l'}^*(k') f_{eB}(k',k,Q) \tag{214}$$

$$\int\limits_0^\infty r^2 \, dr \, j_l(kr) \, j_\infty(Qr) j_{l'}(k'r).$$

When the fine-structure splitting $\Delta E_{l'-1/2,l'+1/2}$ of the final $n'l'$-level with $J' = |l' \pm 1/2|$ can be neglected, we can perform a summation of the cross section $\sigma_{nlJ,\beta}^{n'l'J',\beta'}$ over J' with the use of a known relation for the $6j$-symbols. Then, we obtain the following relation for the sum of squared matrix elements

$$\tfrac{1}{2J+1} \sum_{MM'} \left| \mathsf{f}_{nlJM,\beta}^{n'l'J'M',\beta'}(\mathbf{q}',\mathbf{q}) \right|^2$$

$$= \tfrac{1}{2l+1} \sum_{mm'} \left| \mathsf{f}_{nlm,\beta}^{n'l'm',\beta'}(\mathbf{q}',\mathbf{q}) \right|^2 \tag{215}$$

$$= \left(\tfrac{2\pi\mu}{m}\right)^2 \times \sum_{\infty=|l'-l|}^{l'+l} B_{l'l}^{(\infty)} \left| \Phi_{n_*l',n_*l}^{(\infty)}(Q) \right|^2,$$

The angular coefficients in Eq. (215) are expressed in terms of the $3j$-symbol

$$B_{l'l}^{(\infty)} = (2l'+1)(2\infty+1) \left(\begin{array}{ccc} l' & \infty & l \\ 0 & 0 & 0 \end{array} \right)^2.$$

Thus, calculations of the cross sections $\sigma_{n'l',nl} = \sum_{J'} \sigma_{nlJ}^{n'l'J'}$ of the $nlJ \to n'l'J'$ transitions summed over $J' = |l' \pm 1/2|$ is reduced to the evaluation of the $\sigma_{n'l',nl}$ cross sections of the $nl \to n'l'$ transitions. As directly follows from (215), the general formula of the impulse approximation for the cross section of the $nl \to n'l'$ transition differs from (211) only by some other angular coefficient, i.e. one should replace

$$A_{l'J',lJ}^{(\infty)} \to B_{l'l}^{(\infty)} = \sum_{J'} A_{l'J',lJ}^{(\infty)}.$$

For the hydrogen-like nl-state the radial part of the wave function g_{nl} in the momentum space can be expressed in terms of the Gegenbauer C_{n-l-1}^{l+1}-polynomial

$$
\begin{aligned}
g_{nl}(k) = &\left(\tfrac{a_0}{Z}\right)^{3/2} \left[\tfrac{2}{\pi} \tfrac{(n-l-1)!}{(n+l)!}\right]^{1/2} n^2 2^{2(l+1)} l! \\
&\times \frac{(-ix)^l}{(x^2+1)^{l+2}} C_{n-l-1}^{l+1}\left(\frac{x^2-1}{x^2+1}\right),
\end{aligned}
\tag{216}
$$

where $x = nka_0/Z$, and $Z = 1$ for neutral Rydberg atom.

A particularly simple formula may be obtained from the general expression (211) for pure elastic scattering of a neutral atom by the Rydberg atom in the ns-state. In this case the contribution to the sum over œ is determined only by one term with œ = 0 (when $B_{ss}^{(0)} = 1$) and the integral over r' in (214) is taken analytically. As a result, for slow collisions we have [137]

$$
\sigma_{ns}^{\rm el}(V) = 2\pi \left(\tfrac{\hbar}{mV}\right)^2 \int\limits_0^{2q} \left|\Phi_{n0,n0}^{(0)}(Q)\right|^2 Q\, dQ,
$$

$$
\Phi_{n0,n0}^{(0)}(Q) = \tfrac{1}{2Q} \int\limits_0^{\infty} k\, dk\, g_{n_*s}(k)
\tag{217}
$$

$$
\times \int\limits_{|k-Q|}^{k+Q} k'\, dk'\, g_{n0}^{*}(k')\, f_{\rm eB}(k, k', Q).
$$

Here $q = \mu V/\hbar$, and the radial part of the wave function $g_{n0}(k)$ of the ns-state $(l = 0)$ is given by

$$
g_{n0}(k) = \sqrt{\frac{2}{\pi}} \left(\frac{a_0 n}{Z}\right)^{3/2} \frac{4}{(x^2+1)^2} \frac{\sin(n\beta)}{\sin(\beta)},
\tag{218}
$$

$$
\beta = \arccos\left(\frac{x^2-1}{x^2+1}\right).
\tag{219}
$$

It should be noted that in the special case of the ns-state, a similar expression for the cross section of elastic scattering was obtained in [111].

4.6   Relations Between Cross Sections and Form Factors

If the electron–perturber scattering amplitude depends on the momentum transfer Q alone (i.e. $f_{eB} \equiv f_{eB}(Q)$) or it is a constant (for example, in the scattering length approximation $f_{eB} = -L$), the basic formula (195) for the transition amplitude $f_{\alpha'\alpha}^{\beta'\beta}(\mathbf{q} - \mathbf{Q}, \mathbf{q})$ in slow collisions ($n \ll v_0/V$) of the Rydberg atom with neutral particle can be rewritten as [82]

$$f_{\alpha'\alpha}^{\beta'\beta}(\mathbf{q} - \mathbf{Q}, \mathbf{q}) \approx \left(\frac{\mu}{m}\right) f_{eB}^{\beta'\beta}(Q) \langle \psi_{\alpha'}(\mathbf{r})| \exp(i\mathbf{Q} \cdot \mathbf{r}) |\psi_\alpha(\mathbf{r})\rangle.$$

(220)

Here $\psi_\alpha(\mathbf{r})$ and $\psi_{\alpha'}(\mathbf{r})$ are the initial and final atomic wave functions in the coordinate representation.

Then, the cross section (211) of the $|\alpha, \beta\rangle \to |\alpha', \beta'\rangle$ transition is reduced to a simple form

$$\sigma_{\alpha'\alpha}^{\beta'\beta}(V) = 2\pi \left(\frac{\hbar}{mV}\right)^2 \int\limits_{Q_{min}}^{Q_{max}} \left| f_{eB}^{\beta'\beta}(Q) \right|^2 F_{\alpha'\alpha}(Q) \, Q \, dQ,$$

(221)

where $F_{\alpha'\alpha}(Q)$ is the atomic form factor defined by the following basic relation

$$F_{\alpha'\alpha}(\mathbf{Q}) = \frac{1}{g_\alpha} \sum_{a/\alpha, a'/\alpha'} |\langle a'| \exp(i\mathbf{Q} \cdot \mathbf{r}) |a\rangle|^2,$$

(222)

and $\mathbf{Q} = \mathbf{k}' - \mathbf{k}$ is the momentum transferred to the Rydberg electron. The sum in (222) is taken over all sublevels of the total a, a' sets of quantum numbers except of α and α', respectively, while g_α is the statistical weight of the initial α-level. By using the expression for the form factor through the radial matrix elements of the spherical Bessel function, formula (222) can be rewritten as

$$\sigma_{\alpha'\alpha}^{\beta'\beta}(V) = 2\pi \left(\frac{\hbar}{mV}\right)^2 \sum_{\infty=|l'-l|}^{l'+l} A_{l'J',lJ}^{(\infty)}$$

(223)

$$\times \int\limits_{Q_{min}}^{Q_{max}} \left| f_{eB}^{\beta'\beta}(Q) \right|^2 |\langle n'l'| j_\infty(Qr) |nl\rangle|^2 \, Q \, dQ.$$

Comparison of expression (221) with the basic formula of the Born approximation shows that they are in full agreement with each other. Hence, the impulse approximation in its general form contains the Born approximation as a special case.

In the simplest case of the scattering length approximation $f_{eB} = -L = const$ for the electron–atom scattering amplitude formula (221) may be rewritten as

$$\sigma_{fi}(V) = 2\pi L^2 \left(\frac{\hbar}{mV}\right)^2 \int\limits_{Q_{min}}^{Q_{max}} F_{fi}(Q) \, Q \, dQ \, . \qquad (224)$$

This formula gives the opportunity to obtain simple analytic expressions for the cross sections σ_{fi} of various types of the inelastic and quasielastic $|i\rangle \rightarrow |f\rangle$ transitions with the use of the appropriate approximations for atomic form factors $F_{fi}(Q)$. Equivalent analytic expressions may be derived directly from the general formulae of Sect. 4.5.2 assuming $f_{eB} = -L$ and performing integration over the momentum transfer dQ (or over the scattering angle $d(\cos\theta)$) and over the electron momentum dk.

As it was mentioned in Sect. 3.5, the impulse approximation in the scattering length approximation (224) allows us to derive analytic expressions (132) and (139) for the cross sections of the $nl \rightarrow n'$ and $n \rightarrow n'$ transitions summed over the orbital quantum numbers. For this it should be combined with the binary-encounter approximation for the form factor and with simple semiclassical expressions (206) or (209) for the momentum distribution function. Thus, the binary-encounter approach in the impact-parameter representation [132] and in the momentum representation [150, 146] are equivalent to each other.

4.7   Optical Theorem and Asymptotic Scattering Theory

Now we turn to the behavior of the total cross section $\sigma_i^{tot} = \sum\limits_f \sigma_{fi}$ of the Rydberg-atom–neutral collision at high principal quantum numbers n. It is convenient to present it as the sum of pure elastic scattering cross section σ_i^{el} and the total contribution σ_i^{in} of all inelastic bound–bound and bound–free $|i\rangle \rightarrow |f\rangle$ transitions $(f \neq i)$

$$\sigma_i^{\text{tot}} = \sigma_i^{\text{el}} + \sigma_i^{\text{in}}, \qquad \sigma_i^{\text{in}} = \sum_{f \neq i} \sigma_{fi}, \qquad \sigma_i^{\text{el}} \equiv \sigma_{ii} , \qquad (225)$$

where the initial $i = \{\alpha, \beta\}$ and final $f = \{\alpha', \beta'\}$ states of the composite system $A^* + B$ include the sets of quantum numbers of the Rydberg atom (α) and projectile (β). Here we analyze the contribution of the electron–perturber scattering alone, while the role of elastic and inelastic perturber–core scattering mechanism will be discussed in Sect. 7. The asymptotic expression for the total cross section $\sigma_i^{\text{tot}} = \sum_f \sigma_{fi}$ of the Rydberg-atom–neutral collision, which is applicable for an arbitrary relationship between the relative velocity V of colliding particles A^* and B and the mean orbital velocity of the valence electron $v_{nl} \sim v_0/n$, can be obtained from the basic equation (171). Assuming $\mathbf{q}' = \mathbf{q}$ for the imaginary part of the elastic scattering amplitude in the forward direction $\theta_{\mathbf{q}'\mathbf{q}} = 0$, i.e.

$$\text{Im}\,\{f_{ii}\,(\mathbf{q}, \mathbf{q})\} = (\mu/\mu_{eB}) \int \text{Im}\,\left\{f_{eB}^{\beta\beta}\,(\mathbf{k}, \mathbf{k})\right\} |G_\alpha\,(\boldsymbol{\kappa})|^2 \, d\boldsymbol{\kappa} \qquad (226)$$

and using the optical theorem for the total cross sections of Rydberg-atom–neutral collision $\sigma_i^{\text{tot}}(q)$ and for the electron–perturber scattering $\sigma_{eB}(k)$:

$$\text{Im}\,\{f_{ii}\,(\mathbf{q}, \mathbf{q})\} = \frac{q}{4\pi} \sigma_i^{\text{tot}}\,(q) , \qquad (227)$$

$$\text{Im}\,\left\{f_{eB}^{\beta\beta}\,(\mathbf{k}, \mathbf{k})\right\} = \frac{k}{4\pi} \sigma_{eB}^{\beta\beta}\,(k) , \qquad (228)$$

we finally obtain the well known asymptotic expression [82, 83]

$$\sigma_i^{\text{tot}}(V) = \frac{1}{V} \int v_{eB} \sigma_{eB}^{\text{tot}}\,(k) |G_i\,(\boldsymbol{\kappa})|^2 \, d\boldsymbol{\kappa}, \quad n \gg (v_0/V)^{1/2} . \qquad (229)$$

For fast collisions $V \gg v_0/n$ expression (229) yields the result (193) of *Butler* and *May* [180] discussed in Sect. 4.4. In the opposite case of slow collisions $V \ll v_0/n$ (when the electron velocity $v = v_{eA^+} \approx v_{eB} = \hbar k/m$ and momentum $\boldsymbol{\kappa} \approx \mathbf{k}$), expression (229) directly leads to the following formula

$$\sigma_{nl}^{\text{tot}}(V) \approx \int_0^\infty (\hbar k/mV)\, \sigma_{eB}^{\text{el}}(k)\, W_{nl}\,(k)\, dk = \frac{\langle v\sigma_{eB}^{\text{el}}\,(v)\rangle_{nl}}{V} , \qquad (230)$$

derived by *Alekseev* and *Sobel'man* [154] for collisions of a Rydberg atom with an atomic particle. Here

$$\sigma_{eB}^{el}(k) = 2\pi \int\limits_{0}^{\pi} |f_{eB}(k,\theta)|^2 \sin\theta\, d\theta = \frac{4\pi}{k} \text{Im} \left\{ f_{eB}(k, \theta = 0) \right\} \quad (231)$$

is the integral cross section for elastic electron–atom scattering. The asymptotic formula (230) is applicable in the region of sufficiently high principal quantum numbers $(v_0/V)^{1/2} \ll n \ll (v_0/V)$, i.e. $40 - 60 \ll n \ll 10^3 - 10^4$ at thermal energies. In this range of n the cross section σ_{nl}^{el} for an elastic Rydberg-atom–neutral collision becomes very small (since $\sigma_{nl}^{el} \propto n^{-4}V^{-2}$, see (123) at $l' = l$) as compared to the total contribution σ_{nl}^{in} of all inelastic $nl \to n'$ transitions. Hence, its contribution to the total cross section σ_{nl}^{tot} can be neglected, so that $\sigma_{nl}^{tot} \approx \sigma_{nl}^{in}$. Furthermore, in this range of n the inelastic cross section σ_{nl}^{in} is determined by the contribution of a great number of final n' states with $\Delta n = 0, \pm 1, \pm 2, \ldots$. As was shown in [106], formula (230) proceeds directly from the general equation of the impulse approximation (202) if we approximately replace the summation over all n' values by integration over the transition energy defects $d(\Delta E_{n',nl})$.

In the scattering length approximation $f_{eB} = -L$ formula (230) reduces to simple analytic expression for the total cross section σ_{nl}^{tot} of a Rydberg-atom–neutral-atom collision for a given nl-state with $l \ll n$

$$\sigma_{nl}^{tot}(V) \approx \sigma_{nl}^{in}(V) \approx \frac{8L^2 v_0}{nV}. \quad (232)$$

A similar expression for the total scattering cross section

$$\sigma_n^{tot} = \sum_{l=0}^{n-1} \left[(2l+1)/n^2 \right] \sigma_{nl}^{tot}$$

averaged over all possible values of the orbital angular momentum l takes the form

$$\sigma_n^{tot}(V) \approx \frac{32L^2 v_0}{3nV}. \quad (233)$$

These expressions were first derived [128] by the semiclassical impact-parameter method using the Fermi pseudopotential model and the JWKB-approximation for the atomic wave functions.

It is worthwhile to point out that the general formula of the impulse approximation (202) is applicable within the framework of the quasifree electron model not only in the asymptotic region of high n but also at much lower magnitudes of the principal quantum number. In contrast to the asymptotic expressions (230), (232) it provides reliable quantitative results for the total contribution $\sigma_{nl}^{in} = \sum_{n'} \sigma_{n',nl}$ of all inelastic $nl \to n'$ transitions with the use of appropriate values of the energy defects $\Delta E_{n',nl}$. Therefore, formula (202) can also be used for the calculations of the quenching $\sigma_{nl}^{Q} \equiv \sigma_{nl}^{in}$ cross sections, when the main contribution is determined by an $nl \to n'$ transition to a single (nearest to the initial nl state) or several n' levels.

4.8 Transitions Induced by Resonance Scattering

In the presence of a low-energy resonance on the quasidiscrete level of the perturbing atom B the standard scattering length approximation is certainly inapplicable for the description of collisions involving Rydberg atoms. Due to the strong energy dependence of scattering amplitude $f_{eB}(\epsilon, \theta)$ the cross sections of transitions between highly excited states cannot be expressed in terms of form factors. Hence an appropriate description of the Rydberg-atom–neutral collisions at high principal quantum numbers needs an application of the general impulse-approximation approach [105, 106, 137] presented in Sect. 4.5. The mechanism of the resonance scattering in such collisions have been the subject of intensive theoretical and experimental research (see review [36]). The first theoretical works were devoted to study of the possible influence of extremely narrow ($\Gamma_r \ll E_r$) resonances in electron–molecule [185] and electron–atom [186] scattering upon the behavior of the pressure width and shift of high Rydberg levels. They were based on the asymptotic theory [154] of spectral-line broadening. The next theoretical works on the quenching [105, 111] and broadening [106, 103] of Rydberg states in alkali vapors have demonstrated the major contribution of the ^3P-resonance scattering by alkali-metal atoms (for which $\Gamma_r \sim E_r = 10 - 100$ meV) in these processes. This resonance leads to sharp energy and angular ($\propto \cos \theta$) dependence of the scattering amplitude $f_{eB}(\epsilon, \theta)$ in the range above 10 meV (see Sect. 2.8). As a result, the quenching and broadening cross sections become particularly large. Furthermore, the resonance scattering leads to the appearance of some new phenomena in dependencies of the broadening and shift of spectral lines on the principal quantum number (such as their oscillatory behavior at intermediate $n \sim 15 - 25$ for K, Rb, and Cs). Theoretical description of the identified phenomena were the subject of several papers within the framework of the adiabatic quasimolecular approach [110, 187, 188].

Here we present simple results of calculations [105, 106] for the contributions of the resonance $\sigma_r^{in}(nl)$ and potential $\sigma_p^{in}(nl)$ electron–atom scattering into the total cross section $\sigma_{nl}^{in} = \sum_{n'} \sigma_{n',nl}$ of all inelastic $nl \rightarrow n'$ transitions between highly excited states induced by collisions with the ground-state alkali-metal atoms.

We shall proceed from the general equation (202) for slow collisions. In contrast to the asymptotic expression (230), this equation gives the opportunity to explain quantitatively the essential influence of the quantum defect value δ_l on the magnitude of the cross section σ_{nl}^{in}, which was observed in experiments on the quenching [114, 189] and broadening [112, 113] of high-Rydberg atomic levels by the ground-state alkali-metal atoms. It is convenient to present the sum σ_{nl}^{in} of all inelastic $nl \rightarrow n'$ transitions as

$$\sigma_{nl}^{in} = \sigma_r^{in} + \sigma_p^{in} + \text{interference term} , \qquad (234)$$

in accord with expression (63). Each term in this expression is a result of substitution of the corresponding term in (63) into the general formula (202).

Let us discuss the contribution of the ^3P-resonance scattering to the total inelastic cross section. With the use of the simple expression (66) for the resonance part of the differential cross section, the final result can be written as [105]

$$\sigma_r^{in}(nl) = \frac{63\pi v_0^2}{10V^2} \sum_{n'} \frac{1}{(n')^3} \int_{\epsilon_{min}}^{\infty} \left| f_+^r(\epsilon) \right|^2 D(\epsilon) \, W_{nl}(\epsilon) \, d\epsilon , \qquad (235)$$

where the $D(\epsilon)$ function is given by the relation

$$D(\epsilon) = 1 - \frac{15}{7} \left(\frac{\epsilon_{min}}{\epsilon} \right)^{1/2} + \frac{20}{7} \left(\frac{\epsilon_{min}}{\epsilon} \right)^{3/2} - \frac{12}{7} \left(\frac{\epsilon_{min}}{\epsilon} \right)^{5/2} , \qquad (236)$$

and $\epsilon_{min} = \hbar^2 k_{min}^2 / 2m = \left| \Delta E_{n',nl} \right|^2 / 8mV^2$. The distribution function of the Rydberg electron kinetic energy $\epsilon = \hbar^2 k^2 / 2m$ in (235) is expressed through the momentum distribution function, considered in Sect. 4.5, by the relation

$$W_{nl}(\epsilon) = \frac{\sqrt{2m^3\epsilon}}{\hbar^3} \left| g_{nl}\left[k\left(\epsilon\right)\right]\right|^2, \qquad \int\limits_0^\infty W_{nl}(\epsilon)d\epsilon = 1. \qquad (237)$$

Formula (235) directly follows from (202) by performing a trivial integration of the differential cross section $d\sigma^r_{el}/d\Omega \propto \cos^2\theta$ over the scattering angle $\cos\theta$. Thus, the very appearance of the $D(\epsilon)$ factor in this formula is due to the p-type of the resonance partial scattering wave. A simple asymptotic estimate for the resonance part of the total inelastic cross section can be obtained from (235) as a special case. The final asymptotic expression is given by [106]

$$\sigma_r^{in}(nl) \approx \frac{9\pi\gamma^2\hbar^2}{2^{1/2}m^{3/2}V} \int\limits_0^\infty \frac{W_{nl}(\epsilon)\epsilon^{5/2}d\epsilon}{[(\epsilon-\epsilon_0)^2 + \gamma^2\epsilon^3]}. \qquad (238)$$

As was mentioned in Sect. 2.8.1, for the alkali-metal atoms the energies and widths of the ^3P-resonances prove to be of the same order of magnitude ($\Gamma_r \sim E_r \sim 10^{-1} - 10^{-2}$ eV). Therefore, at sufficiently high n the period of oscillations of the energy distribution function $W_{nl}(\epsilon)$ (described by (204)) is much smaller than the width of such resonances. Therefore, these oscillations practically cancel each other in (235) and (238). Hence in the region of applicability of the quasi-free electron model ($n > 25 - 30$ for thermal collisions with K, Rb, and Cs, see below Sect. 6.3) the oscillations in the dependencies of the broadening cross sections upon n do not arise. Thus, the simple expression (206) for the average energy distribution function can certainly be used in calculations of the broadening and quenching cross sections of Rydberg atoms perturbed by alkali-metal atoms.

At sufficiently high n the resonance $\sigma_r^{in}(nl)$ contribution to the broadening or quenching cross sections falls with an increase of the principal quantum number. It is due to the decrease of the energy distribution function $W_{nl}(\epsilon)$ for the resonance energies $\epsilon \sim E_r$ at $n \gg (Ry/E_r)^{1/2}$. As a result, in the asymptotic region of very high n the main contribution to the total cross section σ_{nl}^{in} of the Rydberg nl level by alkali atoms is due to the electron–perturber potential scattering. Thus, a reliable quantitative description of the quenching and impact broadening processes at high principal quantum numbers needs an account of both the ^3P-resonance and potential triplet and singlet scattering. Since the partial phase shifts $\eta_\ell^{(+)}(k)$ and

$\eta_\ell^{(-)}(k)$ fall strongly near the zero energy $(k \ll a_0^{-1})$ with increasing ℓ $(\ell > 1)$, it is enough to take into account the contribution of
only the first few partial waves to the scattering amplitude (57) and
corresponding differential cross section (58). We will show in Sect.
6 that general impulse-approximation approach [105, 106] combined
with the theory of resonance and potential electron–alkali-atom scattering provides a quantitative description of these processes at high
principal quantum numbers ($n > 15 - 20$ for Li(2s), Na(3s), and
$n > 25 - 30$ for K(4s), Rb(5s), and $n > 30$ for Cs). In this range of
n the quasifree electron model is valid and practically no oscillations
occur. Therefore, the theory presented above allows us to obtain the
values for the ^3P-resonance energies E_r and widths Γ_r, which are in
reasonable agreement with the results [102, 103] of the modified effective range theory (see Table 8). These semiempirical data have been
obtained for K and Rb [105, 106] from the comparison of theoretical
results for the monotone component of the broadening cross sections
of Rydberg levels in alkali vapors with experimental data measured
in [112, 113].

4.9   Analysis of Applicability Conditions

The applicability conditions of the impulse approximation and the
quasifree electron model for quasielastic, inelastic and ionizing collisions between the Rydberg atom A^* and neutral perturbing particle
were discussed in reviews [36, 82, 83]. Here only the physical meaning
of basic approximations is under discussion. In the impulse approximation for an independent treatment of the two possible mechanisms
of the Rydberg-atom–neutral collisions caused by the scattering of
the perturber B on the outer electron and on the ionic core A^+, the
following conditions should be satisfied

$$r_{\mathrm{eA}^+} \gg A = \max \left\{ |f_{\mathrm{eB}}|, \ |f_{\mathrm{A}^+\mathrm{B}}|, \ \lambda_{\mathrm{eB}} \right\}, \qquad (239)$$

$$T_n \gg \tau_{\mathrm{eB}} , \ \tau_{\mathrm{A}^+\mathrm{B}} , \qquad (240)$$

where f_{eB} and $f_{\mathrm{A}^+\mathrm{B}}$ are the amplitudes for the scattering of the perturbing particle B on the outer electron and on the parent core A^+,
respectively; while λ_{eB} is the de Broglie wavelength for the electron–
perturber relative motion. Due to the large characteristic dimension
r_{eA^+} of the interaction region between the ionic core A^+ and Rydberg
electron (which is responsible for the main contribution to the collision process) the interference effects in the e–B and A^+–B scattering

as well as the effects of multiple scattering can be neglected at high enough n. The second condition (240) means that in order for the "impulse" treatment to be applicable, both the electron–perturber τ_{eB} and the core-perturber $\tau_{\text{A+B}}$ collision times should be small compared to the period $T_n \sim 2\pi \left(a_0/v_0\right) n^3$ of the orbital electron motion. As is apparent from (240), the applicability of the impulse approximation in calculations of various elementary processes involving Rydberg atoms and neutral particles is determined not only by the nature of interparticle $e-\text{B}$ and A^+-B interactions but also the relative velocity V of the colliding partners A^* and B as well as the orbital electron velocity v_0/n. It is important to emphasize that the impulse approximation may be used both for fast $V \gg v_0/n$ and for slow $V \ll v_0/n$ collisions, provided the principal quantum number n is large enough.

One more important condition of applicability of the impulse approximation, which clarifies the physical meaning of the quasifree electron model, can be written as

$$\hbar Q_{\text{eB}} \gg |F_{\text{eA+}}|\tau_{\text{eB}} = \left(e^2/r_{\text{eA+}}^2\right)\tau_{\text{eB}} . \qquad (241)$$

Here $\hbar Q_{\text{eB}}$ is the characteristic momentum transferred to the Rydberg electron in its collision with the perturbing particle B; $(|F_{\text{eA+}}|\tau_{\text{eB}})$ represents the impulse of the Coulomb force acting on this electron from the parent core A^+ during the collision time τ_{eB} of e and B particles. The collision time of the quasifree electron with a neutral projectile B is determined by the relation $\tau_{\text{eB}} \sim r_{\text{eB}}/v_{\text{eB}}$, in which the characteristic dimension of the interparticle interaction region may be estimated as $r_{\text{eB}} \sim \max\{\lambda_{\text{eB}}, |f_{\text{eB}}|\}$. Condition (241) can be approximately rewritten as (240) or $\omega_{n,n\pm1}\tau_{\text{eB}} \ll 1$, where $\omega_{n,n\pm1} = 2Ry/\hbar n^3$ is the characteristic transition frequency (see (15)). Thus, it is clear that the validity conditions (240) and (241) of the impulse approximation (or the "quasifree" electron scattering) corresponds to the applicability criterion of the sudden perturbation theory.

Note also that for the impulse approximation to be applicable it is necessary that the cross sections of the Rydberg-atom–neutral collisions do not exceed the geometrical cross section of the highly excited atom

$$\sigma_{fi} \ll \sigma_{\text{geom}} \sim \pi a_0^2 n_*^4 . \qquad (242)$$

This condition is particularly important in the case of elastic and quasielastic collisions, since it is automatically satisfied for inelastic collisions accompanied by sufficiently large energy transfer.

The value of the electron–perturber scattering amplitude in (239) may be estimated as $|f_{eB}| \sim (\langle \sigma_{eB} \rangle /4\pi)^{1/2}$. It is usually reduced to the scattering length $|f_{eB}| \sim |L|$ for potential electron–atom scattering. The perturber–core scattering amplitude $|f_{A+B}|$ is determined by the characteristic dimension of the polarization interaction $V^{l.r.}_{A+B} = -\alpha_B e^2/2R^4$ of the perturbing atom B and the ionic core A^+. Its value may be estimated by the Weisskopf radius

$$\rho_W = (\pi \alpha_B v_0/4V)^{1/3} \qquad (243)$$

or by the impact parameter

$$\rho_{cap} = (4e^2 \alpha_B/\mu V^2)^{1/4} \qquad (244)$$

corresponding to the capture of atom B by the core A^+ of the Rydberg atom (i.e. $|f_{A+B}| \sim \rho_W$ or $|f_{A+B}| \sim \rho_{cap}$ depending on the inelasticity of one or another process). At thermal energies the values of ρ_W and ρ_{cap} are usually greater than the corresponding magnitudes of electron–atom scattering lengths L, so that $A \sim \rho_W$ or $A \sim \rho_{cap}$ in (239). The characteristic radius r_{eA^+} and the de Broglie wavelength λ_{eB} depend on the principal quantum number n, the relative velocity V of colliding particles, and the energy $\Delta E_{fi} = \hbar \omega_{fi}$ transferred to the Rydberg electron. As was shown in [132], their values can be estimated as

$$r_{eA^+} \sim \frac{2n_*^2 a_0}{1 + (n_* a_0 \omega_{fi}/V)^2}, \qquad \lambda_{eB} \sim (a_0 r_{eA^+}/2)^{1/2} \qquad (245)$$

for the $nl \to n'$ and $n \to n'$ transitions. Thus, for quasielastic collisions (when the inelasticity parameter is small $\lambda = n_* a_0 \omega_{fi}/V \ll 1$) the r_{eA^+} value is of the order of the atomic orbital radius $2n_*^2 a_0$ and the de Broglie wavelength of the electron–perturber relative motion is $\lambda_{eB} \sim n_* a_0$. However, for essentially inelastic collisions (with the parameter $\lambda \gg 1$) these values are determined by the relations $r_{eA^+} \sim 2n_*^2 a_0/\lambda^2$ and $\lambda_{eB} \sim n_* a_0/\lambda$, so that $r_{eA^+} \ll 2n_*^2 a_0$ and $\lambda_{eB} \ll n_* a_0$.

Specific estimates show that at thermal collisions of the Rydberg atoms with the ground-state rare gas atoms the impulse approximation is valid for $n \gg 10-20$ (and $n \gg 20-30$ for ionization) depending on the kind of colliding particles and on the energy defect $|\Delta E_{fi}|$

of the transition under consideration. The validity conditions of the impulse approximation in the presence of the ultra-low energy resonance on the quasidiscrete level of a perturbing atom were discussed in [105, 106]. It was shown that in order for it to be applicable, it is also necessary that the resonance width Γ_r be significantly greater than the energy level spacing $\Delta E_{n,n\pm 1}$, so that $\Gamma_r \gg 2Ry/n^3$. This condition is equivalent to (240), since the collision time in this case is determined by the lifetime $\hbar/\Gamma_r$ of the quasidiscrete level. The detailed analysis of conditions (239) and (240) shows that at thermal collisions of the Rydberg atoms with the ground-state alkali-metal atoms the impulse approximation is applicable for $n > n_{\mathrm{low}}$, where $n_{\mathrm{low}} \sim 15-20$ for Li, Na, $n_{\mathrm{low}} \sim 25-30$ for K, Rb, and $n_{\mathrm{low}} \sim 30-35$ for Cs. For collisions with strongly polar molecules (such as NH_3, HF, HCl etc.), the typical magnitudes of scattering amplitudes $|f_{\mathrm{eB}}|$ and $|f_{\mathrm{A+B}}|$ are much greater than for atomic perturbers since they are determined by the long-range dipole potential (78). As a result, the permissible values of n turn out to be shifted toward high principal quantum numbers and the quasifree electron model is justified only for $n > 50 - 60$. Nevertheless, reasonable agreement with the experimental data of the impulse-approximation calculations may be sometimes achieved in the range of n, in which it is formally inapplicable (see below Sect. 6).

5   Inelastic Transitions Involving Rydberg States

In this section our attention will be focused on further development of the theory of collisions between Rydberg atoms and neutral particles (atoms and molecules) and its application to studies of various types of bound–bound and bound–free transitions. We shall consider different processes of excitation, deexcitation and ionization of the highly-excited atomic states in both the weak- and strong- coupling regions. However, as in the preceding section, we discuss situations here in which these processes are due to the mechanism associated with the scattering of a Rydberg electron on the target atom or molecule. The main emphasis will be put on the detailed description of the total and orbital angular momentum transfer processes, the inelastic transitions with a change in the principal quantum number since they were the subject of intensive experimental and theoreti-

cal research. We shall present a careful analysis of the dependence
of cross section behavior on the principal quantum number, relative
velocity of colliding particles, and the inelasticity parameter of one
or another process. The final formulae will be given in a form convenient for specific calculations. Another goal of this section is to give
the self-consistent description of resonance depopulation of highly
excited states in collisions of Rydberg atoms with the ground-state
atoms having small electron affinities. As in the case of ion–pair formation processes, here we shall consider the mechanism of excitation
(deexcitation) of highly excited atoms accompanied by transitions between the Rydberg-covalent and ionic electronic terms of a diatomic
system.

5.1 Main Types of Collisional Transitions

The elementary processes induced by collisions between the Rydberg atoms and neutral targets (atoms or molecules) can be divided
into two main groups. The first one involves the so-called state-
changing processes (i.e. various types of quasielastic and inelastic
bound–bound transitions between highly excited states) and elastic
scattering. The second group involves different ionization processes
(i.e. bound–free transitions of the Rydberg electron). Summary table 9 containing the most important collisional processes, which will
be the subject of this and next sections, is presented below.

Some of these processes (e.g. elastic scattering, quasielastic l-
mixing collisions and inelastic n, l-changing transitions) have already
been considered in the preceding section. However, all theoretical
results on collisions between the Rydberg atoms and neutral particles presented above were obtained in the weak-coupling approximation like the Born approximation or quantal impulse approximation
and its semiclassical version (binary-encounter theory in the impact-
parameter or momentum representations). Such approaches are valid
if the transition probability between the highly excited states is sufficiently small ($W_{fi} \ll 1$). This means that the corresponding cross
section of elementary process should be small compared with the geo-
metrical area of the Rydberg atom, i.e. $\sigma_{fi} \ll \sigma_{\mathbf{geom}} \sim \pi a_0^2 n_*^4$).

At the same time, the behavior of transition probabilities and
cross sections of collisions between a Rydberg atom and a neutral
particle depends drastically on both the principal quantum number
n and the energy defect ΔE_{fi} of the process as well as on a particular

type of colliding particle and the relative velocity V. In particular, the cross sections and rate constants turn out to be quite different for transitions with small and large energy transferred to the highly excited electron from the relative motion of neutral projectile and ionic core of Rydberg atom.

Table 9. Classification of elementary processes in collisions of Rydberg atoms with neutral atoms and molecules

ELASTIC SCATTERING PROCESS
Collision involving Rydberg atom in selectively excited state α:
$\quad$ A (α) + B $\rightarrow$ A (α) + B,$\qquad$ $(\lvert\alpha\rangle = \lvert nl\rangle$ and $\lvert\alpha\rangle = \lvert nlJ\rangle)$
STATE-CHANGING PROCESSES
1. Transitions between the fine-structure components
(J – mixing process):
$\quad$ A (nlJ) + B $\rightarrow$ A (nlJ') + B,$\quad$ $(J' \neq J,\ \ J = \lvert l \pm 1/2 \rvert)$
2. Transitions with change in the orbital angular momentum
(l-mixing process):
$\quad$ A(nl) + B $\rightarrow$ A(nl') + B,$\qquad$ $(l' \neq l)$
3. Transitions with change in the principal quantum number
(n-changing processes):
(a) *Collision with atom:*
$\quad$ A (α) + B $\rightarrow$ A (α') + B,$\qquad$ $(n' \neq n)$
(b) *Rotationally inelastic collision with molecule:*
A (α) + B (j) $\rightarrow$ A (α') + B (j'),$\qquad$ $(n' \neq n,\ j' \neq j)$
(c) *Rotationally elastic collision with molecule:*
A (α) + B (j) $\rightarrow$ A (α') + B (j),$\qquad$ $(n' \neq n,\ j' = j)$
IONIZING PROCESSES
1. Direct ionization:
$\quad$ A (α) + B (β) $\rightarrow$ A$^+$ + B (β') + e,$\quad$ $(\beta' = \beta)$
2. Ionization via rotational deexcitation of molecule:
$\quad$ A (α) + B (β) $\rightarrow$ A$^+$ + B (β') + e,$\quad$ $(E_{\beta'} < E_\beta)$
3. Associative ionization:
$\quad$ A (α) + B (β) $\rightarrow$ BA$^+$ + e
4. Ionization by electron attaching atom or molecule:
$\quad$ A (α) + B (β) $\rightarrow$ A$^+$ + B$^-$ (γ)
5. Ionization accompanied by dissociative electron attachment:
$\quad$ A (α) + CD $\rightarrow$ A$^+$ + C$^-$ + D

Analysis of the inelastic n, l-changing transitions with sufficiently large energy transfer ΔE_{fi} and especially ionizing collisions involving the Rydberg and the ground-state atoms shows that these processes may be reasonably described within the framework of first-order perturbation theory or in the impulse approximation in a wide enough range of the principal quantum number. This means that in this case the range of weak coupling involves simultaneously high, intermediate and sufficiently low principal quantum numbers (provided the value of n is not too small so that the quasifree electron model is broken down). Close coupling arises here only at very low magnitudes of the impact parameters $0 \le \rho \le \rho_0$ (corresponding to large values of the momentum transfer Q), which do not exert any appreciable influence on the final result for the cross section in integration over all possible values of ρ ($0 \le \rho < \infty$).

However, the approaches based on perturbation theory can be applied to calculations of the cross sections of collisions with small energy transfer (e.g. quasielastic l-mixing and J-mixing processes) only at high principal quantum numbers $n \gg n_{\max}$ which correspond to the weak-coupling range. At low $n \ll n_{\max}$ and intermediate $n \sim n_{\max}$ the probabilities W_{f_i} of these processes become of the order of unity and the cross sections are about the geometrical area of the Rydberg atom $\sigma_{\text{geom}} \sim \pi a_0^2 n_*^4$. The use of first-order perturbation theory in the strong-coupling range of $n \lesssim n_{\max}$ leads to an appreciable overestimate of the transition probabilities and to incorrect qualitative behavior of the corresponding cross sections. A consistent quantitative description of such collisions at low and intermediate n requires the inclusion of all couplings between near-degenerate states, but the application of the close-coupling method to collisions involving Rydberg atoms becomes very difficult at high principal quantum numbers due to the presence of a great number of closely-spaced levels.

A simple version of this method at low n (< 10) was used for the quasielastic l-mixing process in thermal Rydberg-atom–rare-gas-atom collisions [161]. Semiclassical calculations [238] of the n-changing process in thermal collisions of $\text{Na}(nS)+\text{He}$ were performed by the close-coupling method at $n = 6$ and 9. The same approach was used [239] for the description of l-mixing in rotationally elastic collisions of Rydberg atoms with strongly polar molecules HF and HCl in a wide range of n. However, reliable semiclassical calculations based on numerical integration of the close-coupling equations for the transition amplitudes were carried out only for a few specific processes involving Rydberg atoms. Furthermore, numerical close-coupling calculations have not yet given a general picture for different processes with regard to their dependence on the quantum numbers of the Rydberg atom, the relative velocity V of colliding particles and the transition energy defect ΔE_{f_i}. On the other hand, the perturbative approaches violate the conservation of transition probability at low and intermediate n if the energy transfer is not sufficiently large.

In the general case, for arbitrary magnitudes of the principal quantum number n, and energy defect ΔE_{fi}, an efficient method of describing Rydberg-atom–neutral collisions is to use the so-called normalized perturbation theory. The simplest way to restore the conservation of probability is to use the unitarized version of first-order perturbation theory in the impact-parameter representation (see Sect. 3.3). It is based on the separation of the whole range of ρ into two regions ($\rho \leq \rho_0$) and ($\rho_0 < \rho$) with qualitatively different behaviors of the transition probability (99) and its integration in accord with (100).

Such an approach combined with the JWKB-approximation for the Rydberg atom wave functions was developed for description of the J-mixing collisions [137] as well as the n, l-changing and l-mixing processes [140, 138, 139] in a wide range of the principal and orbital quantum numbers and transition energy defects. The identified approach allows us to perform most calculations of probabilities, cross sections and rate constants in analytical form. Hence, it turns out to be especially effective for understanding their dependencies on the principal quantum number n, the energy defect of the process ΔE_{fi}, the relative velocity V, and the parameter of the electron–perturber interaction.

5.2 Total Angular Momentum Transfer

The inelastic transitions between the fine-structure $nlJ \rightarrow nlJ'$ components of Rydberg atom induced by collisions with neutral projectiles

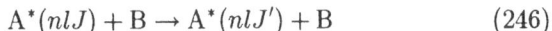

$$A^*(nlJ) + B \rightarrow A^*(nlJ') + B \tag{246}$$

are accompanied by a small energy $\Delta E_{J'J}$ transferred to the highly excited electron from the translational motion of heavy particles. Therefore, they are characterized by large values of the cross sections, which can be of the same order of magnitude as in the process of elastic scattering. Calculations of the J-mixing cross sections in the range of weak coupling (high principal quantum numbers n) has been performed by *Sirko* and *Rosinski* [135], *Lebedev* [136], and *Liu* and *Li* [235] for the $n^2D_{3/2} \rightarrow n^2D_{5/2}$ transitions in the Rydberg Cs and Rb induced by thermal collisions with the rare gas atoms.

An analytic description of the inelastic $nlJ \to nlJ'$ transitions between the fine-structure components in the range of weak and close coupling has been carried out in [137] on the basis of the impulse approximation and semiclassical impact-parameter method.

5.2.1 Inelasticity Parameter of J-Mixing Process

In the range of weak coupling the semiclassical formula (107) for the probability of the $nlJ \to nlJ'$ transition in the impact-parameter representation can be reduced to a rather simple form

$$W_{nlJ}^{nlJ'}\left(\rho, V\right) = \frac{L^2 v_0^2}{\pi^2 n_*^6 V^2} \sum_{s=0}^{l} A_{lJ',lJ}^{(2s)}$$

$$\times \int_{r_{\min}}^{r_{\max}} \frac{\cos\left(\Delta\Phi_R\right) dR}{R k_R (R^2 - \rho^2)^{1/2}} \int_{r_{\min}}^{r_{\max}} \frac{\cos\left(\Delta\Phi_{R'}\right) dR'}{R' k_{R'} (R'^2 - \rho^2)^{1/2}} \qquad (247)$$

$$\times \cos\left[\omega_{J'J}\left(t(R) - t(R')\right)\right] P_{2s}\left(\cos\Theta_{\mathbf{R'R}}\right)$$

if we use the JWKB-approximation for the radial wave functions (112) and the Fermi pseudopotential model for the electron–perturber interaction. Here $r_{\max} = r_2$, $r_{\min} = \max\{\rho, r_1\}$ and $r_{1,2} = n_*^2 a_0(1 \mp \epsilon)$ are the classical turning points, i.e. $r_2 \approx 2n_*^2 a_0$ and $r_1 \approx (l + 1/2)^2 a_0 / 2 \ll 2n_*^2 a_0$ at $l \ll n$. The transition frequency $\omega_{J'J} = |\Delta E_{J'J}| / \hbar$ between the fine-structure components $J = |l - 1/2|$ and $J' = l + 1/2$ (and corresponding quantum defect difference $\Delta\delta_{J'J}$) of the Rydberg atom with one valence electron may be estimated from the relation [69]

$$|\Delta E_{J'J}| = \frac{2Ry\,|\delta_{lJ} - \delta_{lJ'}|}{n_*^3}, \qquad (248)$$

$$|\delta_{lJ} - \delta_{lJ'}| = \frac{\alpha^2 Z_i^2 H_{\mathrm{rel}}(l, Z)}{2l(l + 1)}. \qquad (249)$$

Here $\alpha = e^2 / \hbar c \approx 1/137$; $H_{\mathrm{rel}}(l, Z)$ is the relativistic correction which is relevant only for heavy Rydberg atoms with the spectroscopic symbols $Z \gtrsim 50$ (and $H_{\mathrm{rel}} \approx 1$ for $Z \lesssim 50$); $Z_i e^2$ is the effective charge of the ionic core A^+ for the valence electron which is a short distance from the nucleus ($Z_i \equiv Z_i(l, Z)$). As a rule: $Z - Z_i\,(l = 1) \approx 4 - 6$ and $Z - Z_i\,(l = 2) \approx 10 - 16$ for the p- and d-states of alkali-metal atoms.

One can see that the quantum defect difference of the fine-structure components is very small $\Delta \delta_{J'J} \ll 1$. Hence, the phase difference $\Delta \Phi_R$ of the semiclassical wave functions (116) in formula (247) may be neglected, and we may assume $\cos (\Delta \Phi_R) \approx 1$ and $\cos (\Delta \Phi_{R'}) \approx 1$. The magnitudes of the angular $A_{lJ',lJ}^{(2s)}$ coefficients are presented in [137] for transitions between the fine-structure components of the np-, and nd-levels. In the special case, when the energy splitting of the fine-structure components is very small and can be neglected, the transition probability of the $nlJ \rightarrow nlJ'$ can be approximately presented as

$$W_{nlJ}^{nlJ'}(\rho, V) \approx \frac{C_{J'J}^{(l)}}{2n_*^6} \left(\frac{v_0}{V}\right)^2 \left(\frac{L^2}{\rho a_0}\right), \quad \Delta E_{J'J} \rightarrow 0 . \tag{250}$$

This formula differs from the case of pure elastic scattering of the perturbing atom by the Rydberg atom in the ns-state (see Sect. 3.4) only by a specific value $C_{J'J}^{(l)}$ of the constant coefficient.

Integration of (247) over the $2\pi\rho d\rho$ yields the required result for the cross sections of the J-mixing process $\sigma_{nlJ}^{J-\text{mix}} \equiv \sigma_{nlJ}^{nlJ'}$ (the transition between the fine-structure $nlJ \rightarrow nlJ'$-components with $J' \neq J$) and of pure elastic scattering $\sigma_{nlJ}^{\text{el}} \equiv \sigma_{nlJ}^{nlJ}$ ($nlJ \rightarrow nlJ$ transition) in the range of weak coupling. However, to derive simple result for the dependence of the cross section on the energy of the fine-structure splitting it is more convenient to use expression (221) of the impulse approximation directly. As has been shown in [136], at high enough n the final analytic formulae are given by

$$\sigma_{nlJ}^{J-\text{mix}} = \frac{2\pi C_{J'J}^{(l)} L^2 v_0^2}{V^2 n_*^4} \varphi_{J'J}^{(l)}(\nu) , \tag{251}$$

$$\sigma_{nlJ}^{\text{el}} = \frac{2\pi C_{JJ}^{(l)} L^2 v_0^2}{V^2 n_*^4} \quad \left(n_* \gg n_*^{(0)}\right) , \tag{252}$$

where $\nu = n_*^2 a_0 |\Delta E_{J'J}| / \hbar V$ is the dimensionless inelasticity parameter of the J-mixing process. The $C_{J'J}^{(l)}$ coefficients and the $\varphi_{J'J}^{(l)}(\nu)$ function in (251) may be expressed in terms of spherical Bessel functions

$$C^{(l)}_{J'J} = \xi^{(l)}_{J'J}(0) \, , \qquad \varphi^{(l)}_{J'J}(\nu) = \xi^{(l)}_{J'J}(\nu) / \xi^{(l)}_{J'J}(0) \, ,$$

$$\xi^{(l)}_{J'J}(\nu) = \sum_{s=0}^{l} A^{(2s)}_{lJ',lJ} \int_{\nu}^{\infty} j_s^2(z) J_s^2(z) \, z \, dz, \tag{253}$$

$$\xi^{(l)}_{JJ'}(\nu) = \frac{2J+1}{2J'+1} \xi^{(l)}_{J'J}(\nu) \, .$$

The plot of the $\varphi^{(l)}_{J'J}(\nu)$ function against the value of the inelasticity parameter $\nu_{J'J} = |\delta_{lJ'} - \delta_{lJ}| v_0/V n_*$ is shown in Fig. 11 for the most interesting cases of $n^2 P_{1/2} \to n^2 P_{3/2}$ and $n^2 D_{3/2} \to n^2 D_{5/2}$ transitions. The magnitudes of coefficients for such transitions are as follows: $C^{(p)}_{3/2,1/2} = 0.451$ and $C^{(d)}_{5/2,3/2} = 0.474$ for the $1/2 \to 3/2$ and $3/2 \to 5/2$ transitions, while $C^{(p)}_{1/2,3/2} = 0.225$ and $C^{(d)}_{3/2,5/2} = 0.316$ for the inverse transitions, respectively.

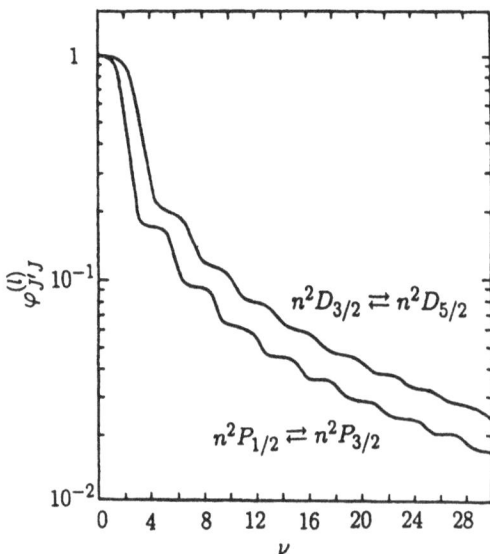

Figure 11. The dependencies of the $\varphi^{(l)}_{J'J}$ functions on the value of the inelasticity parameter $\nu = n_*^2 a_0 \omega_{J'J}/V$ for the $n^2 P_{1/2} \to n^2 P_{3/2}$ and $n^2 D_{3/2} \to n^2 D_{5/2}$ transitions [137].

As is evident from (251) and (253), at small magnitudes of the inelasticity parameter $\nu_{J'J} \ll 1$ the behavior of the J-mixing cross section $\sigma_{nlJ}^{J-\text{mix}}$ is similar to the case of pure elastic scattering ($nlJ \to nlJ$) or quasielastic $nl \to nl'$ transitions ($\omega_{nl',nl} = 0$) with change in the orbital angular momentum alone (123). However, the magnitudes of the constant coefficients differ essentially from each other (e.g. $C_{1/2,3/2}^{(p)}/C_{pp} \approx 0.218$ and $C_{3/2,5/2}^{(d)}/C_{dd} \approx 0.246$). Hence, at $\nu_{J'J} \ll 1$ the formulae (251) and (253) of the weak-coupling approximation leads to the following expression for the sum of the J-mixing and elastic scattering cross sections

$$
\sigma_{nlJ}^{\text{el}} + \sigma_{nlJ}^{J-\text{mix}} \equiv \sum_{J'} \sigma_{nlJ}^{nlJ'} \xrightarrow[\nu \to 0]{} \sigma_{nl,nl}
$$

$$
= \frac{2\pi C_{ll} L^2}{n_*^4 (V/v_0)^2} , \qquad n_* \gg n_*^{(0)},
\tag{254}
$$

$$
C_{ll} = \sum_{J'} C_{J'J}^{(l)} = \sum_{s=0}^{l} B_{ll}^{(2s)} \int_0^\infty j_s^2(z) J_s^2(z)\, z\, dz ,
\tag{255}
$$

where the angular coefficients $B_{ll}^{(2s)}$ are defined by (215).

5.2.2   Collision Strength and Cross Section Behavior

If the energy splitting of the fine-structure components of a Rydberg atom is not too large then the inelasticity parameter of the J-mixing process $\nu_{J'J} = |\delta_{lJ'} - \delta_{lJ}| v_0/V n_*$ turns out to be small enough $\nu_{J'J} < 1$ not only at high n, but also at the intermediate and low n. As is evident from (248), this situation usually occurs for Rydberg atoms with small spectroscopic symbols Z in their thermal collisions with both light and the heavy perturbing atoms. However, even for the $n^2 D_{3/2} \to n^2 D_{5/2}$ transitions in the Rydberg Rb- and Cs- atoms (for which the experimental data are available) the influence of the value $\nu_{J'J}$ on the behavior of the J-mixing collisions with He is not practically important in the whole studied range of n.

Here we show that a simple analytic description [137] of quasielastic J-mixing transitions with $\nu_{J'J} < 1$ and $\varphi_{J'J}^{(l)}(\nu) \approx 1$ may be simultaneously given in the range of weak and close coupling, i.e. at high, intermediate, and sufficiently low principal quantum numbers. This approach is based on the results of first-order perturbation theory for the transition probability $W^{\mathrm{B}} = W_{nlJ}^{nlJ'}$ (250) and their normalization at low and intermediate n in accordance with basic semiclassical formulae of the unitarized perturbation theory (99) and (100).

According to [137], the probability of the $nlJ \to nlJ'$ transition between the fine-structure components with $l \ll n$ may be described by the following expression

$$W_{nlJ}^{nlJ'} = \begin{cases} (\mathrm{g}_{J'}/\mathrm{g}_s\mathrm{g}_{nl})c \,, & 0 \leq \rho \leq \rho_0, \\ \frac{C_{J'J}^{(l)}L^2}{2n_*^6\rho a_0}\left(\frac{v_0}{V}\right)^2, & \rho_0 \leq \rho \leq 2n_*^2 a_0, \\ 0 & 2n_*^2 a_0 < \rho \end{cases} \qquad (256)$$

if we neglect its exponentially rapid decaying tail in the classically forbidden range of the impact parameters ρ out of the right turning point $r_2 = 2n_*^2 a_0$. Here $c < 1$ is the normalization constant whose value may be chosen from comparison with the available experimental data or with the results of the close-coupling calculations at low enough n. It should be emphasized that c is the only empirical parameter of the theory and its value might slightly vary when we switch from one collision system to another. For example, $c = 5/8$ if the cross section $\sum\limits_{J'} \sigma_{nlJ}^{nlJ'}$ is normalized to the geometrical area of the Rydberg atom defined by the relation

$$\sigma_{\mathbf{geom}} = \pi \left\langle r^2 \right\rangle_{nl} \approx \left(5\pi a_0^2/2\right) n_*^4$$

at $l \ll n$. The value of $\rho_0 \equiv \rho_0(V, n_*)$ is determined from the condition

$$W^{\mathrm{B}}\left(\rho_0, V\right) = \left(\mathrm{g}_{J'}/\mathrm{g}_s\mathrm{g}_{nl}\right)c \,,$$

where $\mathrm{g}_{J'} = 2J' + 1$, and

$$\mathrm{g}_{nl}\mathrm{g}_s = \sum_J \left(2J + 1\right) = 2(2l + 1)$$

are the statistical weights of the final nlJ'-sublevel and the total nl-level (involving spin statistical weight $\mathrm{g}_s = 2$), respectively. As is apparent from (256), ρ_0 is given by

$$\rho_0\left(V, n_*\right) = \begin{cases} 2n_*^2 a_0 \,, & n_* \leq n_0 \,, \\ \frac{(2l+1)C_{J'J}^{(l)}L^2}{(2J'+1)cn_*^6 a_0}\left(\frac{v_0}{V}\right)^2, & n_* \leq n_0 \,. \end{cases} \qquad (257)$$

Further, following the paper [137], we introduce the magnitude of the principal quantum number $n_0(V)$, so that the value of the impact parameter ρ_0 becomes equal to the radius $r_2 \approx 2n_*^2 a_0$ of the Rydberg atom at $n_* = n_0(V)$. As a result, we obtain the relation

$$n_0^8 = \frac{(2l+1)C_{J'J}^{(l)}}{2(2J'+1)c}\left(\frac{v_0 L}{Va_0}\right)^2 \tag{258}$$

for the $n_0 = n_0\,(V)$ value, which separates the regions of weak $n_* \gg n_0$ from that of close coupling (where $n_* \ll n_0$ or $n \sim n_0$).

For low principal quantum numbers $n_* \leq n_0(V)$, the whole classically allowed range of integration $(0 \leq \rho \leq 2n_*^2 a_0)$ over the $2\pi\rho d\rho$ in (100) corresponds to the case of close coupling, i.e. $W_{nlJ}^{nlJ'} = const.$ Hence, the value of $\sigma_{nlJ}^{nlJ'}(V)$ is determined by the geometrical cross section σ_{geom} of a Rydberg atom. On the contrary, with $n_* > n_0(V)$ (when $\rho_0(V, n_*) < 2n_*^2 a_0$) it is necessary to take into account the reduction of the transition probability $W^B(\rho, V)$ with increasing ρ in the range of $\rho_0 \leq \rho \leq 2n_*^2 a_0$. As a result, performing the integration over the $2\pi\rho d\rho$ in the classically allowed range of ρ, we derive a general semiclassical expression for the cross section of the $nlJ \to nlJ'$ transition [137]

$$\sigma_{nlJ}^{nlJ'} = \begin{cases} \frac{2J'+1}{2(2l+1)}4c\pi a_0^2 n_*^4\,, & n_* \leq n_0(V)\,, \\ \frac{2\pi C_{J'J}^{(l)}L^2 v_0^2}{n_*^4 V^2}\left(1 - \frac{n_0^8}{2n_*^8}\right)\,, & n_* \geq n_0(V)\,, \end{cases} \tag{259}$$

which is simultaneously applicable at high, intermediate and sufficiently low n_*. As is evident from (259), this result leads to the simple expression (251) of the impulse approximation at high n.

One can also see, that the cross section $\sigma_{nlJ}^{nlJ'}$ reaches its maximum at

$$n_*^{\text{max}}(V) = (3/2)^{1/8} n_0(V)\,,$$

if the inelasticity parameter of the J-mixing process is very small $\nu \ll 1$ (or equals to zero, when $J' = J$), so that we may assume $\varphi_{J'J}^{(l)}(\nu) \approx 1$. Then, the cross section $\sigma_{nlJ}^{nlJ'}$ of quasielastic J-mixing collisions behaves like n_*^4 and n_*^{-4} at low and high principal quantum numbers, respectively. However, the situation becomes quite different at large values of the fine-structure splitting $\Delta E_{J'J}$ and fairly low relative velocities V of colliding A^* and B atoms. In this case, the inelasticity parameter for the $nlJ \to nlJ'$ transition ($J' \neq J$) is large, which leads to an appreciable reduction of the J-mixing cross section as compared to the quasielastic case ($\nu \ll 1$), since $\varphi_{J'J}^{(l)}(\nu) \ll 1$ at $\nu \gg 1$ (see Fig. 11).

The averaged cross sections of excitation ($nlJ \rightarrow nlJ'$, $E_{nlJ} < E_{nlJ'}$) and deexcitation $nlJ' \rightarrow nlJ$ are related through the detailed balance ($g_J = 2J + 1$ and $g_{J'} = 2J' + 1$):

$$\left\langle \sigma_{nlJ}^{nlJ'} \right\rangle_T = \frac{g_{J'}}{g_J} \left\langle \sigma_{nlJ'}^{nlJ} \right\rangle_T \exp\left(-\frac{\Delta E_{J'J}}{kT}\right), \qquad (260)$$

where $\left\langle \sigma_{nlJ}^{nlJ'} \right\rangle_T = \left\langle V\sigma_{nlJ}^{nlJ'} \right\rangle_T / \langle V \rangle_T$. In the special case of pure elastic scattering ($J' = J$ and $\nu = 0$) or the J-mixing of slightly split fine-structure components ($\nu \ll 1$), the averaging expression (259) may be performed analytically [137]

$$\left\langle \sigma_{nlJ}^{nlJ'} \right\rangle_T = \left(\frac{4cg_{J'}C_{J'J}^{(l)}}{g_s g_l}\right)^{1/2} \pi a_0^2 \left(\frac{v_0 |L|}{V_T a_0}\right) \Phi(\eta), \qquad (261)$$

$$\eta = \frac{g_s g_l C_{J'J}^{(l)}}{4cg_{J'} n_*^8} \left(\frac{v_0 L}{V_T a_0}\right)^2, \qquad (262)$$

where $g_{J'} = 2J'+1$, $g_l = 2l+1$, and $g_s = 2$ are the statistical weights.

The dimensionless $\Phi(\eta)$ function in Eq. (261) is given by the following expression

$$\Phi(\eta) = \eta^{1/2} \left\{ E_2(\eta) + \eta^{-1}[1 - \exp(-\eta)] \right\}. \qquad (263)$$

Here $E_2(\eta)$ is the integral exponential function of the second order

$$E_2(\eta) = \int_1^\infty \frac{\exp(-\eta t)}{t^2} dt,$$

$\eta = n_0^8(V_T)/n_*^8$ is the dimensionless parameter characterizing the strength of the collision, and $V_T = (2kT/\mu)^{1/2}$.

It is important to stress that the use of the normalized perturbation theory and the impact-parameter approach for the calculation of the elastic scattering cross section for a given nl state leads to a similar expression. As was shown in [137], the final result $\langle \sigma_{nl}^{el} \rangle$ directly follows from (261) if we put the statistical weights ratio equal to unity ($g_{J'}/g_s g_l \rightarrow 1$). The form of the $\Phi(\eta)$-function is similar to the result obtained previously [131] in the intermediate range of n_* for the elastic scattering cross section in the Rydberg ns-state. As is evident from (263), the $\Phi(\eta)$ function behaves like $\eta^{-1/2}$ and $\eta^{1/2}$ at large $\eta \gg 1$ (strong collision) and small $\eta \ll 1$ (weak collision) values of the collision strength parameter η. The function $\Phi(\eta)$ reaches its maximum $\Phi_{\max} \approx 0.8$ at $\eta_{\max} \approx 0.68$. The plot of $\Phi(\eta)$ function is shown in Fig. 12.

Thus the magnitude and the position of the maximum of the averaged cross section $\left\langle \sigma_{nlJ}^{nlJ'} \right\rangle_T$ for pure elastic $(J' = J)$ or quasielastic $(J' \neq J, \nu_{J'J} \ll 1)$ transitions $nlJ \to nlJ'$ can be determined from the simple expression

$$\left\langle \sigma_{nlJ}^{nlJ'} \right\rangle_T^{\max} = b_{J'J}^{(l)} a_0^2 \left(\frac{|L| v_0}{a_0 V_T} \right) , \tag{264}$$

$$n_*^{\max} = d_{J'J}^{(l)} \left(\frac{|L| v_0}{a_0 V_T} \right)^{1/4} . \tag{265}$$

Here the constant coefficients have the following magnitudes $b_{5/2,3/2}^{(d)} \approx$ 2.12, $d_{5/2,3/2}^{(d)} \approx 0.91$ and $b_{3/2,1/2}^{(d)} \approx 2.18$, $d_{5/2,3/2}^{(d)} \approx 0.89$ for the $n^2 D_{3/2} \to n^2 D_{5/2}$ and $n^2 P_{1/2} \to n^2 P_{3/2}$ transitions, respectively.

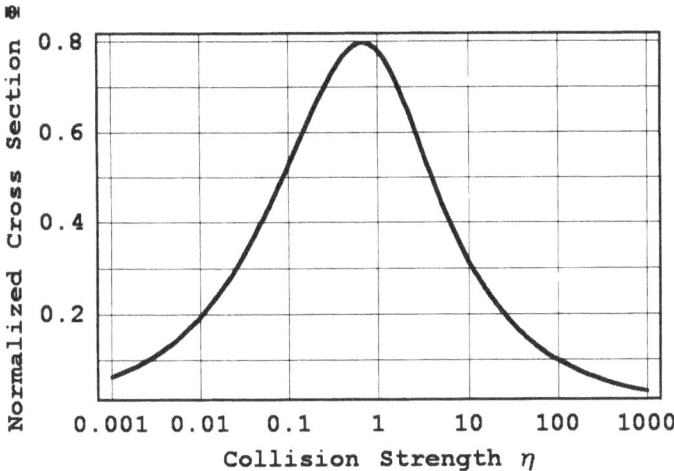

Figure 12. The dependence of the normalized cross section Φ on the value of dimensionless collision strength parameter η.

5.3   Energy and Orbital Momentum Transfer

In the theory of the J-mixing process of the fine–structure components, presented above, we only deal with a transition between two Rydberg states with the given quantum numbers nlJ and nlJ' (or with one state in the case of elastic scattering). Here we shall continue the theoretical analysis of transitions

$$A^* \, (nl) + B \rightarrow A^* \, (n') + B$$

from a state nl to a degenerate manifold n', considered previously in Sects. 3.5.1 and 4.5.2 for weak collisions. As was mentioned above, this transition is characterized by the energy defect $\Delta E_{n',nl} \approx 2Ry \, |\delta_l + \Delta n| \, /n^3$ (where $\Delta n = n' - n$). If the non-integer part of the quantum defect of the nl-state is substantial, the transition energy defect is relatively large even between the nearest nl and n' levels and the process is inelastic. If the value of δ_l is close to zero and the principal quantum number is not changed $n' = n$, we have a quasielastic l-mixing process. The focus of this section is to present recent results of semiclassical theory developed in Refs. [138, 139] and [140] for the simultaneous description of inelastic n, l-changing and quasielastic l-mixing processes in collisions of Rydberg atoms with atomic targets induced by the electron–perturber interaction. As in the preceding Sect. 5.2.2, it is based on the normalized perturbation theory and, hence, may be used at high, intermediate and sufficiently low n.

5.3.1 n, l-Changing Impulse Collisions

We now assume that the interaction between the perturbing atom B and the ion core can be ignored, and the short-range interaction between the Rydberg electron and B can be described by the zero-range Fermi pseudopotential

$$V_{eB} \, (\mathbf{r} - \mathbf{R}) = \left(2\pi\hbar^2/m\right) L_{\text{eff}} \delta(\mathbf{r} - \mathbf{R}) \, ,$$

where L_{eff} is the effective scattering length for electron–perturber scattering. In contrast to the standard scattering length L, L_{eff} depends on collision parameters and n. This dependence will be discussed in Sect. 5.4. Further, starting from the basic formula of first-order perturbation theory (131) in the impact-parameter representation we apply the standard normalization procedure (99) for the probability of the $nl \to n'$ transition. As a result, the final equation for the transition probability can be presented as [140]

$$W_{n',nl}(\rho) = \begin{cases} c, & 0 \le \rho \le \rho_0, \\ \frac{L_{\text{eff}}^2 v_0^2}{2\pi a_0^2 n_*^6 V^2} \sqrt{\frac{a_0}{\rho}} \Lambda(\rho), & \rho_0 \le \rho \le \rho_{\text{max}}, \\ 0, & \rho_{\text{max}} < \rho. \end{cases} \quad (266)$$

Here $\Theta(z) = 1$ for $z \ge 0$ and $\Theta(z) = 0$ for $z < 0$; and the Λ-quantity

$$\Lambda(\rho) = F(\varphi_2, \mathrm{k}) - F(\varphi_1, \mathrm{k}) + +2\Theta(\tilde{\rho} - \rho) F(\varphi_1, \mathrm{k})$$

is expressed in terms of the incomplete elliptic integrals $F(\varphi, \mathrm{k})$ of the first kind

$$F(\varphi, \mathrm{k}) = \int_0^\varphi \left(1 - \mathrm{k}^2 \sin^2 \theta\right)^{-1/2} d\theta, \quad (267)$$

$$\mathrm{k} = \left[\left(1 - \rho/2n_*^2 a_0\right)/2\right]^{1/2}, \quad (268)$$

(see, for example, [144]), while the arguments φ_1 and φ_2 in (266) are

$$\varphi_s = \arcsin\left[\left(\frac{R_s^{(\lambda)}(\rho) - \rho}{\left(1 - \rho/2n_*^2 a_0\right) R_s^{(\lambda)}(\rho)}\right)^{1/2}\right], \quad s = 1, 2. \quad (269)$$

The parameters $R_1^{(\lambda)}(\rho) = 2n_*^2 a_0 x_1^{(\lambda)}(y)$ and $R_2^{(\lambda)}(\rho) = 2n_*^2 a_0 x_2^{(\lambda)}(y)$ are determined from the equation

$$y = \phi_\lambda(x), \quad \phi_\lambda = \frac{(2\lambda)^{1/2} x^{5/4}}{(1-x)^{1/4}}\left[1 - \frac{\lambda}{2}\frac{x^{1/2}}{(1-x)^{1/2}}\right]^{1/2} \quad (270)$$

for a fixed value of the scaled impact parameter $y = \rho/2n_*^2 a_0$. Here $x = R/2n_*^2 a_0$ is the scaled internuclear separation, and

$$\lambda = n_* a_0 |\Delta E_{n',nl}| / \hbar V$$

is the inelasticity parameter for the $nl \to n'$ transition, introduced in Sect. 3.5. The impact parameter ρ_{max} in (266) is the maximum possible value of ρ in the classically allowed region determined by the inelasticity parameter λ of the $nl \to n'$ transition and by the principal quantum number n_*. Within the framework of the semi-classical approach, the transition probability $W_{n',nl}(\rho)$ becomes zero for $\rho > \rho_{\max}(\lambda)$. It corresponds to the maximum value of $\phi_\lambda(x)$.

Figure 13 present the plot of the $\phi_\lambda(x)$ for three magnitudes of the inelasticity parameter $\lambda = 0.2$, 1, and 3 (panel a). One can see, that the ratio $\rho_{\max}/2n_*^2 a_0$ of the classically allowed region of the impact parameter to the characteristic radius $2n_*^2 a_0$ of the Rydberg atom reveals a strong drop with increasing λ. Figure 13 (panel b) demonstrates the positions of the routes $x_1^{(\lambda)}(y)$ and $x_2^{(\lambda)}(y)$ of equation $y = \phi_\lambda(x)$ and the magnitudes of $\tilde{y} = \tilde{\rho}/2n_*^2 a_0$ and $x_{\max}$ for a fixed value of the scaled impact parameter $y \equiv \rho/2n_*^2 a_0 = y_0$ ($\lambda = 1$).

The impact parameter $\tilde{\rho} = 2n_*^2 a_0 \tilde{y}$ in (266) is determined by the relation $\tilde{y} = \phi_\lambda(\tilde{x}) = \tilde{x}$, where $\tilde{x} = \tilde{R}/2n_*^2 a_0$. From (270), we obtain

$$\tilde{\rho}(\lambda) = 2n_*^2 a_0 \tilde{y}(\lambda), \qquad \tilde{y}(\lambda) = 1/\left(1+\lambda^2\right),$$

$$\rho_{\max}(\lambda) = 2n_*^2 a_0 \phi_\lambda^{\max}, \qquad y_{\max}(\lambda) = \phi_\lambda^{\max}. \tag{271}$$

Thus, $\tilde{\rho}$ and $\rho_{\max}$ are determined only by the inelasticity parameter λ and by the principal quantum number n, whereas $\tilde{\rho} < \rho_{\max}$ for given n_* and λ.

The parameter ρ_0 can be calculated from the equation $W_{fi}^{B}(\rho_0) = c$ using expression (266) for the transition probability. It is determined by the principal quantum number n_* and the energy defect $|\Delta E_{fi}|$, as well as by the particular type of colliding partners and their relative velocity V. Notice that the roots $x_1^{(\lambda)}(y)$ and $x_2^{(\lambda)}(y)$ of equation (270) depend significantly on λ. Integration of the general expression (266) over all impact parameters leads to the following result for the cross section of the inelastic $nl \to n'$ transition [140]:

$$\sigma_{n',nl} = c\pi\rho_0^2 + \frac{2\pi L_{\text{eff}}^2 v_0^2}{V^2 n_*^3} \mathcal{F}_\lambda\left(\rho_0/2n_*^2 a_0\right), \quad (\rho_0 \le \rho_{\max}), \tag{272}$$

where the function $\mathcal{F}_\lambda(y_0)$ can be written as

$$\mathcal{F}_\lambda(y_0) = \frac{1}{\pi}\left\{2\Theta(\tilde{y}-y_0)\int_{y_0}^{\xi_1}\left(\frac{x^2-y_0^2}{x-x^2}\right)^{1/2}dx\right.$$

$$\left. + \int_{\xi_1}^{\xi_2}\left(\frac{x^2-y_0^2}{x-x^2}\right)^{1/2}dx + \mathcal{Q}(\xi_2) - \mathcal{Q}(\xi_1)\right\}. \tag{273}$$

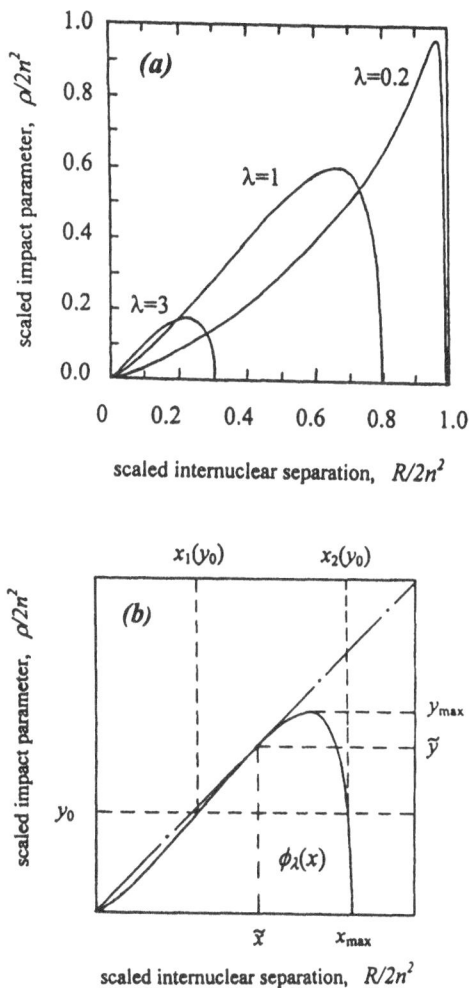

Figure 13. The function $\phi_\lambda(x)$ against the scaled internuclear separation $x = R/2n_*^2 a_0$ (in a.u.) for different values of $\lambda = 0.2$, 1, and 3 (panel a). Panel b shows the scaled parameters of the model: $x_{\max} = R_{\max}/2n_*^2 a_0$ and $y_{\max} = \rho_{\max}/2n_*^2 a_0$, $\tilde{y} = \tilde{\rho}/2n_*^2 a_0$ and $\tilde{x}$, and the routes $x_1^{(\lambda)}(y_0)$ and $x_2^{(\lambda)}(y_0)$ of Eq. (270) for a given value of $y = y_0 = \rho_0/2n_*^2 a_0$.

Here the $Q(z)$-function is given by the following relation

$$Q(z) = \arctan\left[\left(\tfrac{z}{1-z}\right)^{1/2}\right] - [z(1-z)]^{1/2}$$

$$+\lambda\left[z - \ln\left(\tfrac{1}{1-z}\right)\right] , \qquad (274)$$

and $y_0 = \rho_0/2n_*^2 a_0$ is determined from the normalization condition for the transition probability $W_{n',nl}^{\mathrm{B}}(y_0) = c$ calculated by the first-order perturbation theory. The parameters $\xi_{1,2} = x_{1,2}^{(\lambda)}(y_0)$ are the roots of the equation (270) for a given value of the scaled impact parameter $y_0 = \rho_0/2n_*^2 a_0$, and $\tilde{y} = \tilde{y}(\lambda)$ is given by expression (271).

The expression for the total contribution of all inelastic transitions (quenching cross section σ_{nl}^{Q}) for a given nl-level (with $l \ll n$) can be written as

$$\sigma_{nl}^{\mathrm{Q}} \equiv \sigma_{nl}^{\mathrm{in}} = \sum_{n'} \zeta_{l_0,n'}\sigma_{n',nl} , \qquad (275)$$

where the $\zeta_{l_0,n'}$ is a correction factor for a non-hydrogenic Rydberg atom which takes into account the fact that not all final states are hydrogen-like. There are several prescriptions for incorporating this factor into the calculations discussed in [36, 127, 130, 133]. A natural choice of the correction factor is based on a simple physical picture according to which it is determined by the ratio of statistical weights

$$\zeta_{l_0,n'} = (\mathrm{g}_{n'} - \mathrm{g}_0)/\mathrm{g}_{n'} ,$$

where $\mathrm{g}_{n'} = (n')^2$ is the total statistical weight of the final n'-level ($n' = n$ for quasielastic l-mixing process), and

$$\mathrm{g}_0 = \sum_{l'=0}^{l_0-1}(2l'+1) = l_0^2$$

is the statistical weight of the first few non-hydrogenic Rydberg $n'l'$-states (with $l' < l_0$). We assume that all other Rydberg $n'l'$-states with $l' \geq l_0$ with very small values of the quantum defects $\delta_{l'}$ are practically degenerate (so that their energy is equal to $E_{n'} = -Ry/(n')^2$). According to this assumption the correction factor can be written as [36, 133]

$$\zeta_{l_0,n'} = 1 - l_0^2/(n')^2. \qquad (276)$$

As is apparent from (276), $\zeta_{l_0,n'} \approx 1$ in the range of sufficiently large principal quantum number since all Rydberg $n'l'$-states with $l' \geq 2-3$ are practically hydrogen-like (i.e. the l_0 value usually does not usually exceed $2-3$ depending on the specific type of Rydberg atom).

5.3.2   Analytic Formulae for l-Mixing Collisions

In the particular case of a pure quasielastic transition ($\Delta E_{n',nl} = 0$ and hence $\lambda = 0$), when $\tilde{y} = 1$ and $y_{\max} = 1$, we have $x_{1,2}^{(\lambda)}(y) = 1$. Then, the basic expression for the transition probability $W_{n,nl}(\rho, V)$ is significantly simplified and can be written as

$$
W_{n,nl} = \begin{cases} c, & 0 \leq \rho \leq \rho_0, \\ \frac{L_{\mathrm{eff}}^2 v_0^2}{\pi a_0^2 n_*^6 V^2} \sqrt{\frac{a_0}{\rho}} K\left[\mathrm{k}\left(\rho\right)\right], & \rho_0 \leq \rho \leq 2n_*^2 a_0, \\ 0, & 2n_*^2 a_0 < \rho < \infty, \end{cases} \tag{277}
$$

where $K(\mathrm{k})$ is the complete elliptic integral of the first kind, and $\mathrm{k}(\rho)$ is determined by (120). The basic semiclassical formula of normalized perturbation theory for the cross section (272) takes the following form

$$
\sigma_{n,nl} = \begin{cases} c4\pi n_*^4 a_0^2, & n_* < n_0, \\ c\pi\rho_0^2 + \frac{2\pi L_{\mathrm{eff}}^2 v_0^2}{V^2 n_*^3} \mathcal{F}_{\lambda=0}\left(\frac{\rho_0}{2n_*^2 a_0}\right), & n_* \geq n_0. \end{cases} \tag{278}
$$

The function $\mathcal{F}_\lambda(y_0)$ of the scaled impact parameter $y_0 = \rho_0/2n_*^2 a_0$ (see (273)) is reduced to the simple expression at $\lambda = 0$:

$$
\begin{aligned}
\mathcal{F}_{\lambda=0}(y_0) &= \frac{2}{\pi} \int_{y_0}^1 \left(\frac{x^2 - y_0^2}{x - x^2}\right)^{1/2} dx \\
&= \frac{16}{\pi} \int_0^{\mathrm{k}_0} K(\mathrm{k}) \left(1/2 - \mathrm{k}^2\right)^{1/2} \mathrm{k}\, d\mathrm{k},
\end{aligned} \tag{279}
$$

whereas $\mathrm{k}_0 = \left[(1 - y_0)/2\right]^{1/2}$. Using the approximate relation $K(\mathrm{k}) \approx K(0) = \pi/2$ and performing integration over k this formula can be rewritten in the analytical form

$$
\sigma_{n,nl} \approx \begin{cases} c\left(\frac{3\sqrt{2}}{4}\right) 4\pi n_*^4 a_0^2, & n_* \leq n_0, \\ \left\{\pi c\left(\frac{3\sqrt{2}}{4}\right)\rho_0^2 \right. \\ \left. + \frac{2\pi L_{\mathrm{eff}}^2 v_0^2}{V^2 n_*^3}\left[1 - \left(\frac{\rho_0}{2n_*^2 a_0}\right)^{3/2}\right]\right\}, & n_* \geq n_0. \end{cases} \tag{280}
$$

Here ρ_0 is determined from the normalization condition for the Born probability of the quasielastic $nl \to n$ transition (277). Its value is given by

$$\rho_0\left(V, n_*\right) \approx \begin{cases} 2n_*^2 a_0 & n_* \leq n_0, \\ a_0\left(\frac{L_{\text{eff}}^2 v_0^2}{2ca_0^2 V^2 n_*^6}\right)^2 & n_* \geq n_0. \end{cases} \tag{281}$$

The upper relation reflects the fact that ρ_0 can not exceed the maximal possible value of the impact parameter $\rho_{\max}\left(\lambda = 0\right) = 2n_*^2 a_0$, for which the quasielastic transition occurs in the classically allowed range. Thus, the relation $\rho_0\left(V, n_*\right) = 2n_0^2 a_0$ yields the value of the principal quantum number n_0 for which the value of ρ_0 becomes equal to the radius of the Rydberg atom, which is determined by the right turning point $r_2 = n_*^2 a_0\left(1 + \epsilon\right) \approx 2n_*^2 a_0$ at $l \ll n$. As a result, we have

$$n_0\left(V\right) \approx \left(\frac{L_{\text{eff}}^2 v_0^2}{2^{3/2} ca_0^2 V^2}\right)^{1/7}. \tag{282}$$

The maximum of the quasielastic cross section $\sigma_{n,nl} = \sum_{l'} \sigma_{nl',nl}$ appears at $\left(n_{\max}/n_0\right)^{21} = 2$ and its value is given by

$$\sigma_{n,nl}^{\max} \approx \pi a_0^2 \frac{c^{3/7} 7\sqrt{2}}{4} \left(\frac{|L_{\text{eff}}| v_0}{a_0 V}\right)^{8/7}. \tag{283}$$

The cross section of quasielastic transition

$$\langle \sigma_{n,nl}\rangle_T = \frac{\langle V\sigma_{n,nl}\rangle_T}{\langle V\rangle_T}$$

averaged over the Maxwellian velocity distribution can be expressed in terms of the collision strength parameter ς

$$\langle \sigma_{n,nl}\rangle_T = \pi a_0^2 \frac{3c^{3/7}}{2^{5/14}} \left(\frac{|L_{\text{eff}}| v_0}{a_0 V_T}\right)^{8/7} P\left(\varsigma\right), \tag{284}$$

$$\varsigma = \frac{1}{c}\left[\frac{1}{2^{3/2} n_*^7}\left(\frac{v_0 L_{\text{eff}}}{V_T a_0}\right)^2\right], \tag{285}$$

Here the $P(\varsigma)$-function takes the form

$$P(\varsigma) \;=\; \varsigma^{-4/7}\left\{1+\left(\frac{\varsigma}{3}-1\right)\exp(-\varsigma)-\frac{\varsigma^2}{3}E_3(\varsigma)\right\}, \quad (286)$$

$$E_3(z) \;=\; \int\limits_{1}^{\infty}\frac{e^{-zt}}{t^3}\,dt. \quad (287)$$

and $E_3(z)$ is the integral exponential function of the third order. The plot of the $P(\varsigma)$ function against the collision strength parameter ς is presented in Fig. 14.

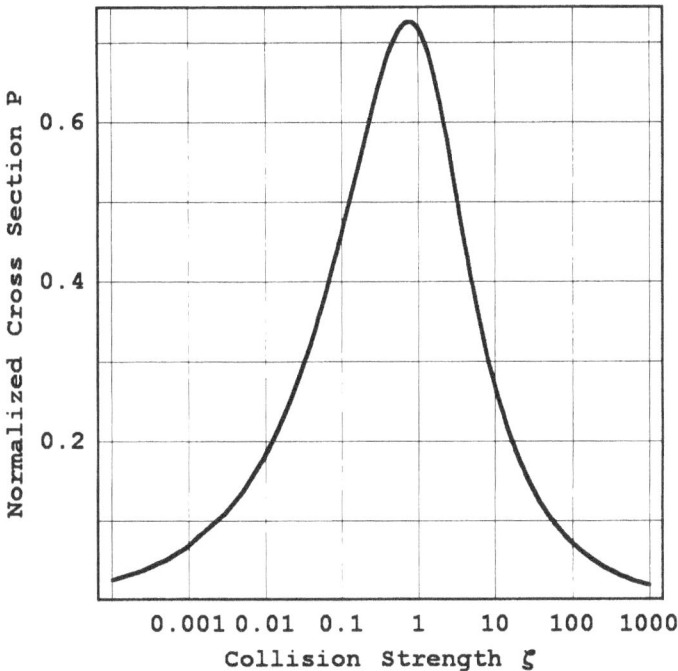

Figure 14. Normalized cross section P (see Eq. (286)) of a quasielastic l-mixing process averaged over the Maxwellian distribution of velocities as a function of the collision strength parameter ς.

The asymptotic behavior of this function is given by

$$P \xrightarrow[\varsigma \to 0]{} P_1 = \frac{4}{3}\varsigma^{3/7}, \qquad P \xrightarrow[\varsigma \to \infty]{} P_2 = \varsigma^{-4/7}. \tag{288}$$

The maximum $P_{\max} \approx 0.73$ appears at $\varsigma_{\max} \approx 0.77$. As follows from (288), the range of weak coupling $n \gg n_{\max}$ and close coupling $n \ll n_{\max}$ correspond to small ($\varsigma \ll 1$) and large ($\varsigma \gg 1$) values of the collision strength parameter, respectively. Using expressions (288) the averaged cross section can be rewritten as

$$\langle \sigma_{n,nl} \rangle_T = \begin{cases} 3\sqrt{2}c\pi a_0^2 n_*^4 & n \ll n_{\max}, \\ 5.37 a_0^2 c^{3/7} \left(\frac{|L_{\text{eff}}|v_0}{a_0 V_T} \right)^{8/7} & n = n_{\max}, \\ 2\pi L^2 v_0^2 / n_*^3 V_T^2 & n \gg n_{\max}. \end{cases} \tag{289}$$

As is evident from (289), the cross section of the l-mixing process is proportional to the geometrical area of the Rydberg atom $\pi a_0^2 n_*^4$ at sufficiently low principal quantum number and reaches its maximum in the range of $n_* \sim (v_0 |L_{\text{eff}}|/a_0 V_T)^{2/7}$. Then, at high $n \gg n_{\max}$, where first-order perturbation theory becomes valid, the total l-mixing cross section behaves like $\sigma_{nl}^{l-\text{mix}} \propto L^2/V^2 n^3$ in accord with the result (135).

5.4   Inclusion of Long-Range Interaction Effects

If the scattering length approximation $f_{\text{eB}} = -L$ does not hold it is necessary to take into account the actual momentum and angular dependencies of the amplitude $f_{\text{eB}}(k,\theta)$ for scattering of free electron by the ground state target atom. This may be achieved by combining the semiclassical impact parameter approach based on the normalized perturbation theory with the results of the impulse approximation at high n. The first semiempirical attempts to include this dependence in the theory of Rydberg-atom–ground-state-atom collisions [161, 162, 130] incorporate the dependence of the cross section for elastic electron–perturber scattering on n. To improve the results of calculations for quasielastic l-mixing processes induced by collisions with rare-gas atoms they include the n-dependent parameter $L_{\text{eff}}(n_*) = \left[\sigma_{\text{eB}}^{\text{el}}(\epsilon_n)/4\pi \right]^{1/2}$ in their models instead of standard scattering length L. According to [162, 130] the effective scattering length $L_{\text{eff}}(n_*)$ is determined by the total elastic scattering cross section $\sigma_{\text{eB}}^{\text{el}}\,(\epsilon_n)$ of a free electron by a perturbing atom for the mean kinetic energy $\epsilon_n = Ry/n_*^2$ of the orbital electron motion [160] (or by its averaged value $\langle \sigma_{\text{eB}}^{\text{el}}\,(k) \rangle_{nl}$ over the momentum distribution function in the nl-state [130]).

A method which also includes dependence on the inelasticity parameter of the process was suggested in [138, 140]. First note that in the strong-coupling region the transition probabilities do not depend significantly on the specific form of the electron-perturber interaction. It can be seen directly from (272) at $\rho_0 \to \rho_{\max}$ (see also (280)) and reflects a general feature of any collisional processes involving a large number of closely-spaced levels. Indeed for strong collisions the quenching cross section for a given energy level is practically the same as the total scattering cross section, which is determined by the unitarity condition. Thus, the required modification of the scattering amplitude can be accomplished in the weak-coupling limit. In accordance with the results of [139], we will do it here within the framework of the quasifree electron model. We proceed from the general semiclassical formula (202) for the cross section of the $nl \to n'$ transition in the momentum representation [105]. In the scattering length approximation $(f_{\text{eB}} = -L = const)$ it directly yields the analytical result (132).

Comparison of (132) with the general formula (202) allows us to introduce a parameter $L_{\text{eff}}^2\,(n_*, \lambda)$ which can be incorporated into all formulae (272)–(289) in order to take into account actual behavior of the electron–perturber scattering amplitude $f_{\text{eB}}\,(k, \theta)$. According to [138] the final result for the square of the effective scattering length $L_{\text{eff}}^2\,(n_*, \lambda)$ can be written as

$$L_{\text{eff}}^2 = \frac{1}{2^{3/2}\mathsf{f}_{n',nl}(\lambda)} \int_{k_{\min}}^{\infty} k^2\, dk\, |g_{nl}(k)|^2$$

$$\times \int_{-1}^{\nu_{\max}(k)} \frac{d\,(\cos\theta)}{(1 - \cos\theta)^{1/2}}\, |f_{\text{eB}}\,(k, \theta)|^2 , \tag{290}$$

where $\nu_{\max}\,(k)$ and $k_{\min}$ are given by (202), and $\mathsf{f}_{n',nl}\,(\lambda)$ is the function (132) of the inelasticity parameter $\lambda = n_* a_0\,|\Delta E_{n',nl}|\,/\hbar V$, introduced in Sect. 3.5, whereas $\mathsf{f}_{n',nl}\,(\lambda = 0) = 1$ for a quasielastic transitions without energy transfer. As is evident from (290), the procedure for calculation of the effective scattering length $L_{\text{eff}}(n_*, \lambda)$ for a given $nl \to n'$ transition includes the average of the differential scattering cross section $d\sigma_{\text{eB}}/d\Omega = |f_{\text{eB}}(k, \theta)|^2$ over the momentum distribution function $W_{nl}\,(k) = k^2\,|g_{nl}(k)|^2$ taking into account its actual momentum and angular dependencies.

The ratio L_{eff}^2/L^2 can be also written as $\sigma_{n',nl}/\sigma_{n',nl}^L$ where $\sigma_{n',nl}$ is calculated in the impulse approximation (202) with the actual scattering amplitude $f_{\text{eB}}(k,\theta)$, while $\sigma_{n',nl}^L$ is the corresponding value obtained for $f_{\text{eB}} = -L = const$ (see (132)). In the scattering length approximation $f_{\text{eB}} = -L$, formula (290) yields $L_{\text{eff}}(n_*,\lambda) = L = const$. In the general case it incorporates both the short and long range parts of the electron–perturber interaction. The parameter $L_{\text{eff}}(n_*,\lambda)$ characterizes the actual electron–atom interaction in the range of the electron momenta which gives the main contribution to the $nl \to n'$ transition for given n, relative velocity V, and energy defect $|\Delta E_{n',nl}|$. Therefore $L_{\text{eff}}(n_*,\lambda)$ depends both on n_* and the inelasticity parameter $\lambda = n_* a_0 |\Delta E_{n',nl}|/\hbar V$. The inclusion of the dependence of L_{eff} on the inelasticity parameter of the $nl \to n'$ transition is the major modification of the method [138] as compared to [162, 130].

To derive an analytic formula for the effective scattering length [140] we use the general expression (290) and the effective range theory [91, 92] for electron scattering by the target atom. In the presence of the short– and long–range (polarization) interaction the first terms of the low-energy expansion of the differential scattering cross section are given by expression (62) of Sect. 2.8.2. According to (62) the first few terms of this expansion are expressed in terms of the standard scattering length L and the polarizability α is the of the target atom B and some constant coefficient B. As noted previously, this expression is applicable if the electron wave number satisfies the condition $\alpha k^2/a_0 \ll 1$. On inserting Eq. (62) into (290) and performing the integration over the scattering angle, we obtain

$$L_{\text{eff}}^2 = \frac{1}{f_{n',nl}(\lambda)} \int_{k_{\min}}^{\infty} \left[L^2 + \frac{\pi\alpha L}{2}(k + k_{\min}) + \frac{8\alpha L^2}{3} k^2 \ln(k) + Bk^2 \right]$$

$$\times \left[1 - \frac{k_{\min}}{k} \right] W_{nl}(k)dk, \qquad W_{nl}(k) = k^2 |g_{nl}(k)|^2 .$$

$$(291)$$

Further we use the well known semiclassical formula (206) for the momentum distribution function of the Rydberg nl–state with small $l \ll n$. Then, the first terms of the expansion for the effective scattering length square can be written as [140]

$$L_{\text{eff}}^2 (n_*, \lambda) = L^2 + \frac{\alpha L}{a_0^2 n_*} \chi_1 (\lambda)$$

$$+ \frac{1}{n_*^2 a_0^2} \left[B\chi_2 (\lambda) - \frac{8\alpha L^2}{3a_0} \ln (n_*) \chi_3 (\lambda) \right].$$

(292)

Here $\chi_i (\lambda) = f_i (\lambda) / f_0 (\lambda)$ and the $f_i (\lambda)$ functions are given by ($i = 1, 2, 3$)

$$f_1 (\lambda) = \frac{2}{\pi} \left[\left(1 - \frac{\lambda^2}{4} \right) \arctan \left(\frac{2}{\lambda} \right) + \frac{\lambda}{2} \right],$$

$$f_2 (\lambda) = \frac{2}{\pi} \arctan \left(\frac{2}{\lambda} \right),$$

(293)

$$f_3 (\lambda) = \frac{4}{\pi} \int\limits_{\lambda/2}^{\infty} \frac{(z - \lambda/2) z \ln (z)}{(1 + z^2)^2} dz,$$

and the $f_0 (\lambda) \equiv f_{n', nl} (\lambda)$ function is determined by (132), whereas $\chi_i (\lambda) \to 1$ at $\lambda \to 0$. Formula (292) for the effective scattering length takes a particularly simple form for a quasielastic transition ($\lambda = 0$)

$$L_{\text{eff}}^2 (n_*) = L^2 + \frac{\alpha L}{n_* a_0^2} + \frac{1}{n_*^2 a_0^2} \left[B - \frac{8\alpha L^2}{3a_0} \ln (n_*) \right]. \quad (294)$$

To demonstrate the inapplicability of the scattering length approximation for the heavy rare gas atoms Ar, Kr, and Xe the modified effective range theory [91, 92] and the basic formula (290) were used in [138]. Calculations of the effective scattering length were performed for transitions to the nearest energy levels which provide the major contribution to the quasielastic or inelastic transitions. The results for the ratio L_{eff}^2 / L^2 are plotted in Fig. 15 for the case of e-Ar scattering. The ratio L_{eff}^2 / L^2 depends significantly on n and turns out to be quite different for transitions with small and large energy defects. In all cases the deep minimum in L_{eff}^2 / L^2 results from the Ramsauer–Townsend effect in free-electron scattering, but it occurs at different values of n_* ($n_* = 10, 15,$ and 23 for $nF \to n$, $nS \to n-3$, and $nD \to n - 1$ transitions, respectively). Simple estimates of the typical electron momentum

$$k \sim k_{\min} = |\Delta E_{n', nl}| / 2\hbar V$$

at $n_* \approx 10, 15,$ and 23 show that they are in full agreement with the values of the free-electron momentum which corresponds to the

Ramsauer–Townsend minimum. Note also that the use of simple relation $[\sigma_{\text{el}}(\epsilon_n)/4\pi]^{1/2}$ instead of the standard scattering length L improves the results. However, more accurate calculations should incorporate the dependence of L_{eff} on the energy defect. Similar results have been obtained [138] for the L_{eff}^2 parameters of the e-Kr, and e-Xe interaction. They are presented in Figs. 16 and 17.

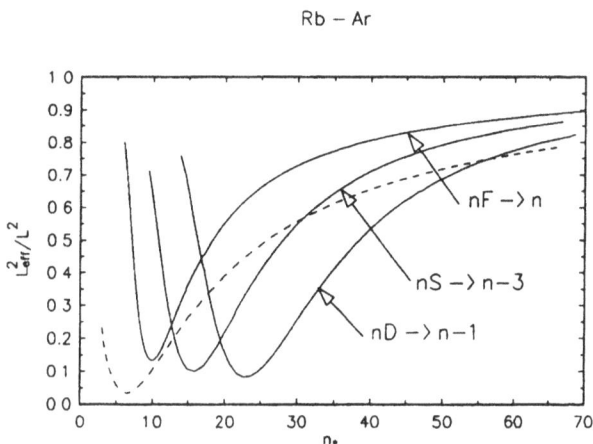

Rb − Ar

Figure 15. The n-dependence of the ratio L_{eff}^2/L^2 for the e-Ar scattering. Full curves are the semiclassical results [138] for the $nS \to n-3$, $nD \to n-1$, and $nF \to n$ transitions in the Rb (nl)+Ar collisions at E=0.026 eV. The dashed curve gives the ratio $\sigma_{\text{el}}(\epsilon_n)/4\pi L^2$, where $\epsilon_n = Ry/2n_*^2$ is the mean kinetic energy of the orbital electron motion.

Thus calculations [138, 139] confirm the previous conclusions [130, 162] that the standard scattering length approximation $f_{\text{eB}} = -L = const$ becomes inapplicable for collisions of Rydberg atoms with the heavy rare gas atoms, in contrast to the case of the He atom. For reliable quantitative results on inelastic and quasielastic transitions induced by collisions with Ar, Kr, and Xe it is necessary to get an appropriate description of both the short– and long–range parts of the electron–perturber interaction. Note that the basic formula (290) for L_{eff} can also be applied to analysis of the effective interaction between the Rydberg electron and the other atoms (e.g. alkali-metal atoms).

Rb − Kr

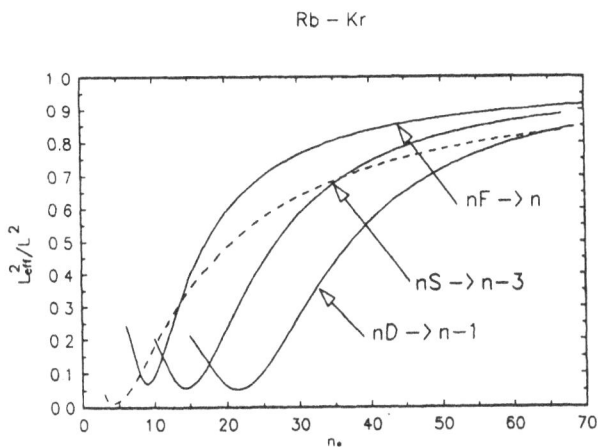

Figure 16. The same calculations as in Fig. 15 for the $Rb(nl)+Kr{\rightarrow}Rb(n')+Kr$ transitions [138].

Rb − Xe

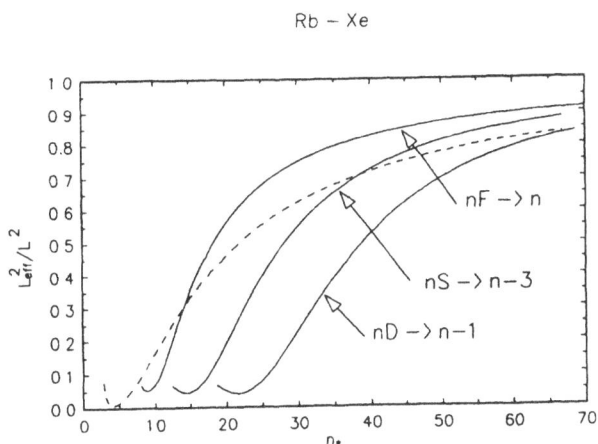

Figure 17. The same calculations as in Fig. 15 for the $Rb(nl)+Xe{\rightarrow}Rb(n')+Xe$ transitions [138].

5.5 Scaling Relations for Probabilities and Cross Sections

It is convenient to rewrite equations (266) and (272) in terms of scaled parameters $x = R/2n_*^2 a_0$ and $y = \rho/2n_*^2 a_0$. Note that the radius $2n_*^2 a_0$ corresponds to the right turning point in the Coulomb potential $r_2 \approx n_*^2 a_0 (1 + \epsilon)$ at $l \ll n_*$, when the eccentricity of the orbit $\epsilon \approx 1$. For the transition probability $W_{\lambda\zeta}(y)$ of the $nl \rightarrow n'$ we obtain [140]

$$
W_{\lambda\zeta}(y) =
\begin{cases}
c, & 0 \leq y \leq y_0, \\
\frac{\zeta}{\pi\sqrt{2}} \left\{ 2\Theta\left(\tilde{y} - y\right) \mathcal{J}\left[y, x_1^{(\lambda)}(y)\right] \right. & \\
\left. + \mathcal{J}\left[x_1^{(\lambda)}(y), x_2^{(\lambda)}(y)\right] \right\}, & y_0 \leq y \leq y_{\max}, \\
0, & y_{\max} \leq y.
\end{cases}
\tag{295}
$$

Here $\zeta = L_{\mathrm{eff}}^2 v_0^2 / 2^{3/2} a_0^2 V^2 n_*^7$ is a scaled parameter characterizing the collision strength (which differs from the dimensionless parameter ς introduced previously in Sect. 5.3.2 only by the value of normalization constant c, i.e. $\varsigma = \zeta/c$); and

$$
\mathcal{J}(u, v) = \int_u^v \frac{dx}{\sqrt{(x - x^2)(x^2 - y^2)}}.
$$

Analysis of the transition probability behavior as a function of the scaled impact parameter $y = \rho/2n_*^2 a_0$ for different values of the inelasticity parameter $\lambda = n_* a_0 |\Delta E_{n',nl}| / \hbar V$ and the collision strength parameter ζ was given in [138]. In the region $0 \leq y \leq y_0$ the first-order probability $W_{\lambda\zeta}^{\mathrm{B}}(\rho)$ becomes large and should be normalized to a constant $c = 0.25$ according to the normalized perturbation theory. The probability $W_{\lambda\zeta}^{\mathrm{B}}(\rho)$ becomes zero at $y > y_{\max} = \rho_{\max}/2n^2 a_0$. This happens because the semiclassical approach neglects the exponentially decaying tail of the electron wave function.

The maximum impact parameter $\rho_{\max}$ depends strongly on the inelasticity parameter λ. To show this we shall proceed from the conservation of energy law for a collisional $nl \rightarrow n'$ transition according to which the minimum possible value of the momentum transfer

$Q = |\mathbf{q}' - \mathbf{q}|$ is determined by the relation $Q_{\min} \approx |\Delta E_{n',nl}|/\hbar V$ if the kinetic energy $\mathcal{E} = \hbar^2 q^2/2\mu \gg |\Delta E_{n',nl}|$. The minimum value of the Rydberg electron momentum k for the $nl \to n'$ transition with the energy transfer $|\Delta E_{n',nl}|$ is $k_{\min} \approx |\Delta E_{n',nl}|/2\hbar V$. It corresponds to the backward scattering $(\mathbf{k}' = -\mathbf{k})$ of the quasifree electron by the perturber B. Substitution of $k_{\min} = a_0^{-1}(\lambda/2n_*)$ into the classical expression

$$\frac{\hbar^2 k^2}{2m} - \frac{e^2}{r} = -\frac{Ry}{n_*^2} \tag{296}$$

for the energy of the Rydberg electron in the Coulomb field of the ion core A^+, yields

$$r_{\max}(\lambda) = \frac{2n_*^2 a_0}{1 + (n_* k_{\min} a_0)^2} = \frac{2n_*^2 a_0}{1 + (\lambda/2)^2}. \tag{297}$$

Within the framework of the Fermi pseudopotential model for the electron–perturber interaction, the collisional transition of the Rydberg electron occurs when its radius r relative to the ion core A^+ is equal to the separation R between A^+ and B. Thus, expression (297) for the radius $r_{\max}(\lambda)$ corresponds to the maximum possible value of the internuclear separation, and therefore can be considered as an upper bound for $\rho_{\max}$. In accord with (271), its value may be estimated as

$$\left(1 + \lambda^2\right)^{-1} < \rho_{\max}(\lambda)/2n_*^2 a_0 < \left[1 + (\lambda/2)^2\right]^{-1}.$$

The results of calculation [138] of the dependence $\rho_{\max}/2n_*^2 a_0$ on λ is shown in Fig. 18 by the dashed curve. The dependence of the impact parameter $\rho_0(\lambda)$ (which separates the regions of weak and strong coupling) is shown by full curves for different values of the collision strength ζ. As is evident from Fig. 8, at large $\zeta > 1$ practically the whole range of classically allowed impact parameters $0 < \rho < \rho_{\max}(\lambda)$ corresponds to the strong coupling of Rydberg states. Therefore, first-order perturbation theory becomes inapplicable for almost all possible values of $y = \rho/2n_*^2 a_0$.

The basic formula (272) for the cross section of the $nl \to n'$ transition can be rewritten in terms of the scaled parameters λ and ζ

$$\sigma(\lambda, \zeta) = \pi n_*^4 a_0^2$$

$$\times \begin{cases} 4cy_{\max}^2(\lambda), & y_0 = y_{\max}, \\ 4cy_0^2(\lambda, \zeta) + 2^{5/2}\zeta\mathcal{F}_\lambda(y_0), & y_0 \le y_{\max}. \end{cases} \tag{298}$$

Here $y_{\max} = \rho_{\max}(\lambda)/2n_*^2 a_0$; and $y_0 = \rho_0(\lambda, \zeta)/2n_*^2 a_0$ should be calculated from equation $W_{\lambda\zeta}(y_0) = c$ in which the first-order transition probability is given (295); and the $\mathcal{F}_\lambda$-function is determined by (273). Note that $\mathcal{F}_\lambda(y_0) \to 0$, when $y_0(\lambda, \zeta) \to y_{\max}(\lambda)$, i.e. for large collision strength ζ. The ratio of the cross section $\sigma_{n',nl}$ to the geometric area of the Rydberg atom depends only on λ and ζ. Thus, it is interesting to analyze the scaled cross section $\sigma_{n',nl}/\pi n_*^4 a_0^2$ as a function of the inelasticity parameter λ and the collision strength ζ.

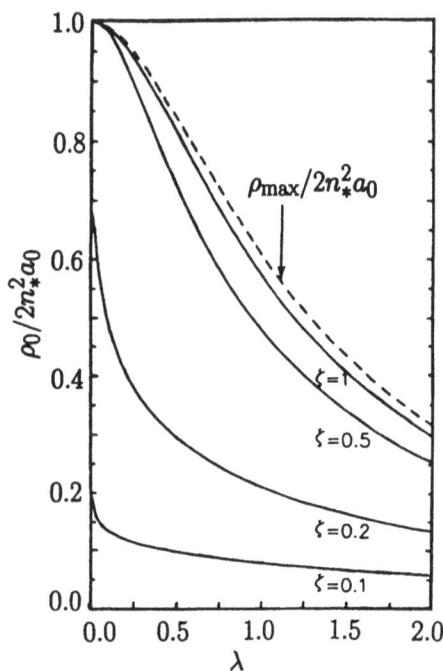

Figure 18. Semiclassical calculations [138] of scaled impact parameter $y_0 = \rho_0/2n_*^2 a_0$, separating the close- and weak-coupling regions, as a function of the inelasticity parameter λ for different collision strengths $\zeta = 0.1$, 0.2, 0.5, and 1 (full curves). All data were obtained for the value of normalized constant $c = 1/4$. The dashed curve represents the maximum possible scaled impact parameter $\rho_{\max}(\lambda)/2n_*^2 a_0$.

The scaling formula (298) has an analytical form in two limiting cases. The first corresponds to quasielastic transitions without energy transfer ($\lambda = 0$) to the Rydberg atom. In this case $y_{\text{max}} = 1$ in (298), the $\mathcal{F}_\lambda (y_0 = 0)$ function is given by simple formula (279) in which the $y_0(\zeta)$ is determined from the equation

$$\left(2\zeta/\pi y_0^{1/2}\right) K\left[k\left(y_0\right)\right] = c . \tag{299}$$

Hence, it is approximately equal to $(\zeta/c)^2$ since $K(k) \approx \pi/2$ for $0 \leq k \leq 2^{-1/2}$, when $0 \leq y \leq 1$. For $\lambda = 0$ there is a certain boundary value of the collision strength $\zeta_0 = c$ (and hence the principal quantum number $n_* = n_0$) for which the scaled parameter y_0 reaches y_{max}. This results from the non-zero value of the transition probability at $y_{\text{max}}(\lambda = 0) = 1$, in contrast to the general case of $\lambda \neq 0$ when $W\left[y_{\text{max}}(\lambda)\right] = 0$. The strong coupling region corresponds to $\zeta \geq c$. In the weak coupling region $\zeta \ll c$ $(n_* \gg n_0)$ the total cross section (298) approaches the asymptotic expression

$$\sigma_{n,nl} = 2^{5/2}\zeta\pi n_*^4 a_0^2 . \tag{300}$$

In the scattering length approximation $L_{\text{eff}} = L$ this limiting expression is in full agreement with the *Omont*'s result (135). In this case the magnitude of the quasielastic cross section is much lower than the geometric area of a Rydberg atom.

The second case corresponds to inelastic $nl \rightarrow n'$ transitions in the range of weak coupling $\zeta \ll 1$. The contribution of the strong-coupling region $0 \leq \rho \leq \rho_0$ can be neglected. Thus, assuming $y_0 = \rho_0/2n_*^2 a_0 \rightarrow 0$ in (273) and (298), we have

$$\mathcal{F}_\lambda (y_0 = 0) \rightarrow f_{n',nl}(\lambda) ,$$

where $f_{n',nl}(\lambda)$ is defined by (132). Hence, for the scaled cross section we obtain the result

$$\sigma (\lambda, \zeta) = \left(\pi n_*^4 a_0^2\right) 2^{5/2}\zeta f_{n',nl}(\lambda), \tag{301}$$

which is in full agreement with the general semiclassical formula (202) of the weak coupling limit [105] in the momentum representation. It is reduced to the simple analytical formula (132) of first-order perturbation theory [132, 146] if the scattering length approximation is applicable. For pure quasielastic transitions the $f_{n',nl}(\lambda)$-function in (301) becomes equal to one, while the collision strength parameter $\zeta_0 \approx c$. Hence, the aforementioned *Omont*'s formula [128] for the l-mixing cross section in the range of weak coupling may also be derived from (301). It is important to stress that in the weak coupling limit the cross section is independent of the specific choice of the normalized constant c.

The general case, given by the scaling formula (298), is presented in Figs. 19 and 20. Figure 19 demonstrates the scaled cross section $\sigma_\lambda(\zeta)/\pi n_*^4 a_0^2$ as a function of ζ for different values of λ. Note that the inelastic cross section falls strongly with increasing inelasticity parameter λ both in the range of weak and strong coupling.

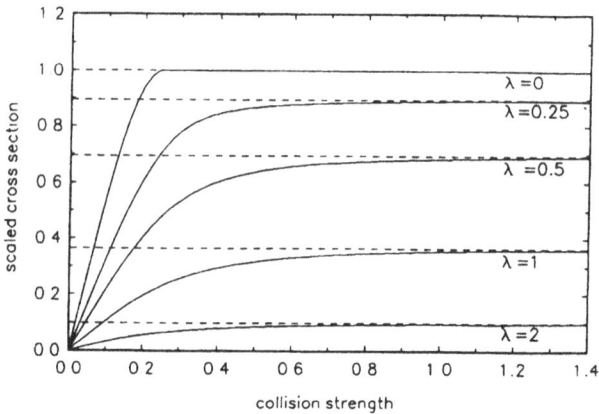

Figure 19. The ratio $\sigma_\lambda(\zeta)/\pi n_*^4 a_0^2$ of the cross section to geometric area of the Rydberg atom as a function of the collision strength ζ. Full curves were calculated in [138] at $c = 0.25$. Dashed curves represent the limiting value of this ratio at $\zeta \to \infty$ for $\lambda \neq 0$ and $\zeta \geq \zeta_0 \approx c$ for $\lambda = .0$ in the close-coupling range. Numbers near the curves mark the values of the inelasticity parameter λ.

To demonstrate the failure of the perturbative approach in the strong-coupling region, we make a comparison between the semiclassical calculations [138] and first-order perturbation theory in Fig. 20. The results are presented for $\zeta = 0.1$, 0.25, and 1. For each couple of curves, the lower one has been calculated using the scaling formula (298), while the upper curves have been obtained in first-order perturbation theory. First-order perturbation theory gives satisfactory results at $\zeta < 0.1$. In this case the cross sections are very close to those calculated by the normalized perturbation theory. The difference between these two methods becomes particularly small for large λ. However, at intermediate ζ (full curves for $\zeta = 0.25$) both methods give close results only for inelastic transitions with large λ. The difference becomes very large at small λ and large $\zeta > 1$. In the strong-coupling case first-order perturbation theory (or the impulse approximation in the momentum representation) leads to significantly overestimated magnitudes and to qualitatively incorrect behavior of the cross section (see long-dashed curves for $\zeta = 2$).

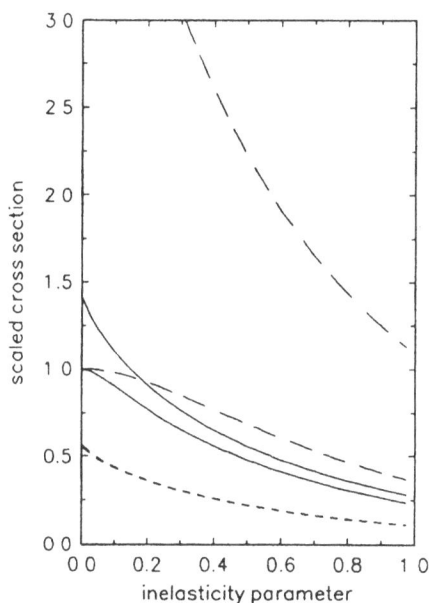

Figure 20. The scaled cross section $\sigma_\zeta (\lambda) / \pi n_*^4 a_0^2$ as a function of the inelasticity parameter λ. Long-dashed, full, and short-dashed couples of curves are the semiclassical results [138] for three values of the collision strength parameter $\zeta = 0.1$, 0.25, and 1, respectively. The lower curves for each couple were calculated using the scaling formula, while the upper curves correspond to first-order perturbation theory.

5.6 Simple Formulae for Direct Ionization

The direct ionization of Rydberg atom by neutral atom

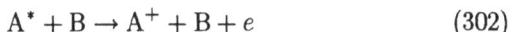

$$A^* + B \rightarrow A^+ + B + e \qquad (302)$$

is accompanied by a large value of the energy transferred to highly excited electron from the translational motion of colliding particles as compared to all inelastic bound-bound transitions considered above. If the principal quantum number is not too small the cross section of this process can be evaluated in the quasifree electron model using the general formula (203) of the impulse approximation. In order to clarify the major features of this process in its dependence on the principal quantum number and relative velocity we restrict our analysis here to the simplest case of the scattering length approximation $f_{eB} = -L$. In this approximation analytic expressions for the ionization cross sections $\sigma_{nl}^{ion}(\mathcal{E})$ and $\sigma_n^{ion}(\mathcal{E})$ has been derived in [171].

In the range of $n \ll v_0/V$, corresponding to slow collisions, one can use the following approximation $na_0 k \geq na_0 k_1 \approx v_0/4nV \gg 1$ for all possible values of the electron momenta. Then, the use of the semiclassical expression (206) for the momentum distribution function with $Z = 1$, yields a simple formula

$$\sigma_{nl}^{ion}(\mathcal{E}) = \frac{8n_* \sigma_{el}^{eB}}{3\pi v_0} \left(\frac{2\mathcal{E}}{\mu}\right)^{1/2} \left(1 - \frac{|E_{nl}|}{\mathcal{E}}\right)^{3/2} \qquad (303)$$

for the ionization cross section at $l \ll n$ (where $\sigma_{el}^{eB} = 4\pi L^2$). Averaging the value $V\sigma_{nl}^{ion}(V)$ over the Maxwellian distribution function in the range of $|E_{nl}| \leq \mathcal{E} = \mu V^2/2 < \infty$, we have the corresponding analytic expression for the ionization rate constant [171]

$$K_{nl}^{ion}(T) = \langle V\sigma_{nl}^{ion} \rangle_T = \frac{8kTn_*}{\pi\mu v_0} \sigma_{el}^{eB} \exp\left(-\frac{Ry}{kTn_*^2}\right). \qquad (304)$$

Here T is the gas temperature. Formulae (303) and (304) describe the case of selectively excited nl levels with given values of the principal n and orbital l ($l \ll n$) quantum numbers. Given equally populated nlm sublevels of the Rydberg $A(n)$ atom at a definite value of n, similar calculations lead to the following expression for the ionization cross section [171]

$$\sigma_n^{\text{ion}}(\mathcal{E}) = \frac{2^{10} n^3 \sigma_{\text{el}}^{\text{eB}}}{15 \pi v_0^3} \left(\frac{2\mathcal{E}}{\mu} \right)^{3/2} \left(1 - \frac{|E_n|}{\mathcal{E}} \right)^{1/2}$$

$$\times \left[1 - \frac{7}{6} \frac{|E_n|}{\mathcal{E}} + \frac{1}{6} \left(\frac{|E_n|}{\mathcal{E}} \right)^2 \right] . \tag{305}$$

The ionization rate constant averaged over all lm-sublevels ($0 \leq l \leq n-1$) for a given value of the principal quantum number n takes the form

$$K_n^{\text{ion}}(T) = \frac{2^{10} n^3 (kT)^2 \sigma_{\text{el}}^{\text{eB}}}{\pi \mu^2 v_0^3} \left(1 + \frac{Ry}{3kTn^2} \right) \exp\left(-\frac{Ry}{kTn^2} \right) . \tag{306}$$

As can be seen from (304) and (306), the ionization rate constants calculated in the quasifree electron model are changed exponentially K_{nl}^{ion}, $K_n^{\text{ion}} \propto nT \exp\left(-Ry/n^2 kT\right)$ for low $n \ll (Ry/kT)^{1/2}$ (when $|E_{nl}| \gg kT$). In the range of $(Ry/kT)^{1/2} \ll n \ll v_0/V_T$ the K_{nl}^{ion} and K_n^{ion} values are proportional to the first and third powers of the principal quantum number, i.e. $K_{nl}^{\text{ion}} \propto n$ and $K_n^{\text{ion}} \propto n^3$.

Analysis of the validity conditions of the impulse approximation (see Sect. 4.9) for a direct ionization process shows that the formulae presented above are applicable for a sufficiently wide range of the principal quantum numbers

$$(A/32a_0)^{1/4} (v_0/V)^{1/2} = n_{\text{low}} \ll n \ll v_0/V, \tag{307}$$

where $A = \max\{|L|, \rho_W\}$ and $\rho_W = (\pi \alpha v_0/4V)^{1/3}$ is the Weisskopf radius, and α is the polarizability of the perturbing atom.

For very high principal quantum numbers of $n > n_{\text{up}} \sim v_0/V$ (where $n_{\text{up}} \sim 5 \cdot 10^2 - 5 \cdot 10^4$ for thermal collisions) expressions (303)–(306) prove inapplicable due to the condition (307). For such values of n, the increase of the ionization σ_{nl}^{ion} and σ_n^{ion} cross sections (or the rate constants) with growing n would become slower compared with their behavior for $n \ll v_0/V$. According to (193), the limiting values of σ_{nl}^{ion} and σ_n^{ion} are determined by the total cross section $\sigma_{\text{el}}^{\text{eB}} = 4\pi L^2$ for elastic scattering of the ultra-low energy electron e on the perturbing atom B:

$$\sigma_{nl}^{\text{ion}} \xrightarrow[n \to \infty]{} \sigma_{\text{el}}^{\text{eB}} = 4\pi L^2, \qquad K_{nl}^{\text{ion}} \xrightarrow[n \to \infty]{} \langle V \rangle_T \sigma_{\text{el}}^{\text{eB}} . \tag{308}$$

Here $\langle V \rangle_T = (8kT/\pi\mu)^{1/2}$ is the average thermal velocity of colliding atoms.

Thus, at thermal velocities of the Rydberg atoms with neutral atomic particles, the traditional mechanism of the perturber–quasifree-electron scattering can be effective for the direct ionization (302) only for high enough n. As a rule, in the range of $n < 20 - 40$ other physical mechanisms of ionization induced by the perturber-core scattering prove to be predominant at thermal collisions of Rydberg atoms with neutral atomic particles (see below Sect. 7).

5.7 Transitions Induced by Ion–Covalent Coupling

5.7.1 Destruction of Rydberg States by Atoms with Small Electron Affinities

Collision processes accompanied by transitions between covalent and ionic states has been the subject of intensive theoretical and experimental works for many years (see [190, 191]). The well known processes caused by the ion–covalent coupling of electronic terms are the ion-pair formation and mutual neutralization reactions. A number of recent experimental works was devoted to studies of ion-pair $(A^+ + B^-)$ formation processes in thermal collisions of Rydberg atoms A^* (nl) with the ground-state atoms and polar molecules B having small electron affinity (*McLaughlin* and *Duquette* [192], *Desfrançois* [193-195], *Compton* et al. [196], and *Reicherts* et al. [197]).

Experimental studies of these processes and measurements of their cross sections and rate coefficients as regards their dependence on the principal quantum number n provide information about the electron affinities of the ground-state atoms and molecules. In particular, this method was used to obtain electron affinities of several polar molecules. A recent application of this method for collisions with the ground-state Ca [197] atom confirmed the most accurate values of electron affinity of Ca [116]. There is also a series of recent theoretical works by *Fabrikant* and coworkers [198-201] and *Desfrançois* [202] aimed at a description of ion-pair formation processes in thermal collisions between the highly excited atom and neutral targets.

It is evident that collisions of Rydberg atoms A^* (nl) with such neutral targets can also lead to inelastic transitions between the highly excited states $nl \to n'l'$, i.e. to the destruction of the initial Rydberg nl-level via the excitation and deexcitation processes. One of the possible mechanisms of such transitions is connected with the formation of a temporarily negative B_t^- and positive A^+ ions [203]. Thus, the collisional depopulation of Rydberg atomic states by the ground-state atoms with small electron affinities can be performed in the following two ways

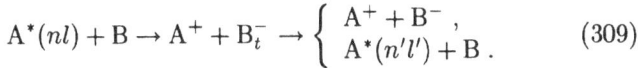

$$A^*(nl) + B \rightarrow A^+ + B_t^- \rightarrow \left\{ \begin{array}{l} A^+ + B^- , \\ A^*(n'l') + B . \end{array} \right. \qquad (309)$$

The theoretical description of these two channels of reaction (309) can be given on the basis of various modifications of the theory of non-adiabatic transitions between the Rydberg covalent A^*+B and ionic A^++B^- electronic terms of diatomic system. We shall present below the main idea and basic equations of an approach developed by *Fabrikant* [200] for the process of ion-pair formation and extended to the case of resonance quenching of Rydberg states in Ref. [203]. The ion-pair formation (the first channel of reaction (309)) occurs when the temporarily negative ion B_t^- survives all pseudocrossings with covalent terms. Resonance depopulation of nl-state via the inelastic transitions (the second channel of reaction (309)) takes place when one of the covalent $A^*(n'l')+B$ terms turns out to be populated.

As noted in Ref. [203], experimental observations of the second channel of reaction (309) might be helpful for negative-ion spectroscopy [204], particularly when the electron affinity is very small. For example experimental investigations [205] confirmed by recent calculations [206] have shown that if a stable or long-lived metastable Yb^- ion does exist, its binding energy will be below 3 meV. In fact, theory [206] predicts that the Yb^- ion "is most certainly unbound". However, we still cannot completely rule out the possibility that the $p_{1/2}$ component is very slightly bound. Such a weakly bound state is easily destroyed by external electric fields, therefore its direct observation could be very challenging. The detection of the excitation (deexcitation) reactions might be quite helpful in this case.

However, interpretation of experimental data in this case might be complicated by the existence of another excitation (deexcitation) mechanism which is due to scattering of a weakly-bound electron by the target atom or molecule B. The theory of such processes induced by the electron–projectile interaction is presented in Sect. 5.3. This scattering leads to an energy exchange between the Rydberg electron and B which causes inelastic or quasielastic transitions. This mechanism is more efficient at high n where it can be described in terms of the impulse approximation (therefore it will be called the impulse mechanism below). The question appears if the resonant and impulse contributions can be separated. If they interfere substantially, it might cause difficulties in interpreting experimental data.

The situation here is similar to the process of quenching occurring through the formation of a quasistationary negative-ion state. The intermediate state B_t^- can decay in this case not only due to interaction with the ion A^+, but also spontaneously. A typical example is the process of destruction of Rydberg states in collisions of Rydberg atoms with ground-state alkali-metal atoms. Experimental data [98, 112] on collisional broadening of Rydberg states by alkali-metal perturbers exhibit very large broadening cross sections and their oscillatory dependence on n in the region $n = 25 - 35$. Whereas the large magnitude of the cross sections is explained by the impulse approximation [103, 105], the oscillatory structure can be explained only through the mechanism of formation of a very low-energy ^3P resonance state of the alkali-metal anion [187].

A quantitative description involves the curve-crossing model where the coupling parameter is proportional to the wave function of the Rydberg electron. The important difference between this case and the quenching through formation of a stable (with respect to spontaneous decay) anion is that in the former case the crossing point lies in the classically allowed (for the Rydberg electron) region, whereas in the latter it lies in the classically forbidden region.

Figure 21 illustrates this feature schematically [203]. In the case of a quasistationary intermediate anion the coupling parameter oscillates with n, and this is what causes oscillations in the cross sections. In the case of a stable intermediate anion the coupling parameter decays exponentially with n. This causes the resonant region to be rather narrow in n. In contrast, in the case of a quasistationary anion the region of n where resonant mechanism is important is rather broad and overlaps strongly with the impulse region. The simplest way to account for both mechanisms is just to add two contributions to the cross sections incoherently. In this way the overall behavior of the quenching cross sections was explained [110, 188] qualitatively and semiquantitatively. However, this approach is not rigorous, and it is not clear how to add these two contributions coherently.

The resonance quenching through the formation of a stable (with respect to spontaneous decay) anion might be advantageous in this regard. As follows from the results of the recent paper [203], the region of resonant quenching and the impulse quenching in some cases can be completely separated, therefore observation of this process can be very useful for studying properties of weakly-bound negative ions. The theory presented below will be illustrated in Sect. 6.7 by two examples: Ca^- whose binding energies are well established [116] and the hypothetical Yb^- with binding energy 2 meV.

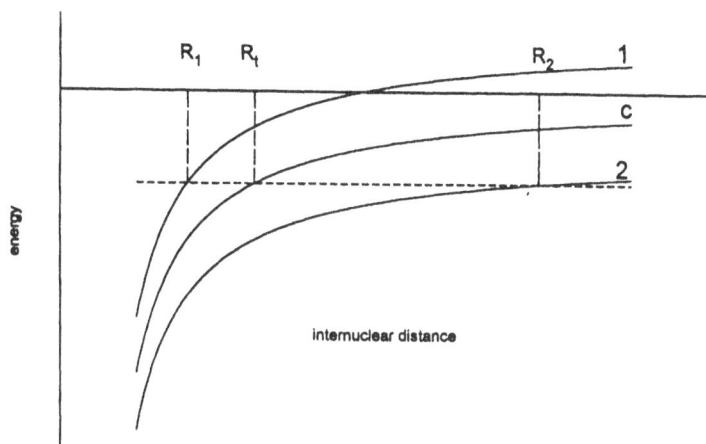

Figure 21. Schematic potential energy curves representing ion–covalent coupling [203]. Ionic curves are given for the case of a quasistationary negative-ion state (curve 1) and a bound state (curve 2). Curve c represents the Coulomb potential for the Rydberg electron with R_t being the classical turning point, and the horizontal dashed curve is the covalent state. The crossing point R_1 lies in the classically allowed region for the electron motion, and R_2 in the classically forbidden region.

5.7.2   Ion-Pair Formation and Resonance Quenching

Most theoretical treatments of ion-pair formation processes involving excited atoms in the initial channel of reaction are based on the two-state Landau–Zener formula [37] and the multi-state Demkov–Osherov model [207] (see [190, 191] and references therein). Recent experimental observations [197] of the peak in Ca^- production in thermal collisions of $Ne(ns)$, $Ne(nd)$ atoms with the ground-state Ca atoms were explained by three theoretical models: the multi-state Landau–Zener model [202], the decay model [200] combined with the distorted-wave theory [199] and semiclassical close coupling method [201]. The decay model was also used for a description of the quenching processes for the ns and nd states of neon by Ca and Yb atoms in Ref. [203].

An important common feature of all identified models is that the probabilities and cross sections of processes (309) are determined by the coupling between the covalent Rydberg state $A^*(nl)+B$ and the ionic state $A^+ + B^-$ of a diatomic system. The matrix elements $\langle \Psi_f(\mathbf{r}) | V(|\mathbf{r}|) | \Psi_i(\mathbf{r}') \rangle$ of such transitions for the first and second channels of reaction are expressed in terms of an integral of the LCAO type (see, for example, [190]). This integral in the coordinate space contains the final wave function $\Psi_f(\mathbf{r})$ of the weakly-bound electron of unperturbed negative ion B^-, the spherically symmetric interaction potential $V(|\mathbf{r}|)$ of this electron with the atom B, and the initial wave function $\Psi_i(\mathbf{r}')$ of a highly excited electron relative to the ion core A^+ of the Rydberg atom $A^*(nl)$. It is important to stress that the $\Psi_i(\mathbf{r}')$ and $\Psi_f(\mathbf{r})$ functions are referred to different atomic centres (i.e. to the Coulomb centre A^+ and to the nucleus of the B atom because of $\mathbf{r}' = \mathbf{r} - \mathbf{R}$, where the $\mathbf{R}$ vector is directed from B to A^+). Therefore, the exact calculation of such matrix elements in the general form is a rather complicated problem even if the specific form of the interaction potential $V(r)$ and the wave functions $\Psi_i(\mathbf{r}')$ and $\Psi_f(\mathbf{r})$ are well known.

Basic Formulae of Decay Model. Here we will follow the approach developed by *Fabrikant* [200] for the process of ion-pair formation. Briefly, for a given impact parameter ρ, the probability W_i of ion-pair formation is given by

$$W_i = p(1 - p)(S_1 + S_2), \tag{310}$$

where $S_{1,2}$ are the probabilities for the ion pair to survive if it is formed before (transition point 1) or after (transition point 2) the closest approach, and p is the Landau–Zener probability of transition from the initial covalent state $A(nl) + B$ to the ionic state [190]

$$p = 1 - \exp\left[-\frac{2\pi}{V(\rho)} \mathcal{V}_{\mathrm{cp}}^2 R_s^2 \right], \tag{311}$$

where R_s is the internuclear separation at which the transition occurs, $V(\rho)$ is the radial relative $A - B$ velocity at the transition point, and $\mathcal{V}_{\mathrm{cp}}$ is the coupling parameter.

Similarly, the probability of quenching the initial covalent state is

$$W_q = p(1 - S_{12}) + pS_{12}(1 - p)(1 - S_2)$$
$$+(1 - p)p(1 - S_2), \tag{312}$$

where S_{12} is the probability for the ion pair to survive between the points 1 and 2. The first term in Eq. (312) describes the transition to the ionic state at point 1 with the following decay between points 1 and 2. The second term describes transition to the ionic state at point 1 with the following evolution of the ionic state between points 1 and 2 and decay after passing point 2. Finally, the third term describes the transition to the ionic state at point 2 with the following decay. These three options exhaust all possible routes for destruction of the initial state above the ion-pair formation threshold. However, below the threshold, the route which initially forms the ion pair cannot lead to infinite separation of ion pairs because there is not sufficient kinetic energy. In this case the ions are captured into an elliptic orbit and eventually decay into a covalent channel. Therefore the ion-pair formation probability, Eq. (310), should be added to the quenching probability (312). The resulting probability is

$$W_q = p(2 - p - S_{12}p) . \tag{313}$$

The cross sections for the ion-pair formation σ_i and resonance quenching σ_q processes are then calculated on integration of Eqs. (310) and (312) over the $2\pi\rho\,d\rho$

$$\sigma = \int_0^\infty W\left(\rho\right) 2\pi\rho\,d\rho . \tag{314}$$

Note that oscillations of the transition probability due to the phase difference between the wave functions in ionic and covalent channels are very fast and can be completely neglected in the calculation of the cross section [199].

The most important parameters in the above equations are the coupling parameter $\mathcal{V}_{cp}$ and the survival probabilities. The standard method of calculation of the coupling parameter [190, 208, 209] involves two approximations: the approximation of fixed internuclear axis and the Landau–Herring asymptotic method [210, 211] for evaluation of the exchange integral. The first approximation is reasonable for slow collisions [212], since due to the strong Coulomb interaction the electron angular momentum follows the internuclear axis adiabatically. (More discussion on this can be found in Ref. [199]). Note also that the approximation of the fixed nuclear axis is necessary for employing the decay model where the decaying states are classified according to projection on the internuclear axis.

The Landau–Herring method for evaluation of the coupling parameter might lead to substantial errors in the case of collisions with highly-polarizable atoms [200]. The authors of Ref. [203] have employed the approximation of fixed internuclear axis but calculated the coupling parameter according to the distorted-wave approximation [199, 200].

Coupling Parameter in the Dipole Approximation. In the distorted-wave theory the coupling between the covalent and the ionic states is given by an integral of the LCAO type [199, 200]

$$\mathcal{V}_{\rm cp}(\mathbf{R}) = \int d\mathbf{r}\, \Psi_f^*(\mathbf{r}) V_{\rm eB}(\mathbf{r})\psi_i(\mathbf{r} - \mathbf{R}), \qquad (315)$$

where $\mathbf{r}$ is the electron radius-vector relative to the atom B , $V_{\rm eB}(\mathbf{r})$ is the interaction of electron with the atom B, ψ_i is the initial wave function of the Rydberg electron, and Ψ_f is the electron wave function in the anion B$^-$. The vector $\mathbf{R}$ is directed from B to A$^+$.

In the laboratory frame the matrix element $\mathcal{V}_{\rm cp}(\mathbf{R})$ can easily be calculated if one of the states (i or f) is spherically symmetric. This was done previously for transitions from a state with arbitrary l_c to a state with $l_i = 0$ [199] and for $s - p$ transitions [200]. Since the weakly-bound p-states (like Ca$^-$ and Yb$^-$) are of most interest in current experimental studies, in the present paper we calculate $\mathcal{V}_{\rm cp}$ for transitions from arbitrary l_c into $l_i = 1$ using the fixed-nuclei approximation.

For high principal quantum number n, ψ_i varies much slower with $\mathbf{r}$ than $V_{\rm eB}$ and Ψ_f. Therefore one can use the Taylor expansion

$$\psi_i(\mathbf{r} - \mathbf{R}) = \psi_i(-\mathbf{R}) + \nabla\psi_i|_{-\mathbf{R}} \cdot \mathbf{r}. \qquad (316)$$

We will assume that the $e - $B interaction $V_{\rm eB}$ is spherically symmetric. Then the first term in Eq. (316) does not contribute to $\mathcal{V}_{\rm cp}$ if $l_f > 0$, and we will concentrate on the second term. Using the spherical coordinates r, θ, φ it is convenient to rewrite the expression for $\nabla\psi_i(\mathbf{r})$ in terms of the radial part $\mathcal{R}_i(r)$ of the wave function $\psi_i(\mathbf{r})$ and its derivative dR_i/dr. Thus, it is clear that there are two contributions to $\mathcal{V}_{cp}$: $\mathcal{V}_{cp}^{(1)}$, which is proportional to dR_i/dr, and the other $\mathcal{V}_{cp}^{(2)}$ proportional to $\mathcal{R}_i$. The first contribution is given by [203]

$$\mathcal{V}_{cp}^{(1)}(R) = -\delta_{m_i 0}\delta_{m_f 0}(-1)^{l_i} \left.\frac{dR_i}{dr}\right|_R \sqrt{\frac{2l_i+1}{3}}$$

$$\times \int\limits_0^\infty \phi_f(r)\mathsf{V}_{eB}(r)\,r^3\,dr, \tag{317}$$

where $\phi_f(r)$ is the radial part of $\Psi_f(\mathbf{r})$.

The evaluation of the second contribution

$$\mathcal{V}_{cp}^{(2)} = \nabla\psi_i|_{-\mathbf{R}} \cdot \int \Psi_f^*(\mathbf{r})\mathsf{V}_{eB}(r)\,\mathbf{r}\,d\mathbf{r} \tag{318}$$

leads to the following result [203]

$$\mathcal{V}_{fi}^{(2)} = -(\delta_{m_i,-1}\delta_{m_f,-1} + \delta_{m_f 1}\delta_{m_f 1})(-1)^{l_i}$$

$$\times \frac{\mathcal{R}_i(R)}{R}\sqrt{\frac{(2l_i+1)l_i(l_i+1)}{6}} \int\limits_0^\infty \phi_f(r)\mathsf{V}_{eB}(r)\,r^3\,dr. \tag{319}$$

As can be expected, in the approximation of the fixed internuclear axis $\mathcal{V}_{fi}$ is non-zero only for $m_i = m_f$.

Survival Probability. The probability for the ion pair to survive between instants t_i and t_f is given by [213, 214]

$$S_{fi} = \exp\left(-\int_{t_i}^{t_f} \Gamma(t)\,dt\right), \tag{320}$$

where Γ is the probability of negative ion decay in the Coulomb field of the A^+ core. The method of calculation was discussed in detail by *Fabrikant* [200]. Note that in the calculation of Γ as a function of time the relative A^+-B motion along a classical path is to be assumed. The Coulomb distortion of the classical path affects the survival factor very little even at low velocities [200]; therefore the rectilinear approximation can safely be used.

The decay model assumes that depopulation of the ionic state occurs continuously. Although in general this assumption is reasonable because of the high density of Rydberg states, in specific applications we have to distinguish between the capture into an ionic state from a Rydberg state with a negligible quantum defect and from a Rydberg state with a substantial quantum defect. Indeed, in the first case, due to interaction with a highly degenerate Rydberg manifold, the ionic state starts to decay almost immediately after its formation. In the second case the ionic state has to evolve until the ionic state crosses a Rydberg manifold. The neglect of this effect underestimates the survival factor and leads to an underestimated ion-pair formation cross section. In particular the rates for the charge transfer from d states of Ne (small quantum defect) calculated by the decay model agree with calculations by the curve crossing model [197] whereas for s states with a large fractional part of the quantum defect the curve crossing model gives results which are about 57 % higher [200]. To correct this drawback of the decay model, we have assumed in Ref. [203] that decay is forbidden in the region between the crossing of the ionic state with the initial isolated covalent state and crossing of the ionic state with the nearest Rydberg manifolds.

This limited-decay model with the asymptotic coupling parameters yields rates whose absolute value is in almost perfect agreement with the multiple-crossing Landau–Zener model employed in Ref. [197]. However, using the distorted-wave coupling parameter leads to a further increase of the rates. Limitation of the decay is not important for the d states whose quantum defect is small. On the other hand, a small discrepancy is still observed there for the shape of the rate dependence on n which might be caused by using the asymptotic coupling parameter for intermediate states in the Landau–Zener model. Note that results of the decay model for d states agree better with experiment than the results of the Landau–Zener model [197].

Note that applications of the theory [203] of resonant excitation and quenching processes to concrete atomic systems involving Rydberg states and ground-state atoms with small electron affinities will be discussed in Sect. 6.7.

6   Quenching and Ionization at Thermal Energies

6.1   Introductory Remarks

The cross section σ_i^Q of collisional quenching (depopulation) of a given Rydberg state $|i\rangle$ is determined by the total contribution of all inelastic bound-bound and bound-free $|i\rangle \rightarrow |f\rangle$ transitions, i.e. $\sigma_i^Q \equiv \sigma_{nl}^{in} = \sum_f{}' \sigma_{fi}$. Here the prime on the summing symbol cancels the contribution of pure elastic scattering (when $f = i$) from the summation over all final states f. Thus, for a Rydberg atom with one valence electron the quenching cross section σ_{nl}^Q of the nl-state can be represented as the sum of the total cross section $\sigma_{nl}^{n,l\text{-ch}} = \sum_{n'l'}{}' \sigma_{n'l',nl}$ of the bound–bound $nl \rightarrow n'l''$ transitions with a change both in the principal and orbital quantum numbers (including the contribution of the l-mixing $nl \rightarrow nl'$ transitions with $n' = n$) and the ionization cross section , i.e.

$$\sigma_{nl}^Q \equiv \sigma_{nl}^{in} = \sigma_{nl}^{n,l\text{-ch}} + \sigma_{nl}^{ion} \ . \tag{321}$$

The quenching cross section σ_{nl}^Q of selectively excited Rydberg nlJ-state with given magnitudes of n, l, and the total angular momentum $J\ (= |l - 1/2|$ or $l + 1/2)$ is determined by the relation

$$\sigma_{nlJ}^Q \equiv \sigma_{nlJ}^{in} = \sigma_{nlJ}^{J\text{-mix}} + \sigma_{nlJ}^{n,l\text{-ch}} + \sigma_{nlJ}^{ion} \ , \tag{322}$$

i.e. it additionally includes the J-mixing cross section of the fine-structure components. In thermal collisions between the Rydberg atoms and neutral atoms the influence of the direct ionization process on the value of total depopulation cross section is usually very small and can be neglected in calculations of the quenching cross sections.

In this section we shall present a large number of theoretical results for quite different types of inelastic and quasielastic quenching processes and ionization of highly excited atoms in thermal collisions with the ground-state atomic and molecular particles. Our major goal is to demonstrate the characteristic features of one or another process and to make a detailed comparison of theory with experiment. It may be worthwhile to remember that in most of the available experiments on the quenching and ionization of Rydberg states by neutral targets the measured value is the cross section

$$\langle \sigma \rangle_T = \langle V\sigma \rangle_T / \langle V \rangle_T \,, \qquad \langle V \rangle_T = (8kT/\pi\mu)^{1/2} \qquad (323)$$

averaged over the Maxwellian velocity distribution for a given gas temperature T, where $\langle V \rangle_T$ is the mean relative velocity of colliding partners.

We shall start the analysis of the quenching processes from a detailed consideration of recent theoretical works on thermal collisions of Rydberg atoms with the rare gas atoms. In this case there is an extensive experimental material needed for comparison with theoretical calculations. The behavior of the cross sections and their magnitudes depend essentially on the value of the quantum defect δ_l. The quenching of Rydberg nl-states with very small quantum defects $\delta_l \approx 0$ (for example, $n^1 P$ -states of helium; nd and nf states of sodium; nf states of rubidium and xenon) is due, mainly, to quasielastic $(\Delta E_{nl',nl} \approx 0)$ transitions with a change in the orbital angular momentum $nl \to nl'$, but not in the principal quantum number. Therefore, the calculation of the total quenching cross section $\sigma_{nl}^{\mathrm{Q}} = \sum_{n'l'}' \sigma_{n'l',nl}$ is usually reduced to the evaluation of the contribution of all quasielastic l-mixing collisions $\sigma_{nl}^{l\text{-mix}} = \sum_{l'}' \sigma_{nl',nl}$ if the principal quantum number is not too large. The analysis of previous theoretical and experimental results in this field was given in reviews [36, 81].

In the case of thermal collisions of Rydberg atoms with the ground-state rare gas atoms there is also a series of experimental works (see [2, 36] and references therein) on the quenching of Rydberg nl levels of alkali-metal atoms with low values of the orbital angular momentum l (for example, ns -levels of sodium, np-levels of potassium, and ns-, np-, and nd -levels of rubidium). Due to the large quantum defects δ_l the quenching process of these levels is accompanied by a substantial energy transfer from the Rydberg atom to the kinetic energy of the relative motion of the colliding particles . Therefore, quenching occurs preferentially with $nl \to n'l'$ transitions to the $n'l'$ states which are closest in energy to the initial nl level (i.e., with minimum energy defects $\Delta E_{n'l',nl}$).

As a result, it gives rise to a change both in the orbital angular momentum and principal quantum number not only at high enough n, but also at low and intermediate n. The presence of appreciable energy defects means that the quenching process can no longer be regarded as quasielastic.

Another aim of this section is to present a number of theoretical results on quenching processes of high-Rydberg atoms $A(nl)$ in thermal collisions with the ground-state alkali-metal atoms and to make a comparison with experimental data. As noted previously, an important feature of such processes is the presence of the ^3P-resonance on the quasidiscrete level of negative alkali-metal B^- ions temporarily formed during the scattering of a quasifree electron by the perturbing ground-state atom B. Then we shall discuss the results of recent calculations of the depopulation cross sections for Rydberg ns and nd states of neon in thermal collisions with the ground-states of Ca and Yb atoms having small electron affinities. Further we shall consider a mechanism of the quasiresonant energy exchange between the quasifree electron and rotational degrees of freedom of molecular projectiles and present some results on quenching and ionization of Rydberg states by some polar and non-polar molecules. At the end of this section we shall consider some experimental and theoretical data on the ion-pair formation processes in thermal collisions of Rydberg atoms with neutral targets.

6.2   Quenching of Rydberg States by He

We start the theoretical analysis of different orbital angular momentum transfer processes by considering the quenching of the Rydberg $He(n^1P)$ and $Na(nD)$ states by $He(^1S_0)$ atom, for which there are reliable experimental data at high, intermediate and low magnitudes of n. The identified quenching processes with small quantum defects $\delta_{1P}^{He} = 0.01$ and $\delta_{nD}^{Na} = 0.015$ of the initial states are primarily due to the l-mixing transitions $nl \rightarrow nl'$ to a degenerate manifold of the final nl' sublevels of the same n-level with $l' \geq l_0$, where $l_0 = 2$ and 3 for nP- and nD-states, respectively (see Eqs. (275) and (276) in Sect. 5.3). The contribution of transitions with $n' \neq n$ turns out to be very small and can be neglected if the principal quantum number is not too large.

$He(n^1P) + He(^1S_0)$.   The results of theoretical calculations [140] of the Maxwell-averaged cross sections for the l-mixing collision process $He(n^1P) + He(^1S_0)$ at thermal energy ($T = 300$ K, $V_T = 7.16 \cdot 10^{-4}$ a.u.) are plotted in Fig. 22. The dashed curve represents the results obtained using the simple formula (284) (see also (286)) which corresponds to the limiting case of quasielastic transitions without energy transfer $\lambda = 0$. The full curve indicates the calculations by basic formulae of normalized perturbation theory (272) and (273) taking into account the energy defect $\Delta E_{n,n^1 P} \approx \delta_{1 P}/n^3$ of the $n^1 P \to n$ transition. The corresponding inelasticity parameter is equal to $\lambda \approx 14/n^2$ and, hence, it increases from 0.083 to 0.87 when the principal quantum number n decreases from 13 to 4.

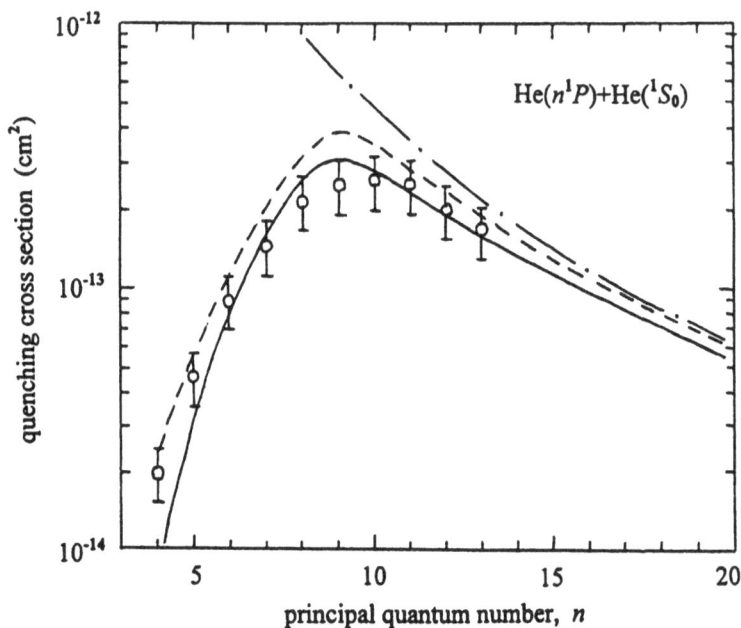

Figure 22. Quenching of $He(n^1 P)$ states induced by the l-mixing collisions with $He(^1 S_0)$ at $T = 300$ K. Full curve, calculations [140] of the Maxwell-averaged cross section $\langle \sigma_{n^1 P}^{l\text{-mix}} \rangle_T$ by the basic formula (272) of normalized perturbation theory $(c = 0.25)$ containing the energy defect ΔE of the $n^1 P \to n$ transition. Dashed curve, calculation by the simple formula (284) for quasielastic transitions with $\Delta E = 0$. Dashed-dotted curve, first-order perturbation theory. Open circles, experimental data [218].

It can be seen that the basic formula (272) provides a good agreement with experimental data of *Pendleton* et al. [218] in the whole of the studied range of n. The simple formula (284) correctly describes the cross section behavior at high, intermediate and sufficiently low n; however, it gives slightly overestimated values. On the other hand, the asymptotic Omont's formula of first-order perturbation theory (dashed-dotted curve) yields reasonable results only at high n. It significantly overestimates the cross sections in the vicinity of the maximum $n \sim n_{\mathrm{max}}$ and leads to incorrect qualitative behavior of the l-mixing process at the intermediate and low n.

Na(nD) **+He**$(^1S_0)$. Analogous calculations [140] of the Maxwell-averaged cross sections for the l-mixing Na(nD)+He$(^1S_0)$ collisions obtained using the basic formula (272) are shown in Fig. 23 by full the curve. This process has been experimentally studied by *Gallagher* et al. [219] at the gas temperature $T = 430$ K. Due to sufficiently close quantum defect values of the Na(nD)- and He(n^1P)-states the energy defect of the $nD \to n$ transition in sodium is only 1.5 times greater than for the $n^1P \to n$ transition in helium, while the relative velocity $V_T = 6.57 \cdot 10^{-4}$ a.u. of the Na+He collisions is close to the previous case. Therefore, the inelasticity parameter $\lambda = n\,|\Delta E_{n,nD}|\,/V_T$ varies from 0.10 to 0.91 at of $15 \geq n \geq 5$. Hence, it is natural to expect similar influence of the transition energy defect on the cross section magnitudes. Indeed, as in the previous case the simple formula (284) corresponding to the quasielastic limit ($\lambda = 0$) reasonably reproduces the behavior of the l-mixing process Na(nD)+He$(^1S_0)$ but it gives somewhat overestimated results.

For example, for $n = 20$, 10, and 6 this formula gives the cross section values which exceed present results taking into account the energy defect of the process for 1.25, 1.45, and 1.8 times, respectively. The agreement between the calculations by the basic formula (272) and the experimental data is good in the whole of the studied range of n except for $n = 9$ (However, the measured cross section for Na (nD) +He exhibits a jump at $n = 9$ which has not been observed in other experimental works on the l-mixing collisions and is not reproduced by any of the theoretical calculations).

In Fig. 23 we also make a comparison of our theory with the results of a purely classical model (dashed-dotted curve) of *Burkhardt and Leventhal* [220], with calculations (dotted curve) by the semi-empirical scaling formula of *Hickman* [162], and with the impulse approximation results (dashed curve) of *Petitjean* and *Gounand* [150] combined with the binary-encounter approximation for the form factors. One can see that classical model for angular momentum mixing (developed for $\Delta E_{n,nl} = 0$) leads to significantly underestimated cross sections at high n and to overestimated values at low n. On the other hand our results (full curve) agree well with the calculations of Ref. [162]. Since at low enough n they were performed by the close coupling method this means that in the range of strong coupling the present theory is in correspondence with the results obtained using more sophisticated models of quasielastic collisions. It is evident from Fig. 23 that for the l-mixing collisions with small energy transfer, the impulse approximation does not hold in the strong coupling region. The corresponding theoretical values significantly exceed the experimental data near the maximum of the l-mixing cross section and especially at low $n \ll n_{\max}$. However, the impulse approximation provides a successful description of quasielastic collisions at high n.

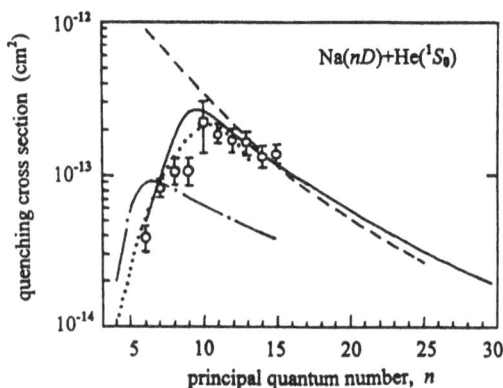

Figure 23. Comparison of the l-mixing cross sections for Na(nD)+He collisions with experimental and theoretical data ($T = 430$ K). Full curve, calculations [140] of $\left\langle \sigma_{nD}^{l-\mathrm{mix}} \right\rangle_T$ by the basic formula (272) of normalized perturbation theory ($c = 0.25$). Dotted curve, calculations by scaling formula [162]. Dashed-dotted curve, classical model [220]. Dashed curve, impulse approximation results [150]. Open circles, experimental data [219].

$\mathrm{Li}(nD, nP, nS) + \mathrm{He}(^1S_0)$. To demonstrate quite different behavior of the l-mixing processes in the dependence on the value of the energy transferred to the Rydberg atom (or the inelasticity parameter λ) we consider below thermal collisions of the $\mathrm{He}(^1S_0)$ atom and Rydberg lithium in the nD-, nP-, and nS-states. A vivid example of a quasielastic process, in which the energy transfer can practically be neglected, is the l-mixing collision process of $\mathrm{Li}(nD) + \mathrm{He}$. The results of calculations [140] of the Maxwell-averaged quenching cross section $\left\langle \sigma_{nD}^{\mathrm{Q}} \right\rangle_T = \left\langle V \sigma_{nD}^{\mathrm{Q}}(V) \right\rangle_T / \left\langle V \right\rangle_T$ for this process are plotted in Fig. 24 by the full curve together with available experimental data [221] ($T = 700$ K, $V_T = 9.71 \cdot 10^{-4}$ a.u.). Due to an extremely small value of quantum defect $\delta_D^{\mathrm{Li}} = 0.002$ the energy defect of the $nD \to nl'$ transition in lithium turns out to be for 7.5 times lower than for the corresponding transition in sodium considered above. Therefore, the weak inelasticity of this l-mixing process ($\lambda \ll 1$) practically does not affect the cross section values. As a result, calculations [140] performed by the basic formula (272) of normalized perturbation theory and by the simple formula (284) for quasielastic transitions without energy transfer practically coincide.

As is apparent from Fig. 24 these calculations are in good agreement with the measured quenching cross sections for the nD-levels in the whole of the studied range of n. Thus, the quasielastic processes of the orbital angular momentum mixing with very small energy transfer can be successfully described by a simple formula which is simultaneously valid in the weak- and close-coupling regions.

In Fig. 24 we also present calculations [140] of the Maxwell-averaged cross sections for quenching of $\mathrm{Li}(nP) + \mathrm{He}$ at $T = 700$ K obtained using the basic formula (272). Despite the quantum defect of $\mathrm{Li}(nP)$ states remaining considerably smaller than unity, its value $\delta_p^{\mathrm{Li}} = 0.053$ exceeds the quantum defect of the nD-state by about 25 times. Therefore, the inelasticity parameter $\lambda_T = \delta_p / n^2 V_T$ of the $nP \to nl'$ ($l' > 1$) transitions ($n' = n$) is sufficiently large at the intermediate and sufficiently low n (e.g. $\lambda \approx 0.53$, 1.1, and 3.4 at $n = 10$, 7, and 4, respectively). As a result, its influence on the l-mixing cross section values in the whole of the studied range of n becomes more pronounced than that for $\mathrm{He}(n^1P) + \mathrm{He}$ and $\mathrm{Na}(nD) + \mathrm{He}$ collisions. As is evident from the comparison of full and dashed curves in Fig. 24, the quenching cross sections for nP-states turn out to be about 2, 3, and 6 times lower than that for the nD-states for $n = 10$, 7, and $n = 5$, respectively. Thus, the $\text{Li}(nP)$+He collisions clearly demonstrate that orbital angular momentum mixing processes can be significantly affected by the energy transferred to the Rydberg atom even for quantum defect values $\delta_l \ll 1$. Nevertheless, the qualitative behavior of the quenching cross section for the nP-levels remains similar to the case of the nD-levels and the position of the maximum n_{max} is not changed appreciably.

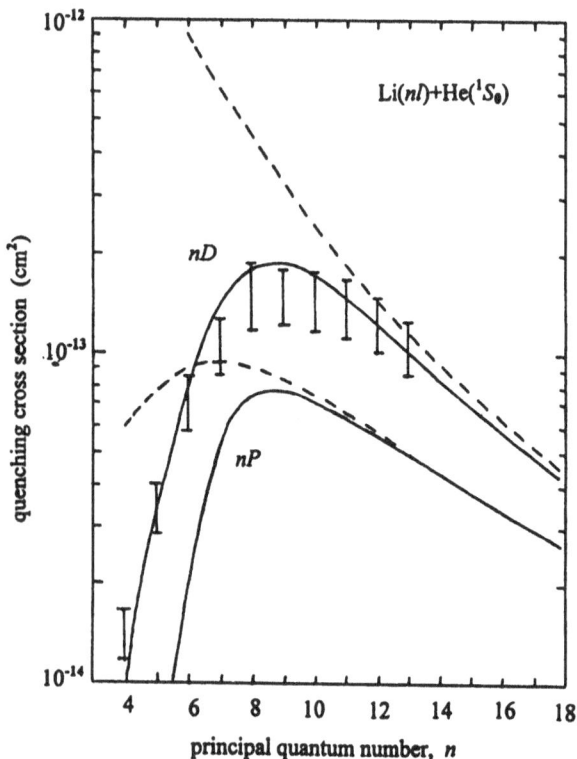

Figure 24. Quenching cross sections $\left\langle \sigma_{nl}^{l\text{-mix}} \right\rangle_T$ for the l-mixing collisions $\text{Li}(nD)$+$\text{He}\left({}^1S_0\right)$ and $\text{Li}(nP)$+$\text{He}\left({}^1S_0\right)$ at $T = 700$ K. Full curves, calculations [140] by the basic formula (272) of normalized perturbation theory ($c = 0.25$). Dashed curves, results of the first-order perturbation theory obtained by simple formula [128] for quasielastic $nD \to n$ transitions and by formula [132] for the inelastic $nP \to n$ transitions. Experimental data for the nD-levels are those from [221].

It is important to stress that the available weak-coupling models for quasielastic [128]) and inelastic [132, 146] collisions do not provide an appropriate description of both l-mixing processes $Li(nD)$ +He and $Li(nP)$+He. It can be seen from Fig. 24 that first-order perturbation theory does not hold near the maximum of the cross section and in the range of the principal quantum number $n < n_{max}$.

The situation is changed radically for collisions involving Rydberg states with a substantial quantum defect as, for example, for quenching $Li(nS)$ +He, when $\delta_s^{Li} = 0.399$. This process is accompanied by especially large values of the energy transfer even for the l-mixing $nS \to nl'$ transitions without change in the principal quantum number. As a result, the cross sections of this process measured by *Harnafi* and *Dubreil* [221] at $n = 4-7$ turn out to be $2 \cdot 10^{-16} - 10^{-15}$ cm^2 (see Fig. 25), i.e. about two orders of magnitude lower than the corresponding values for the quasielastic l-mixing thermal collisions involving the Rydberg $Li(nD)$-states.

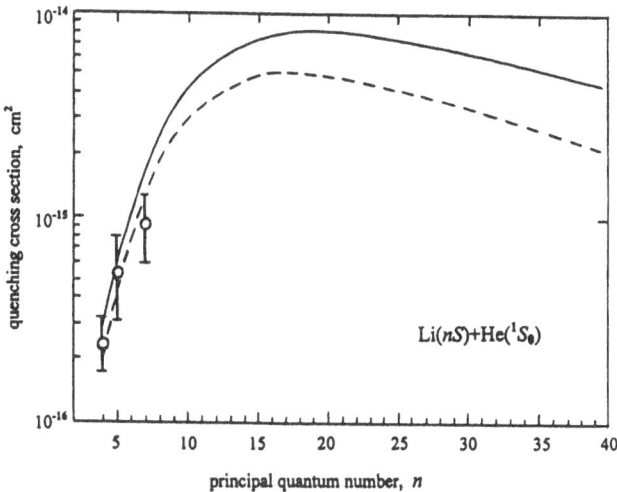

Figure 25. Quenching of $Li(nS)$ states by $He(^1S_0)$ at $T = 700$ K. Full curve, results of Ref. [140] for the total quenching cross section $\langle \sigma_{nS}^Q \rangle_T$ including the contributions of the l-mixing process ($n' = n$) and the n, l-changing $nS \to n'$ transitions with $n' \neq n$. Dashed curve, contribution $\langle \sigma_{nS}^{l\text{-mix}} \rangle_T$ of the l-mixing process alone. Experimental data are those from [221].

This effect cannot be described within the framework of standard theoretical models for the orbital momentum transfer (see review [81]) because they certainly do not hold when the inelasticity parameter of the $nl \rightarrow nl'$ transitions is large $\lambda \approx \delta_l/n_*^2 V \gg 1$.

Calculations [140] for $\mathrm{Li}(nS) + \mathrm{He} \rightarrow \mathrm{Li}(nl') + \mathrm{He}$ collisions clearly show (see Fig. 25) that the orbital angular momentum mixing collisions with large energy transfer behave substantially like inelastic n, l-changing processes. Therefore, if the quantum defect becomes substantial the maximum of the l-mixing cross section appears in the range of $n \sim n_{\mathrm{max}} \approx 0.9 \, (\delta_l/V_T)^{1/2}$ (where $\lambda \approx 1$) and its value may be estimated as $\sigma_{nl}^{l-\mathrm{mix}} \approx 0.1 \left(4\pi L_{\mathrm{eff}}^2 / \delta_l^{3/2} V^{1/2} \right)$ in accord with the simple formula (138) corresponding to the weak-coupling limit. For the l-mixing collisions of the $\mathrm{Li}(nS) + \mathrm{He}$ atoms we have $n_{\mathrm{max}} \approx 18$ and $\langle \sigma_{ns}^{l-\mathrm{mix}} \rangle_T \approx 5 \cdot 10^{-15}$ cm^2. Thus, it turns out to be considerably shifted towards higher magnitudes of n and its value is much lower than for quasielastic $\mathrm{Li}(nD) + \mathrm{He}$ collisions.

In the range of $n \gg n_{\mathrm{max}}$ the cross section of the inelastic l-mixing process $\mathrm{Li}(nS) + \mathrm{He}$ tends to quasielastic limit $\sigma_{nl}^{l-\mathrm{mix}} \propto L_{\mathrm{eff}}^2 / V^2 n^3$. At the same time, at $n \ll n_{\mathrm{max}}$ (when $\lambda \gg 1$) the analytic formula of Ref. [132] predicts the following asymptotic behavior of the l-mixing cross section $\sigma_{nl}^{l-\mathrm{mix}} \propto L_{\mathrm{eff}}^2 V n^3 / \delta_l^3$. Whereas the increase of the cross section with growing n takes place in the whole experimentally studied range of n. As is apparent form Fig. 25, the results [140] for the total quenching cross section of $\mathrm{Li}(nS) + \mathrm{He}$ are in quite reasonable agreement with available experimental data at $n = 4, 5$, and 7. One can see that the contribution of the orbital angular momentum mixing process $(n' = n)$ is predominant in the experimentally studied range of n. The contribution of the inelastic n, l-changing $nS \rightarrow n'$ transitions with $n' \neq n$ increases with growing n and should be taken into account (particularly with $n' = n-1$) to obtain reliable theoretical data.

Some comments should be made concerning the choice of the specific value of the normalization constant c in all calculations presented above. This is an empirical parameter of our theory and its value has been evaluated from the comparison of the calculated cross sections with the available experimental data for quasielastic l-mixing processes. As noted previously, at high $n \gg n_{\mathrm{max}}$ the values of the quasielastic cross sections obtained using the unitarized perturbation theory and standard Born approximation practically coincide.

(That is, they are practically independent of the specific choice of the normalization constant c). The linear dependence on the c value $(\sigma_{nl}^{l-\text{mix}} \propto c)$ appears only at low $n \ll n_{\text{max}}$, while in the close vicinity of the maximum $n \sim n_{\text{max}}$ the l-mixing cross section is proportional to $c^{3/7}$ (see Sect. 5.3.2). Thus, the specific value of c should be obtained from the comparison of theory with experiment or with the close coupling calculation at low or intermediate n (whereas it is enough to have this comparison at only one point of n). As follows from the results obtained, the optimal magnitude of c for the case of collisions of Rydberg $\text{Li}(nD, nP)$, $\text{Na}(nD)$, and $\text{He}(n^1P)$ atoms with $\text{He}(^1S_0)$ atom is equal to $c = 0.25$. The results for substantially inelastic $\text{Li}(nS)$ +He collisions are independent of the choice of the normalization constant. They can be obtained in the first-order of perturbation theory or in the impulse approximation not only at high n but also at the intermediate and sufficiently low n in accord with the discussion presented above.

6.3   Rydberg Atom–Neon Collisions: Polarization Effect

The case of the ground-state Ne atom is special due to a very small value of the electron–Ne scattering length ($L=0.21$ a.u.). The free-electron–Ne scattering cross section grows very sharply at low energies. As follows from the results of the *Lebedev* and *Fabrikant* theory [138, 139] this increase makes the scattering-length approximation completely unreliable, even at relatively large values of n. In order to demonstrate this fact we compare the results of calculations of the quenching cross sections obtained with the use of the actual electron–neon scattering amplitude $f_{eB}(k, \theta)$ with those calculated in the scattering length approximation $f_{eB} = -L = const$. Since the relative role of the short-range and long-range polarization electron-perturber interactions in collisions involving Rydberg atoms depends on both the principal quantum number n and the inelasticity of reaction we consider two different types of processes here: the quasielastic l-mixing collisions of $\text{Na}(nd)$+Ne and $\text{Li}(nd)$+Ne as well as the inelastic n, l-changing collisions of $\text{Rb}(ns)$+Ne.

$\text{Na}(nd)$, $\text{Li}(nd)$ + Ne Collisions.   In Fig. 26 we present cross sections for l-mixing of $\text{Na}(nd)$ states in collisions with the ground-state Ne atoms. The results incorporating the energy and angular dependence of the e–Ne scattering amplitude are about a factor of 4 greater than those obtained in the scattering-length approximation, for n between 7 and 12 (the region of the maximum in the cross section).

Na(nd)–Ne

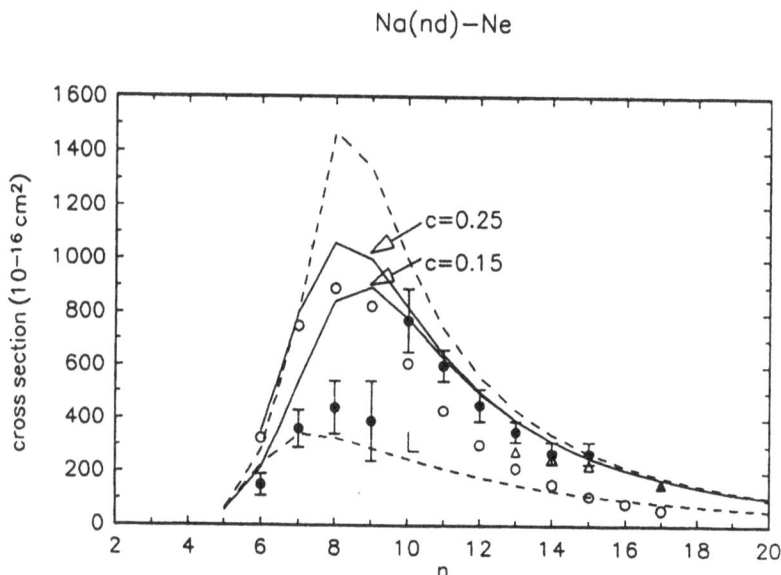

Figure 26. Quenching of Na(nd) states by Ne [139]. Full curves: Maxwell-averaged cross sections (T=430 K) calculated with the normalization constants c =0.25 and c=0.15. Dashed curves, non-averaged cross sections for $V = 3.73 \cdot 10^{-4}$ a.u., calculated (with $c = 0.25$) in the scattering-length approximation (lower curve "L") and with the "exact" scattering amplitude (upper curve). Full triangle, impulse approximation [166]. Empty triangles, semiclassical model [130]. Empty circles, close-coupling method [222]. Full circles with error bars, experiment [219].

As is apparent from the comparison of the full and dashed curves in Fig. 26, the Maxwell average has an appreciable effect on the l-mixing cross section near its maximum. In particular the peak at n=8 is significantly reduced and agreement with experiment becomes better at $n \geq 10$. The results are still too high as compared to the experiment [219] if we employ the value of the normalization constant $c = 0.25$. However, this constant might slightly vary when we switch from one collision system to another. Somewhat different values of the normalization constant c for various perturbers might mean that the parameter L_{eff} characterizing the Rydberg-electron–perturber interaction affects the l-mixing cross sections in the strong-coupling region to some extent. On the whole, however, this influence is considerably smaller than that in the weak coupling region where the cross section is proportional to L_{eff}^2. On the other hand, the variation of c does not affect the cross sections in the weak-coupling region where the quasifree electron model is valid. In Fig. 26 we also give the Maxwell-averaged cross sections for c=0.15, (the value recommended in Ref. [162] for quasielastic collisions). This choice of c slightly reduces the cross sections at $n \leq 9$.

We also present several other calculations [130, 166, 222], which are all in reasonable quantitative agreement with experimental data at $n \geq 10$. However, the measured cross section for Na(nd)+Ne exhibits a jump at $n = 10$ which has not been observed for other colliding atoms and is not confirmed by any of the calculations. Using $c = 0.15$ slightly reduces our results below n=10 but the discrepancy remains large. We suggest that perhaps our "exact" results somewhat overestimate the cross sections in the region between n=7 and n=10, but the agreement between the scattering-length approximation and experiment at $n \leq 9$ is certainly fortuitous.

Another example involving Ne-atom is shown in Fig. 27 where we present the quenching cross sections for quasielastic Li(nd)-Ne collisions ($\delta_d^{\text{Li}} = 0.002$) and compare them with experimental data [221]. The solid lines give the quenching cross sections at the collision velocity $V = 6.61 \times 10^{-4}$ a.u. (mean velocity for T=650 K) and dashed lines give the Maxwell averaged cross sections. Calculations [139] were performed for two values of c in order to analyze the dependence of the results on the normalization constant in different regions of n. For each pair of curves the upper was calculated at c=0.25 and lower at $c = 0.15$. As in the case of Na(nd)-Ne collisions, for $c = 0.25$ and $c = 0.15$ our cross sections are somewhat higher than experimental at low and intermediate n but are much closer to them at $n \geq 11$.

Note that to get a reliable quantitative results for the quenching cross sections involving Ne, Ar, Kr, and Xe in a wide range of n (particularly for the inelastic transitions with large energy defect) it is necessary to take into account higher-order terms in the partial

scattering amplitudes. The values of the scattering lengths, polariz-
abilties, and other parameters (see Sect. 2.8.1) have been taken in
the theory [138, 139] from the experimental works of *Gulley* et al.
[95] (for Ne) and *Weyhreter* et al. [94] (for Ar, Kr, and Xe).

$$\text{Li}(nd)-\text{Ne}$$

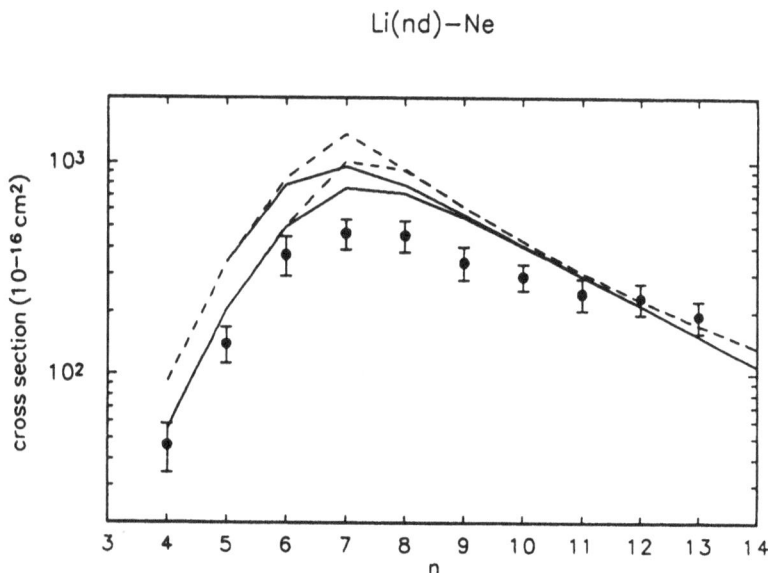

Figure 27. Quenching of Li(nd) states by Ne [139]. Full curves, Maxwell-
averaged cross sections for T=650 K; dashed curves, non-averaged cross
sections for $V = 6.61 \cdot 10^{-4}$ a.u. For each pair of curves the upper is
calculated for $c = 0.25$, and the lower for $c = 0.15$.

Rb(ns) + Ne Collisions. To show how the theory [138, 139] works
for the inelastic n, l-changing processes involving Ne atoms we present
the quenching cross sections for Rb(ns)-Ne collisions in Fig. 28. The
results [139] for the total quenching cross sections are based on the
summation of separate contributions of the $nl \rightarrow n'$ transitions calculated using the normalized perturbation theory. Inclusion of a large
number of the $nl \rightarrow n'$ transitions from the initial Rydberg ns-state
is particularly important at high n. As is evident from the figure
we have good agreement of Lebedev and Fabrikant's theory with the
experimental results of *Hugon* et al. [165] at n=32, 36, and 38. The

model incorporating the exact energy dependence of the scattering amplitude works much better than the scattering-length approximation. In the region $n = 15 - 20$ near the maximum of the inelastic quenching process the calculated cross sections exceed those obtained in the scattering length approximation by an order of magnitude. Calculations [167] at $n=32$ and $n=38$ only include $n \rightarrow n - 3$ transitions; therefore they are substantially lower than [139]. It is also important to note that the results of calculations [139] for the inelastic Rb(ns)+Ne collisions are practically independent of the choice of the normalization constant c since the whole of the studied range of n corresponds to the weak coupling of Rydberg states.

Rb(ns)−Ne

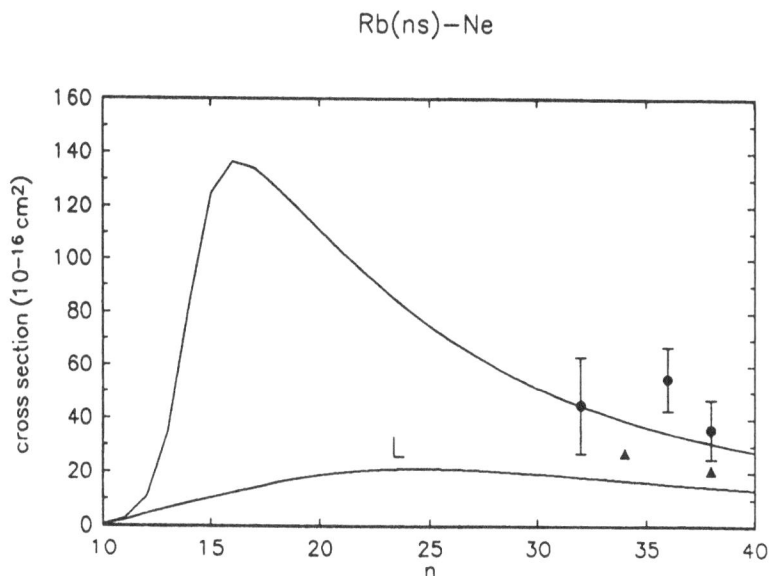

Figure 28. Quenching of Rb(ns) states by Ne. Full curves: theory for $V = 3.41 \cdot 10^{-4}$ a.u. Lower curve "L", scattering length approximation; upper curve, theory [139]. Triangles, impulse approximation results [167]. Full circles with error bars, experimental data [165].

6.4   Quenching by the Heavy Rare Gas Atoms

6.4.1   Na(ns, nd) + Ar, Kr, and Xe

For thermal collisions of a Rydberg sodium atom with the heavy rare gas atoms the experimental data are available for the isolated ns-levels (*Gallagher* and *Cooke* [223], *Boulmer* et al. [224]) with $\delta_s = 1.34$ as well as for quasidegenerate nd-levels (*Gallagher* et al. [219], *Chapelet* et al. [225], *Kachru* et al. [226]) with a small quantum defect $\delta_d = 0.015$. The theoretical analysis of these two cases allows us to demonstrate the influence of inelasticity on the magnitudes and qualitative behavior of the quenching cross sections.

Quenching of Rydberg nd-states.   In Fig. 29 we present the l-mixing cross sections for Na(nd)-Ar collisions and compare them with available experimental and theoretical data. Since it is not very clear what is the relative collision velocity distribution in the experiment [225], we present our data for a fixed collision velocity $3.2 \cdot 10^{-4}$ a.u. corresponding to the mean velocity in experiment [219] (temperature T=430 K), and the results of the Maxwell average for two temperatures: T =430 K and T=296 K. The Maxwell average of the cross section changes the results very slightly (not more than by 10 %), which is consistent with the studies of Refs. [130, 138].

The quenching cross sections describe well both the strong-coupling region corresponding to low n, and the weak-coupling region of high n. At $n \leq 15$ they agree quite well with the experimental data [219] and close-coupling calculations [222], but are lower than the normalized-perturbation-theory results [127]. There are two reasons for this disagreement. First, *Gersten* [127] uses the scattering-length approximation which overestimates the cross section for e–Ar scattering. Second, his value of the normalization constant is close to 1 which is substantially higher than our value c=0.25. At n=5 the cross section obtained in [139] seems to be too low, which demonstrates that reliable quantitative calculations of the sodium-atom-rare-gas-atom collisions for the first few levels $n = 3 - 5$ need the application of physical approaches and theoretical models developed for the description of transitions between the ground and low excited states (see, for example, [227]).

At higher n recent results [139] agree very well with those of the impulse approximation [166] obtained for $17 \leq n \leq 45$. This is not surprising as the general theory [138, 139] incorporates all features of the quasifree-electron model at high n corresponding to the weak coupling of Rydberg states. However, starting from the region $n=45$ and above *Sasano 's* et al. [166] quenching cross sections become lower than those obtained in Ref. [139]. We explain this by the contribution of n, l-changing collisions which were not included in the calculations [166].

Na(nd)−Ar

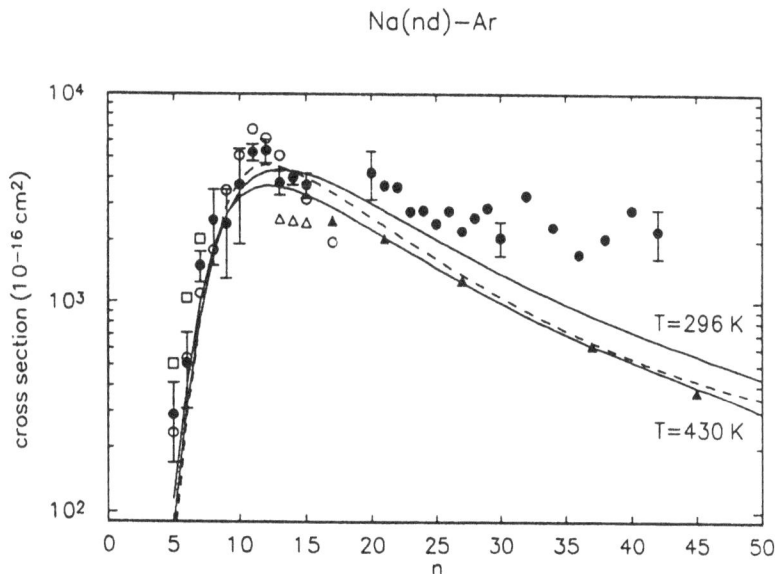

Figure 29. Quenching cross sections of Na(nd) states by Ar. Full curves, the Maxwell-averaged cross sections [139] for temperatures $T=296$ and 430 K. Dashed curve, the results [139] calculated for relative collision velocity $V =3.2 \times 10^{-4}$ a.u. Empty circles, close-coupling calculations [222]. Empty squares, normalized perturbation theory calculations [127]. Empty triangles, semiclassical model [130]. Full triangles, impulse approximation calculations [166]. Full circles, experimental data [219] ($n \leq 15$) and [185] ($n \geq 20$).

To demonstrate this, we present cross sections for the $nd \to n'$ transitions ($\Delta n = 0, \pm 1, \pm 2, ...$) in Fig. 30. One can see that the contribution of the n, l-changing transitions exceeds 30%, starting from $n \approx 55$ and turns out to be about 100 % for $n \approx 80$. Further increase of n leads to growth of the relative contribution of inelastic n-changing transitions.

$$\text{Na(nd)} - \text{Ar}$$

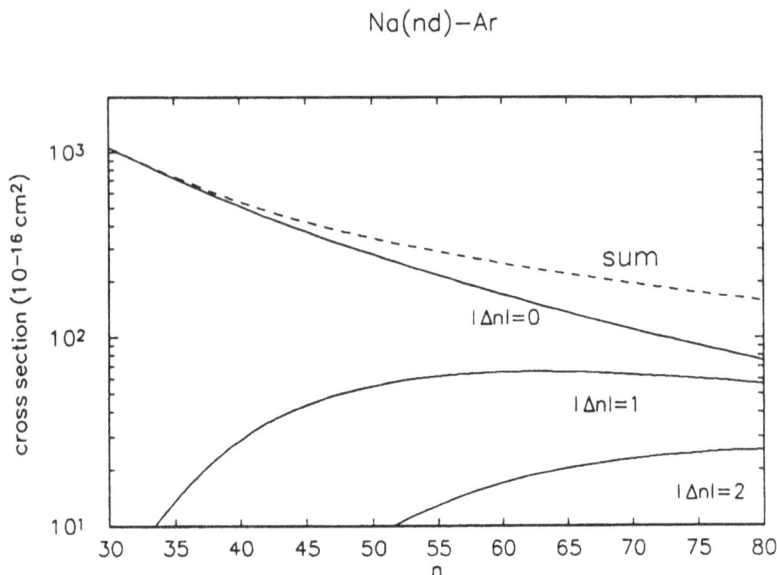

Figure 30. Contribution of different $nd \to n'$ transitions to the quenching of the Na(nd) levels by Ar [139]. Dashed curve, the sum of $|\Delta n|$=0, 1, and 2 contributions.

Since the quasifree-electron model should work quite well at high n, the substantial disagreement with the experiment of *Chapelet* et al. [225] for Ar-perturber in Fig. 1 is surprising. The disagreement between theoretical and experimental data for the quenching Na(nd)+Ar collisions at high n was also pointed out in [167, 228]. Moreover, *Gounand* and *Petitjean* [228] made an attempt to attribute this disagreement to the additional contribution of the perturber–core scattering mechanism considered previously by several authors (see Chap. 6 in the review article [36] and references therein). *Lebedev* and *Marchenko* [106] showed that the contribution of the perturber–core scattering associated with the "shake-up" mechanism grows with n and it becomes about 10-15% at $n \approx 50$. However, the perturber–core scattering effect is small in the experimentally studied range of $n \leq 45$ and can be neglected. Thus, new detailed experimental measurements at high n with a definite value of the gas temperature would be very desirable.

In Fig. 31 we present comparisons of recent theoretical results [139] incorporating the effects of the long-range polarization interaction with the experimental data of *Chapelet* et al. [225] and quasifree electron model calculations of *Sato* et al. [167] for Na(nd)–Kr collisions.

Na(nd)−Kr

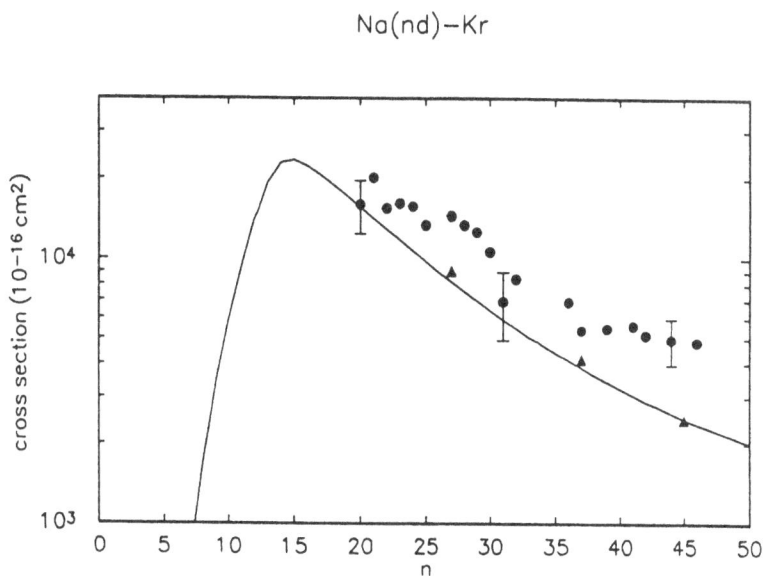

Figure 31. Quenching cross sections of Na(nd) states by Kr. Full curve, the results of Ref. [139] for $V = 2.87 \cdot 10^{-4}$ a.u. Triangles, impulse approximation results [167] for nD states. Full circles, experimental data [225].

Once again, at higher principal quantum numbers n the theoretical calculations [139] agree with the quasifree electron model but are lower than the experimental data.

In Fig. 32 we show that the situation is somewhat better for Na(nd)–Xe collisions. In this case the theory [139] is in good agreement with the experimental data of *Kachru* et al. [226] at low and intermediate magnitudes of the principal quantum number n. At high principal quantum numbers theoretical results [139] for Xe are more consistent with the experimental data ($n = 20-46$) of *Chapelet* et al. [225] than in the case of Ar perturber.

$$Na(nd)-Xe$$

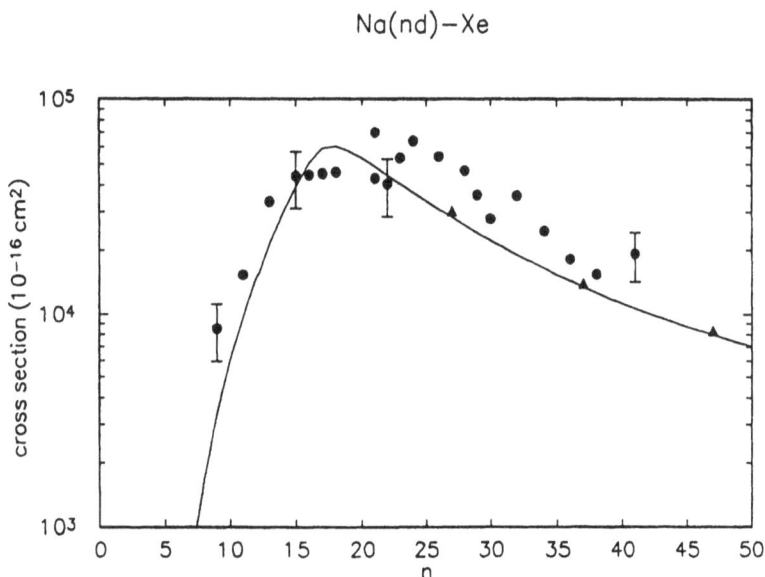

Figure 32. The same as in Fig. 31 for quenching of Na(nd) by Xe atoms ($V = 2.74 \cdot 10^{-4}$ a.u.) [139]. Experimental data for $n \leq 20$ are those of Ref. [226].

Quenching of Rydberg ns-States. The quantum defect of Na(ns) states is 1.34, therefore Na(ns)–rare-gas-atom collisions are essentially inelastic. The cross sections at $n < 50$ are substantially lower than those for l-mixing of Na(nd) states. The major contribution to quenching is due to $nd \rightarrow (n-2)l'$ and $nd \rightarrow (n-1)l'$, $l' \geq 2$, processes, but at higher n other n-changing transitions are also important. Our cross sections take into account the contributions of all final n'.

In Fig. 33 we present the cross sections for the quenching of

Na(ns) states by Ar, Kr, and Xe for a fixed collision velocity corresponding to T=430 K. An interesting feature of the cross sections is their non-monotonic dependence on n in the region of n from 15 to 25. This feature was discussed in Ref. [138] in connection with collisions of Rb Rydberg atoms with the rare-gas atoms, and appeared to be a manifestation of the Ramsauer–Townsend effect in free-electron scattering by rare-gas atoms. It is important to note that the effect becomes apparent only in inelastic collisions. In quasielastic collisions the typical electron velocity is much lower than that corresponding to the Ramsauer–Townsend minimum. In contrast, inelastic collisions are produced by electrons whose velocity is much higher and corresponds to a closer distance between the Rydberg electron and the ion core in the coordinate space. This higher range of velocities covers the Ramsauer-Townsend region for n between 15 and 25.

Theoretical results [139] on Na(ns)–Ar collisions are in reasonable general agreement with the experimental data [224] obtained at $n \geq 20$ if we note the same uncertainty in the precise choice of the relative collision velocity distribution for calculations as in the experiment [225]. There is some disagreement in the position of the global maximum (in contrast to the local maximum at $n = 16$ due to the Ramsauer–Townsend effect): the experimental data exhibit the maximum at $n = 30$ whereas our cross sections approach it only at $n = 50$ before they start to decrease slowly. This feature of the theoretical cross section is due to the contribution of a large number of n-changing $ns \rightarrow n'$ transitions at higher n. Note that *Sato and Matsuzawa's* [167] calculations include only $n \rightarrow n - 1$ transitions. For these transitions their cross sections agree with ours. However the results [139] presented in Fig. 33 include the contribution of all essential n-changing transitions. Therefore the better agreement of the results [167] with the data of *Chapelet* et al. [225] is fortuitous. Unfortunately, the experimental points are too rare to make any conclusion about experimental evidence of the Ramsauer–Townsend effect in inelastic collisions involving Rydberg atoms.

As follows from the analysis of the validity conditions for the independent treatment of the Rydberg-electron–perturber and core–perturber scattering mechanisms, the theory developed in [138, 139] should work for the inelastic Na(ns)+Ar collisions in the range of the principal quantum numbers $n > 15$ (see Sect. 5.4 in the review [36]). However, in order to demonstrate the qualitative behavior of the cross section for quenching due to the Rydberg-electron–perturber

scattering we have extended the calculations to the region of low
n. The theoretical cross sections drop rapidly in this region. At
$n \leq 11$ the theoretical cross sections are substantially lower than the
available experimental data of [223] for collisions of Na(ns) with Ar
and Xe.

Na(ns)–R

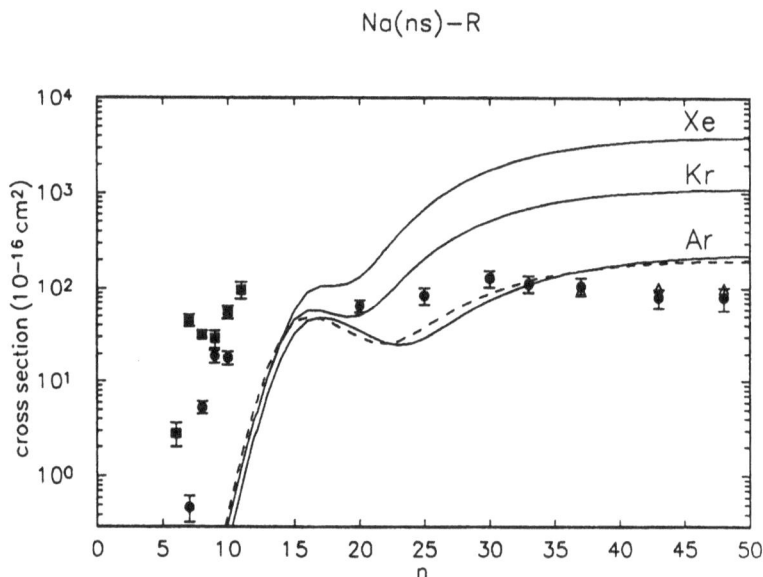

Figure 33. Quenching of Na(ns) states by the heavy rare gas atoms [139].
The collision velocity equals the mean velocity at temperature 430 K for
each Na–R pair. Experimental data for $n \leq 11$, [219]: full circles, Ar;
full squares, Xe. Experimental data for $n \geq 20$, [224]: full circles, Ar.
Triangles, impulse approximation results [167] for $n \rightarrow n - 1$ transitions in
Na(ns)–Ar collisions.

Gallagher and Cooke [223] suggested that at such low n the ion–
core effects become important and discussed a mechanism according
to which the ion core polarizes the incoming neutral atom, and the
interaction of the Rydberg electron with the induced dipole moment
causes the quenching transition. As was shown by *Lebedev* et al. [230]
the dipole interaction of the outer electron with a quasimolecular
ion, temporarily formed during the scattering of a perturbing atom

from the atomic core, can produce inelastic n-changing transitions between the Rydberg states which are more efficient (at $n < 10 - 15$) than those induced by the perturber–electron scattering. This situation takes place for collisions of atoms (for example, $H(n)$+He) with small reduced mass and a large depth of the potential well of the quasimolecular ion when the collision time τ of the neutral perturber and the ionic core of the Rydberg atom is small. As a result, the inelastic n, l-changing processes occur in the non-adiabatic regime $(\omega_{n'l',nl} < \tau^{-1})$ even at low n corresponding to a substantial energy transfer from the translational motion of colliding particles to the Rydberg electron. The theory of these transitions (taking into account both the contribution of the induced dipole and the dipole moment of the positive ionic core relative to the centre of mass of quasimolecule) was developed [231] and will be discussed in Sect. 7. The analysis of similar perturber–core effects at low n in the case of $Na(ns)$+Ar, Kr, Xe collisions is beyond the scope of the present review. It needs a further extension of the identified approach to systems for which the depth of the potential energy curve of the quasimolecular ion is of the same order of magnitude as the thermal energy of colliding atoms and new detailed calculations of dipole-induced transitions in the non-adiabatic $(\omega_{n'l',nl} < \tau^{-1})$ and adiabatic $(\omega_{n'l',nl} \gg \tau^{-1})$ regions.

Moreover, it is possible that an appropriate quantitative description of the identified inelastic quenching processes at low and intermediate n may be given only on the basis of theoretical approaches taking into account the non-impulsive corrections and coherence effects in the electron-perturber and core-perturber scattering as was previously mentioned by *Hahn* [168], *de Prunelé* [229] and *Fabrikant* [111] for quasielastic and slightly inelastic processes. In any case the calculations presented in Fig. 33 clearly demonstrate that the core effects become important in inelastic collisions at $n < 10 - 15$.

It should be worthwhile to clarify why the range of validity of the theory [138, 139] is quite different for the inelastic and quasielastic l-mixing transitions. First, for a given n the typical Rydberg-electron-ion-core separation, $r_n \sim 2n_*^2 a_0/[1 + (n_* a_0 |\Delta E|/\hbar V)^2]$, leading to inelastic collisions is considerably smaller than that for quasielastic transitions with $\Delta E = 0$ [138, 140]. Therefore the influence of the core effects on the inelastic n, l-changing transitions induced by the Rydberg-electron–perturber scattering is important for higher values of n as compared to those for l-mixing transitions.

Another reason is the different influence of the normalization procedure on the behavior and magnitudes of the cross sections for inelastic and quasielastic transitions. For substantially inelastic processes theoretical results [139] are practically independent of the specific value of the normalization constant c and can be obtained in first-order perturbation theory in the whole of the considered region of n. In this case the range of applicability of the approach [139, 140] corresponds fully to the validity condition of the semiclassical impulse approximation [106]. A similar situation occurs for quasielastic collisions at high $n \gg n_{max}$. Here the major contribution to the cross section is determined by the range of weak coupling $\rho \geq \rho_0$ in which only the electron–perturber interaction is taken into account.

On the other hand, the behavior and the cross section for the l-mixing process in the range of low and intermediate n can be determined only on the basis of the unitarized version of perturbation theory. Within the framework of the approach [138, 140] the value of the normalization constant c is chosen from the comparison of calculated cross sections with the experimental data which implicitly contain information about the total contribution of all interactions (whereas it is enough to make this comparison at only one point $n < n_{max}$). Therefore, despite the fact that the perturber–core effects are not incorporated in our model explicitly we can interpret the normalization procedure in the strong-coupling region as relevant to the total probability of quasielastic transition induced by both the perturber–electron and perturber–core scattering. Therefore we obtain reasonable physical results for the l-mixing cross sections in the range of the principal quantum number $n < n_{max}$ where the quasifree electron model is certainly inapplicable.

6.4.2   Rb(ns, nd, nf) + Ar, Kr, and Xe

Here our main goal is to demonstrate the significant difference between the inelastic n, l-changing and quasielastic l-mixing processes for the quenching of a Rb(nl) state by heavy rare gas atoms. In this case experimental data are available for both the nF-level [163] (which has a small quantum defect, $\delta_f = 0.02$), and for isolated nS-, nD-levels [165] (for which $\delta_s = 3.13$, and $\delta_d = 1.34$). The calculations [138] presented below are based on the semiclassical approach and normalized perturbation theory, described in Sect. 5.3.1. The cross sections include the summation over all possible values of the final principal quantum number n'. This summation leads to $1/n$ dependence of the cross section in the high-n limit, in contrast to the $1/n^3$ behavior of $\sigma_{n',nl}$. The cross section σ_n is averaged then over the Maxwellian distribution of relative velocities.

The n-dependencies of the averaged quenching cross sections for nS, nD, and nF-states of Rb atoms in thermal collisions with Ar, Kr, and Xe are shown in Figs. 33–35.

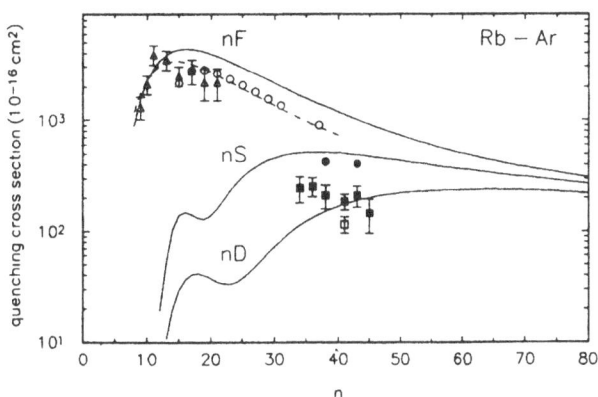

Figure 34. Cross sections for quenching of the nS-, nD-, and nF- states of the Rydberg Rb atom by Ar averaged over the Maxwellian velocity distribution. Full curves, the theoretical results of Ref. [138] (T=296 K). Dashed curve, the same calculation for the nF-level at T=520 K. Full squares, empty squares and triangles are the experimental data for the nS-, nD-levels (Ref. [163], T=296 K), and nF-levels (Ref. [165], T=520 K), respectively. Full circles and empty circles are the free-electron-model calculations [167, 150] for nS and nF states, respectively.

They are compared with experimental data of *Hugon* et al. [163, 165]. For the quenching of the nF and nS levels by Ar and nF levels by Kr we also show the previous theoretical data of Refs. [150, 167] obtained in the impulse approximation. Overall, calculations of *Lebedev* and *Fabrikant* [138] reproduce the experimental observations [163, 165] quite well, although there is a substantial disagreement (about a factor of 2) for the quenching of nS states by Ar. Sato and

Matsuzawa's results [167] are slightly lower than [138] in this region
of n, but they also substantially exceed the experimental data.

The quenching cross sections reveal a strong dependence on the
quantum defect δ_l of the initial Rydberg nl-state in the whole of the
experimentally studied range of n. This behavior is reasonably reproduced by semiclassical calculations [138] for quasielastic l-mixing
quenching of nF-levels as well as for inelastic n, l-changing quenching
of nS and nD-levels.

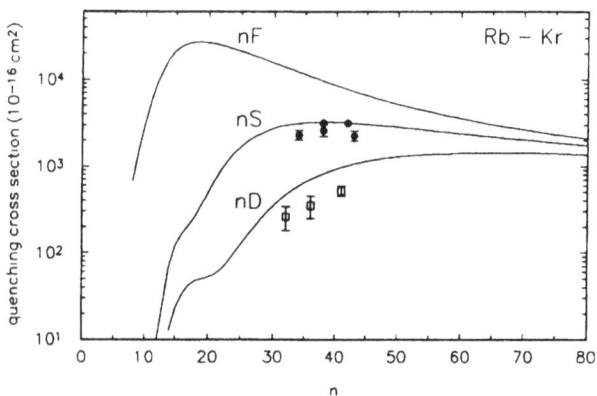

Figure 35. The same as in Fig. 34 for quenching by Kr atoms [138]. Full
circles, calculations [167] for nS states.

Note that the impulse approximation does not describe the observed cross sections for the quenching of nF–states in the range
$n < 25 - 30$, where it is mainly determined by quasielastic l-mixing
collisions. On the other hand, for $n > 40 - 50$ the cross sections, corresponding to different values of the initial orbital momentum, start
to merge. This occurs due to the contribution of a large number
of different $nl \rightarrow n'$ transitions ($\Delta n = 0, \pm 1, \pm 2, ...$), which makes
the total quenching cross section independent of quantum defect in
accordance with the results of the asymptotic Omont's theory [128].

Another interesting feature of the results is non-monotonic dependence of the calculated cross sections for nS and nD-levels on
n in the region of n from 15 to 25. This feature is a manifesta-

tion of the Ramsauer–Townsend effect which is clearly represented in the n-dependence of the effective scattering length, Figs. 34–36. Although averaging over the Maxwellian distribution in relative velocities makes this influence somewhat less pronounced than that for a fixed relative velocity, the Ramsauer–Townsend minimum significantly affects the quenching process in Rb–Ar collisions at $n = 15 - 30$. Unfortunately there is no experimental data available in this region.

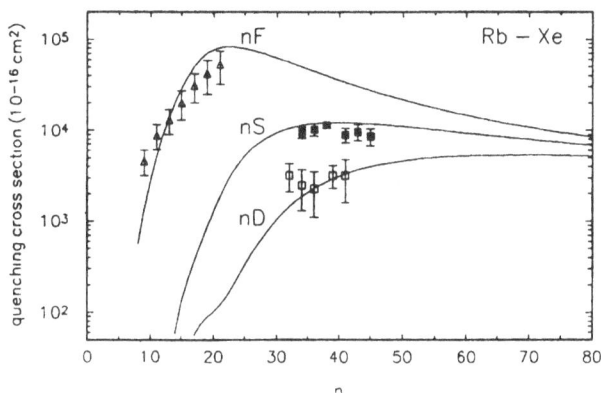

Figure 36. The same as in Fig. 35 for quenching by Xe atoms [138].

For the quenching by Kr the Ramsauer–Townsend minimum is less pronounced since it occurs at higher energies in the free–electron scattering. The Maxwell average washes out this effect completely for Xe. To demonstrate the Ramsauer–Townsend effect for a fixed collision velocity, we present the non-averaged cross sections for quenching of Rb(nD) states in Fig. 37. Now the effect becomes visible even for Xe but, as before, it is most pronounced for Ar.

It is important to stress that actual energy and angular dependencies of the amplitude $f_{eB}(\epsilon, \theta)$ for electron scattering by the heavy rare gas atoms were incorporated to calculations [138] using expression (290) and the modified effective range theory (60). The results obtained clearly demonstrate that the Ramsauer–Townsend effect significantly affects the values of the cross sections, especially, for inelastic transitions with large energy transfer. This means that the

scattering length approximation (which provides quite good results
for the He atom as a perturber) does not allow one to get a reliable
quantitative description of collisions between the Rydberg atom and
the ground-state heavy rare gas atoms.

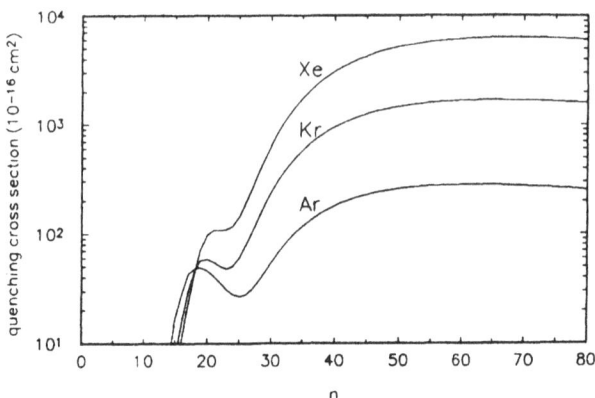

Figure 37. Non-averaged cross sections for quenching of nD states by Ar,
Kr, and Xe [138]. The collision velocity equals the mean velocity at the
temperature 296 K for each Rb–B pair (B=Ar, Kr, and Xe).

6.4.3   Xe(nf) + Ar, Kr, and Xe

An interesting example of thermal collisions between the Rydberg
atoms and the heavy rare gas atoms is the quenching process of
Xe(nf) states by Ar, Kr, and Xe studied experimentally [232] at
$T = 300$ K. The quantum defect of the Xe(nf) states is sufficiently
small ($\delta_f = 0.055$). For such values of quantum defect the quenching
of Rydberg states in thermal collisions with the light ground-state
atoms (He and Ne) is usually a result of purely quasielastic orbital
angular momentum transfer $nl \to nl'$ (i.e., the energy defect ΔE
of reaction is not important). The other situation takes place for
Xe(nf)+Ar, Kr, and Xe collisions. Due to the large reduced mass of
colliding partners their relative velocity V is substantially less than
in the case of collisions with He and Ne targets. As a result, the
magnitudes of the inelasticity parameter $\lambda = na_0\omega_{n,nf}/V$ turn out

to be sufficiently large and strongly affect the cross section behavior even for the l-mixing transitions.

We illustrate the main features of the Maxwell-averaged quenching cross section behavior by presenting *Lebedev* and *Fabrikant's* calculations for thermal Xe(nf)+Xe collisions in Fig. 38. They are based on the theory [138, 139] presented in Sects. 5.3 and 5.4.

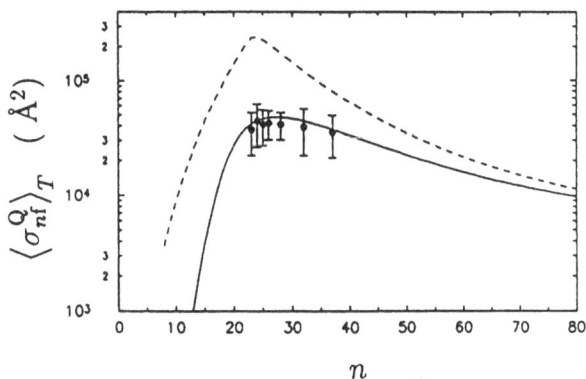

Figure 38. Maxwell-averaged cross sections for Xe(nf)+Xe$(^1S_0)$ quenching ($T = 300$ K). Full curve, calculations with account of the energy defects $\Delta E_{n',nf} = 2Ry\,|\delta_f + \Delta n|\,/n^3$ for all $nf \to n'$ transitions ($\delta_f = 0.055$). Dashed curve, quasielastic approximation ($\delta_f = 0$ and $\Delta E_{n,nf}$) for l-mixing $nf \to n$ transitions with $n' = n$ and $l' \neq 3$. Full circles, experiment [232].

The full curve was obtained taking into account the energy defects of all $nf \to n'l'$ transitions (including $nf \to nl'$ transitions to all degenerate hydrogen-like nl' sublevels of the same level $n' = n$). One can see that these calculations are in very good agreement with experimental data [232]. The dashed curve in Fig. 38 is the result of the analogous calculation of the l-mixing process without using the energy defect $\Delta E_{nl',nf}$ of reaction (i.e., assuming that the quenching Xe(nf)+Xe collisions behave like a purely quasielastic process). As is evident from the comparison of full and dashed curves, the difference between the results of these two calculations is about one order of magnitude near the maximum of the cross section. This fact

clearly demonstrates that collision $Xe(nl)+Xe$ quenching of Rydberg
nl-levels $(l = 3)$ with small quantum defect $\delta_f \ll 1$ (associated,
primarily, with the orbital angular momentum transfer $l' \neq 3$ and
$n' = n$) is a substantially inelastic process. Whereas its cross section
can not be evaluated within the framework of standard models of
quasielastic l-mixing collisions. In Fig. 39 we present the analogous
calculations of *Lebedev* and *Fabrikant* and the experimental data [232]
for quenching $Xe(nf)+Ar$ and $Xe(nf)+Kr$ collisions.

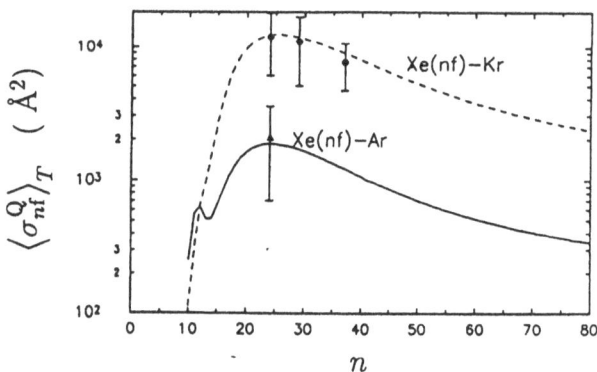

Figure 39. Maxwell-averaged cross sections for quenching $Xe(nf)+Ar$
(full curve) and $Xe(nf)+Kr$ (dashed curve) at $T = 300$ K. Calculations
take into account the energy defects for all $nf \rightarrow n'$ transitions. Full triangles and full circles, experimental data [232] for Ar and Kr, respectively.

Apparently, the agreement between theory and experiment is excellent for both targets. It is seen, that there is a local minimum
(at $n = 12 - 16$) in the n-dependence of cross section for Ar. This
reflects the appreciable influence of the Ramsauer–Townsend effect in
electron–argon scattering on the values of quenching cross sections.

In summary, the semiclassical theory [138–140], presented in Sects.
5.3 and 5.4, incorporates some important features of the Rydberg-
electron–perturber interaction: the correct energy and angular dependencies of the electron–perturber scattering amplitude, the conservation of transition probability in the strong-coupling region, and

explicit dependence on the inelasticity parameter. The numerical data obtained for the cross sections describe a wide region of the principal quantum number n, the relative velocity V, and the energy defect $|\Delta E_{n',nl}|$. Thus, a reasonable description of quasielastic l-mixing and inelastic n, l-changing processes was simultaneously given from a common point of view independently of a particular value of the energy transferred to the Rydberg atom.

For collisions of the Rydberg atom with a He atom the theory [140] is valid in the scattering length approximation. However, calculations [139] for the quasielastic and inelastic Rydberg-atom–neon collisions clearly demonstrate that the scattering length approximation is certainly inapplicable for Ne. In contrast to the He atom as a perturber, a reliable description of the quenching processes in a wide range of n can only be given in the model incorporating actual angular and energy dependencies of the amplitude for elastic scattering of slow electrons by neon. Moreover, the theory [138, 139] confirms the previous conclusions about the invalidity of the scattering length approximation for the Rydberg-atom–heavy-rare-gas-atom collisions. It clearly shows how to incorporate the energy and angular dependencies of the free-electron scattering amplitude in a more general way, valid for both quasielastic and inelastic processes. As follows from the examples presented above, the strong dependence of the electron–heavy-rare-gas-atom scattering amplitude on the momentum and scattering angle significantly affects the cross section values for quasielastic and inelastic transitions. In particular, the electron momenta in the vicinity of the Ramsauer–Townsend minimum contribute significantly to the cross sections of the inelastic processes. Thus, the theory described in Sects. 5.3 and 5.4, provides a quite reasonable quantitative description of major phenomena in inelastic and quasielastic collisions of Rydberg atoms with the rare gas atoms.

6.5   Mixing of Fine-Structure Components

The quenching of selectively excited Rydberg nlJ-states with the given principal n and orbital l quantum numbers and the total angular momentum $J = |l - 1/2|$ (or $J = l + 1/2$) has some specific features. In this case the depopulation of highly excited state may be accounted for not only by the n, l-changing processes, but also by the $nlJ \rightarrow nlJ'$ transitions between the fine-structure components. As follows from the results of experimental [163, 233, 234] and theoretical

[135, 136, 137, 235] studies, the depopulation of the nlJ-states with relatively small n $(n < 15 - 20$ at thermal collisions of Rb $(n^2 D_{3/2})$ and Cs$(n^2 D_{3/2})$ with the rare gas atoms) is primarily the result of total angular momentum transfer $(J' \neq J)$ without change in the orbital l and principal n quantum numbers. However, at high enough n the n, l-changing processes predominate. Therefore, the total cross section for quenching of selectively excited nlJ-level has a qualitatively different dependence on the principal quantum number n than for nl-states. In particular, two clearly pronounced maxima can occur here at low and high values of n.

The available experimental data [163, 233] and theoretical results [136, 137] for the quenching of Rb$(n^2 D_{3/2})$ states by helium $(T = 520$ K) are shown in Fig. 40. The dashed curve shows the n-dependence of the Maxwell-averaged J-mixing cross section $\left\langle \sigma_{nD_{3/2}}^{J-\text{mix}} \right\rangle$. It was calculated by the simple formula (261) corresponding to the quasielastic limit of the J-mixing process. It is justified because of the inelasticity parameter $\nu = \left| \delta_{D_{3/2}} - \delta_{D_{5/2}} \right| v_0 / V_T n_*$ for the $n^2 D_{3/2} \to n^2 D_{5/2}$ transitions of rubidium is small $(\nu_{3/2,5/2} \ll 1)$ in the entire considered range of n (the quantum defects difference is $\delta_{D_{3/2}} - \delta_{D_{5/2}} = 0.002$). As is evident from Fig. 40, the J-mixing cross section reaches its maximum $\left\langle \sigma_{nD_{3/2}}^{J-\text{mix}} \right\rangle_{\text{max}} \approx 1 \cdot 10^{-13}$ cm^2 at $n_{\text{max}}^{J-\text{mix}} = 8$. With increase of $n > n_{\text{max}}^{J-\text{mix}}$ the cross section of the $n^2 D_{3/2} \to n^2 D_{5/2}$ transition rapidly tends to the asymptotic limit of weakly coupled states $\left\langle \sigma_{nD_{3/2}}^{J-\text{mix}} \right\rangle \propto n_*^{-4}$. However, at low principal quantum numbers the J-mixing cross section behaves like $\propto n_*^4$ in accord with (259).

The averaged cross sections $\left\langle \sigma_{n',nD_{3/2}} \right\rangle$ of the inelastic $n^2 D_{3/2} \to n'$ transitions with a change in the principal and orbital quantum numbers $(2 < l' \leq n' - 1, J' = 3/2, 5/2)$ and their total contribution $\left\langle \sigma_{nD}^{n,l-\text{ch}} \right\rangle$ calculated in [137] are shown in Fig. 40 by dashed-dotted and dashed curves, respectively. One can see that the maximum $\left\langle \sigma_{nD_{3/2}}^{n,l-\text{ch}} \right\rangle_T^{\text{max}} \approx 8.8 \cdot 10^{-15}$ cm^2 of the n, l-changing cross section is reached at $n_{\text{max}} \approx 22$. The main contribution is determined by $n^2 D_{3/2} \to n - 1$ and $n^2 D_{3/2} \to n - 2$ transitions, which have the smallest energy defects $\Delta E_{n-1,nD}$ and $\Delta E_{n-2,nD}$ $(\delta_D = 1.34)$. At large $n > 40 - 50$, the inelastic $n^2 D_{3/2} \to n'$ transitions to other degenerate hydrogen-like sublevels $n'l'J'$ $(l' > 2)$ with $n' \neq n-1, n-2$ become important.

The total quenching cross section, which is determined by the sum $\left\langle \sigma^Q_{nD_{3/2}} \right\rangle = \left\langle \sigma^{J\text{-mix}}_{nD_{3/2}} \right\rangle + \left\langle \sigma^{n,l\text{-ch}}_{nD_{3/2}} \right\rangle$ of the J-mixing and n, l-changing contributions, is shown by a full curve in Fig. 40. Thus, for high $n > 20$ the quenching of selectively excited $n^2 D_{3/2}$ states is primarily due to inelastic $n^2 D_{3/2} \rightarrow n'$ transitions with a change of the principal and orbital quantum numbers. On the other hand, at low $n < 15$ it is mainly determined by the $n^2 D_{3/2} \rightarrow n^2 D_{5/2}$ transitions with change a of only the total angular momentum. These processes make comparable contributions in the region $n \sim 15 - 20$.

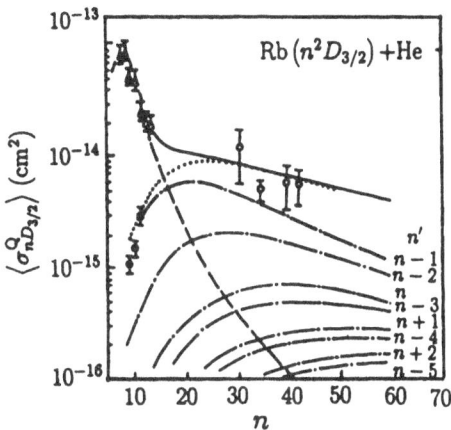

Figure 40. Cross sections of $\mathrm{Rb}\left(n^2 D_{3/2}\right)$ +He collisions averaged over the Maxwellian velocity distribution at $T = 520$ K. Dashed curve, J-mixing $(n^2 D_{3/2} \rightarrow n^2 D_{5/2})$ cross sections from [136, 137]. Dotted curve, the depopulation cross section of the nD-level out of the doublet. Empty circles $(T = 520$ K) and full circles $(T = 296$ K), experimental data [163] and [164, 165]. Dashed-dotted curves, contribution of individual $n^2 D_{3/2} \rightarrow n'$ transitions. Full curve, total quenching cross section of $n^2 D_{3/2}$ level. Empty $(T = 520$ K) and full $(T = 380$ K) triangles, experimental data [163] and [233], respectively.

In Fig. 41 we present theoretical results [137] for the $n^2 D_{3/2} \rightarrow n^2 D_{5/2}$ transitions (the dashed curves) induced by thermal collisions $(T = 353$ K) of Rydberg Cs* and the ground-state He atoms. It is seen that calculations (259) are in good agreement with experimental data [234] for $n^2 D_{3/2} \to n^2 D_{5/2}$ transitions in both the range of weak coupling $(n > 10-12)$ and strong coupling $(n \sim 8-11)$, i.e., the close vicinity of the maximum $\langle \sigma_{nD}^{J-\mathrm{mix}} \rangle \approx 1.03 \cdot 10^{-13}$ cm^2 at $n_{\max} = 9$.

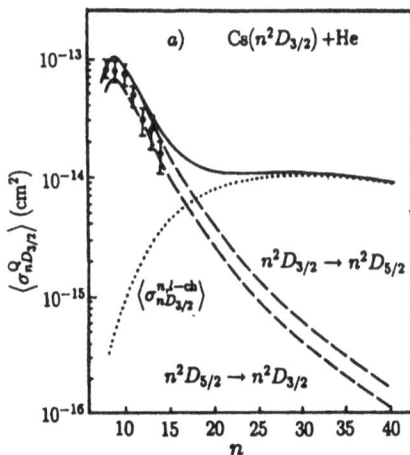

Figure 41. Depopulation cross sections $\mathrm{Cs}(nlJ)$+He averaged over the Maxwellian velocity distribution at $T = 353$ K. Panel a shows the semiclassical results [137] for the $n^2 D_{3/2}$ states. Dashed curves, the J-mixing cross sections for direct $n^2 D_{3/2} \to n^2 D_{5/2}$ and inverse $n^2 D_{5/2} \to n^2 D_{3/2}$ transitions. Dotted curve, the sum of all inelastic $n^2 D_{3/2} \to n'$ transitions. Full curves, the total quenching cross section. Empty circles, experimental data [234]. Panel b presents the same calculations for the $n^2 P_{1/2}$ states.

The quantum defects difference of these states is $\delta_{D_{3/2}} - \delta_{D_{5/2}} = 0.009$, while $\delta_D = 2.47$ for Cs (see Table 2). Thus, the influence of the inelasticity parameter $\nu_{J'J} = |\delta_{J'} - \delta_J| v_0/Vn_*$ on the cross section values of the J-mixing process $(\mathrm{Cs}(n^2 D_{3/2})$ +He) is already sufficiently small at $n \geq 12$. Indeed, in this region the value of $\nu_{3/2,5/2}^{(d)} \leq 1.7$ and, hence, we can put $\varphi_{3/2,5/2}^{(d)} \approx 1$ (see Fig. 41). Small deviations in the behavior of the J-mixing process from the pure quasielastic case $(\nu = 0)$ only arises here at low $n \sim 8-11$.

The cross sections $\langle \sigma_{nlJ}^{n,l-\mathrm{ch}} \rangle$ of the n, l-changing processes for collisions of $\mathrm{Cs}(n^2 D_{3/2})$ + He taking into account the contribution of many inelastic $nlJ \rightarrow n'$ transitions, are shown in Fig. 41 by dotted curves. The total quenching cross section $\left\langle \sigma_{nlJ}^Q \right\rangle$ of the selectively excited $n^2 D_{3/2}$ states of Cs is shown by the full curve in Fig. 41. It is evident that in the region of $n < 20$ the major role is played by the J-mixing transitions without a change of both the principal and orbital quantum numbers. However, at high principal quantum numbers the contribution of the n, l-changing processes into the total depopulation cross section becomes predominant.

As follows from the experimental studies [234], the behavior of the J-mixing cross sections for the $\mathrm{Cs}\left(n^2 D_{3/2}\right) \rightarrow \mathrm{Cs}\left(n^2 D_{5/2}\right)$ transitions induced by thermal collisions with heavy rare gas atoms remains qualitatively similar to the case of a He atom as the perturber. However, their magnitudes increase appreciably in the range of $n > 11 - 12$ (see also review [36] and [135, 235]).

6.6 Quenching by Alkali-Metal Atoms

We now discuss the quenching processes of high-Rydberg states in alkali vapors. The main qualitative features in the experimentally observed behavior of the cross sections and the rate constants of thermal collisions between the Rydberg atoms and the ground-state alkali atoms remain the same as those considered above. However, the maxima of the quenching cross sections become much greater in magnitudes and are shifted toward higher n as compared to the collisions with the ground-state rare gas atoms, particularly so, for the heavy elements K, Rb, and Cs as the perturbers. This is the result of the extremely large value of the cross section for elastic scattering of the ultra-low-energy electron by the alkali-metal atom and, in particular, of the ^3P-resonance on the quasidiscrete level of the corresponding negative ion (see Fig. 8 and Table 9). We shall illustrate some features in the quenching of high-Rydberg states by the ground-state sodium and rubidium atoms below .

Let us first consider thermal collisions ($\mathcal{E} = \mu V^2/2 = 0.037$ eV) of Rydberg sodium in the nS and nD states with the ground-state parent atoms Na($3S$). The impulse-approximation results [105] in the total quenching cross section $\sigma_{nl}^Q = \sum_{n'}{}' \sigma_{n',nl}$, obtained using the general formula (202), are shown in Fig. 42 by the full curves. These calculations were made in the range of applicability of the quasifree electron model on the basis of the resonance and potential scattering

theory, presented in Sects. 2.8.1 and 4.8. The parameters of elastic
scattering of an ultra-slow electron by Na(3S) were taken from [104]
(see Table 9). It can be seen that the relative role of the potential and
resonance electron–perturber scattering in the quenching of Rydberg
levels depends essentially on both the principal quantum number n
and the value of the transition energy defect $\Delta E_{n',nl}$. Because of the
small quantum defect $\delta_d^{Na} = 0.0145$, the quenching of the Na(nD)
levels is mainly determined by quasielastic $nD \to nl'$ $(l' > 2)$ transitions with a change in the orbital angular momentum if the principal
quantum number is not too large. In this case the contributions of the
resonance and potential scattering are of the same order of magnitude
in a wide range of n.

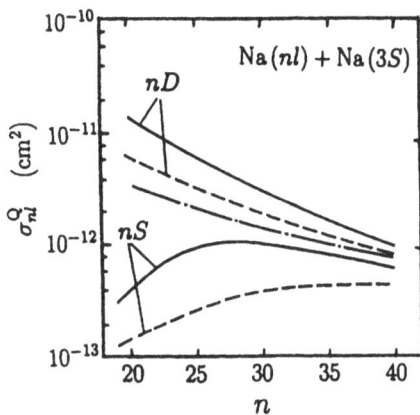

Figure 42. The impulse-approximation results [105] for the quenching
cross section of Na(nD, nS)+Na(3S) ($\mathcal{E} = 0.037$ eV). Dashed curves, contribution $\sigma_{n',nl}^{pot}$ of the potential electron–perturber scattering. Full curves,
total quenching cross section $\sigma_{nl}^{Q} = \sigma_{nl}^{pot} + \sigma_{nl}^{res}$ including the contribution
of the ^3P-resonance scattering. Dashed-dotted curve, the cross section σ_{nl}^{Q}
obtained by the asymptotic formula [154].

The quenching of the Rydberg Na(nS) states ($\delta_s^{Na} = 1.35$) is due
to the inelastic $nS \to n'l'$ transitions with a change in the orbital
and principal quantum number (primarily, with $n' = n - 1, n - 2$).
In this case the resonance scattering plays a much greater relative

role near the maximum of the quenching cross sections σ_{nS}^Q. As is apparent from the comparison of full and dashed curves in Fig. 42, the contribution of resonance scattering to the quenching of the nS levels at $20 < n < 35$ is considerably greater than that of the potential scattering. As the principal quantum number increases further $n > 35 - 40$, the role of the ^3P-resonance begins to decrease. This is a result of a substantial drop in the value of the energy distribution function $W_{nl}(\epsilon)$ for the resonance energies $\epsilon \sim E_r$ at high principal quantum numbers $n \gg (Ry/E_r)^{1/2}$ (see [105, 106] for more details). Note that at high enough n the total quenching cross section $\sigma_{nl}^Q = \sum_{n'}$ $\sigma_{n',nl}$ is due not only to transitions to the nearest energy levels, but also to a large group of final levels n'.

As is evident from Fig. 42, the quenching cross sections σ_{nS}^Q and σ_{nD}^Q calculated in [105] for nS and nD levels (full curves) reveal a strong dependence on the quantum defect values in the region of $n \sim 20-30$ (the impulse approximation is valid for Na^*-Na collisions at $n > 20$). This dependence cannot be described on the basis of the asymptotic theory [154] supplemented by the ^3P-resonance contribution of the electron–perturber scattering, which yields practically the same values $\sigma_{nS}^Q = \sigma_{nD}^Q$ (dashed-dotted curve) for the quenching cross sections. The most important difference between the values of σ_{nS}^Q and σ_{nD}^Q arises, as expected, in the range $n \sim 20 - 30$ in which the main contribution to the quenching processes is determined by the inelastic $nS \to n'$ or quasielastic $nD \to n$ transitions to several of the nearest energy levels ($n' = n - 1$, $n - 2$ for nS and $n' = n$ for nD levels). On the other hand, the calculations within the framework of the asymptotic theory agree reasonably with the results of the quenching theory developed in [105] for growing n ($n > 40$). This is due to the increasing role of the inelastic $nl \to n'$ transitions to a large group of final n' levels.

Since the energy E_r of the ^3P-resonance in electron scattering by the ground-state $Rb(5S)$ atom is considerably smaller than that for $Na(3S)$ (see Fig. 8), this resonance should play an even greater role in the quenching of high-Rydberg $Rb(nS)$ states in its own gas compared with the case of Na^*+Na collisions, considered above. As follows from the impulse-approximation analysis [105, 111], the available experimental data [114] on the quenching process $Rb(nS)+Rb(5S)$ can not be explained within the framework of the quasifree electron model taking into account only the potential electron–perturber scattering. Simple estimates show that its contribution σ_{nS}^{pot} to the total quenching cross section turns out to be appreciably smaller than the values measured [114] in the range of $34 \leq n \leq 43$.

This was confirmed by calculations [105], which indicate (see Fig. 43) that the experimental data for this process can be reasonably described by the general formula (202) taking into account the ^3P-resonance scattering contribution (235) with the use of the semiempirical resonance parameters presented in Table 9.

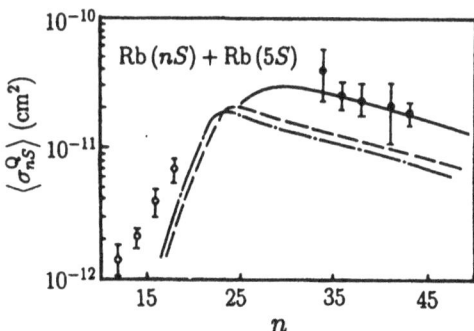

Figure 43. The quenching cross sections of the $\text{Rb}(nS) + \text{Rb}(5S)$. Full curve, the impulse-approximation results [105] calculated at $T = 400$ K with semiempirical ^3P-resonance parameters $\epsilon_0 = 1.2 \cdot 10^{-3}$ a.u. and $\gamma = 18$ a.u.. Dashed-dotted curve, a similar calculation [111] at $T = 530$ K with the resonance parameters obtained using effective range theory [103]. Dashed curve, the same result for $T = 420$ K. The open and filled circles, experimental data [189] and [114] at $T = 520$ K and 400 K, respectively.

The dominant contribution to the collision quenching cross section $\sigma_{nS}^{Q} \equiv \sum_{n'} \sigma_{n',nS}$ at $n \sim 25 - 50$ is made by the $nS \to n - 3$, $l' > 2$ transition with the minimum energy defect ($\delta_s^{\text{Rb}} \approx 3.13$), although all other $nS \to n'$ transitions have also been taken into account in [105]. Similar calculation of the quenching cross section σ_{nS}^{Q} for $\text{Rb}(nS) + \text{Rb}(5S)$ collisions based on the general formula (202) of the impulse approximation has been performed [111] using the data [103] for the resonance and potential amplitudes of electron–alkali-atom scattering. In the range of $n > 25$ the curve of the quenching cross section obtained in [111] shows the same dependence on n as in [105]. However, the magnitudes of the former cross section are one half as large as those in [105] because the values of the ^3P-resonance parameters used in these papers slightly differ from each other.

All results for the quenching processes of Rydberg alkali-metal atoms by the ground-state parent atoms presented above were obtained for high enough values of n, which correspond to the range of weak coupling. Hence, the magnitudes of the calculated cross sections σ_{nl}^{Q} for both inelastic and quasielastic processes are considerably smaller than the geometrical cross section $\sigma_{\mathbf{geom}} \propto \pi a_0^2 n_*^4$ of the Rydberg atom. Therefore, the use of the impulse approximation in calculations [105, 111] provides reasonable agreement with the experimental data [114] at high n. Nevertheless, a major part of the experimental results on quenching of the Rydberg states by the ground-state alkali atoms have been obtained for sufficiently low or intermediate n, where the impulse approximation does not hold. For instance, there are experimental data on quenching of Rydberg states of rubidium (nS, nP, nD, nF with $n \leq 22$) and caesium (nS, nD with $n \leq 16$) in alkali–alkali collisions. Moreover, several authors have also measured cross sections for the fine-structure mixing collisions of Rb*+Rb and Cs*+Cs (see [36] and references therein).

6.7   Quenching of Rydberg States by Ca and Yb

Here we present recent theoretical calculations of *Fabrikant* and *Lebedev* [203] on the quenching of Rydberg ns- and nd-states of neon by atoms with small electron affinities. As follows from the results of this work for Ne(nl)+Ca thermal collisions, there are two contributions to the quenching cross section: the resonance contribution and the impulse contribution. The first is very sensitive to the electron affinity of the projectile and may be used for studies of weakly-bound anions. The resonance contribution is dominant for the process of quenching Rydberg states with large quantum defects (e.g., s states) at relatively low principal quantum numbers n, whereas the impulse contribution dominates in the process of quenching of states with small quantum defects (e.g., d states of Ne). We also present calculations [203] below of quenching of the Rydberg Ne(ns)-states by Yb atoms with a hypothetical electron affinity of 2 meV. One of the key points of this work consists in the important conclusion that experimental studies of quenching of Rydberg states by Yb might help in determining the electron affinity of Yb even if it is negative. In the latter case the quenching cross section as a function of n should exhibit oscillations.

Calculations of the impulse quenching mechanism in [203] are based on the theory developed in Refs. [138–140]. As noted in Sect. 5.3 this theory combines two approaches: the impulse approximation for sufficiently large n where coupling between Rydberg states is weak, and the normalized perturbation theory with the Fermi pseudopotential at lower n providing the conservation of probability. The important feature is a modification of the Fermi pseudopotential by using the impulse approximation with the energy- and angular-dependent scattering amplitude [105, 106]. This allows us to introduce an effective scattering length L_{eff} (see Sect. 5.4) characterizing the actual interaction between the Rydberg electron e and the perturbing atom B. In general, L_{eff} contains contributions of both short- and long-range interactions and depends on n, the collision velocity V and the reaction energy defect ΔE.

For calculation of L_{eff} it is necessary to know the low-energy $e-$B scattering amplitude. It was calculated [203] with the help of the model-potential approach of *Fabrikant* [198, 200] which uses a potential well for the description of the short range interaction and polarization potential. The well parameters were used to reproduce the binding energy for a weakly-bound negative ion for $l = 1$ and *ab initio* low-energy scattering phase shift for $l = 0$. For $l \geq 2$ the modified effective-range theory [91, 92] for the polarization potential was used safely in the low-energy region (see Sect. 2.8.1).

The two identified approaches have been applied in Ref. [203] to collisions involving the Rydberg atom and the ground-state Ca and Yb atoms. Recent R-matrix calculations [215] of $e-$Ca scattering phase shifts provide a good feedback for calculations of the impulse quenching. However, at present there is no available scattering phase shifts for $e-$Yb scattering. Therefore the following procedure was used to describe the $e-$Yb phase shifts. The description of the s-wave scattering is based in [203] on the experimental value for the polarizability [100] of Yb ($\alpha = 142$ a.u.) and on the same well parameters as for Ca [200, 198]. At the same time, for $l = 1$ *Fabrikant* and *Lebedev* used the well parameters yielding a hypothetical Yb$^-$ with the binding energy 2 meV. (As was mentioned above the experimental upper bound [205] for the electron affinity of Yb is 3 meV).

Such an approximate way of calculating phase shifts does not

strongly affect the quenching cross sections in the low and intermediate n (up to $n = 20$) regions. Previous calculations [138, 139] show
that the cross sections are weakly sensitive to the phase shifts in the
strong-coupling regime, where the magnitude of the cross section is
mostly controlled by the unitary limit $c\pi n_*^4 a_0^2$ (c is a constant of the
order of 1). Since the most interesting problem is the region of com-
petition between the resonance mechanism and the impulse mechanism for quenching corresponding to low n, the model described in
[203] should be sufficient for these purposes. Besides, when fitting
the $l = 1$ phase shifts to the binding energies we should be aware
that a bound p-state has two fine-structure components. (It is possible, though, that one of the components for Yb$^-$ is bound and
the other lies in the continuum). Therefore the quenching cross sections were calculated in Ref. [203] for two sets of the interaction
parameters corresponding to two values of the electron affinities, and
the results averaged over statistical weights of the two fine-structure
components, as in the case of the resonance quenching. However, according to the *Fabrikant's* paper [198], p-wave scattering phase shifts,
in contrast to s-wave shifts, are very insensitive to the value of the
binding energy [198]. Hence, the result of this average is practically
indistinguishable from any of the two cross sections.

The choice of Ca and Yb as the target ground-state atoms was
motivated in Sect. 5.7. The specific choice of the Rydberg projectile
does not affect the results qualitatively, so that it is convenient to use
the same projectile as in the experiment [197] on ion-pair formation in
collisions of Rydberg atoms with ground-state Ca atoms. The quantum defects of the Rydberg neon atom were obtained in Ref. [217]
for the ns-state ($\delta = 1.32$) and for the nd-state ($\delta = 0.02$). It should
be worthwhile to point out that the results of calculations turn out to
be substantially different for Rydberg states with a negligible quantum defect and a quantum defect with a substantial non-fractional
part. Therefore the choice of two examples (s series and d series) is
important for the discussion.

The most accurate experimental values [116] for the Ca$^-$ binding
energies (19.73 meV for the $^2P_{3/2}$ state and 24.55 meV for the $^2P_{1/2}$
state) were used in calculations [203]. The results for the resonance
quenching were summed over the fine-structure contributions as discussed by *Fabrikant* [200]. The impulse contribution, as discussed
above, is very insensitive to the value of the binding energy.

The results [203] of a comparison of resonance and impulse con-

tributions to the quenching cross sections for collisions of Rydberg
Ne(ns) and Ne(nd) atoms with Ca are shown in Fig. 44 and 45.

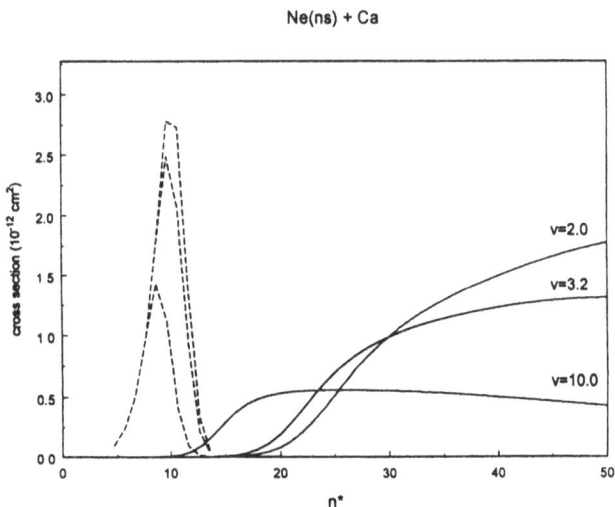

Figure 44. Quenching of Ne(ns) states by Ca($4s^2$) atoms [203]. Full curve,
impulse contribution; dashed curve, the ion-covalent-coupling contribution.
Collision velocity V is given in 10^{-4} a.u. Dashed curves are plotted for
the same magnitudes of V=2.0, 3.2 and 10.0 (from top to bottom).

One can see that the resonance contribution for ns states is somewhat larger than for nd states because of the larger coupling parameter for ns states and higher survival probability for nd states.
The impulse contribution for ns states is substantially lower than
the resonance contribution since the energy exchange between the
nuclear motion and the electron motion required for quenching is
suppressed significantly in this case. Therefore two contributions to
the quenching cross section are very well separated. Moreover, the
impulse contribution is practically negligible in the region of the peak
of the resonance contribution. Both processes turn out to be strongly
velocity-dependent and are more efficient at lower collision velocity,
although the impulse process is suppressed at lower velocities for
$n_* < 25$. In contrast to the process of ion-pair formation, there is no
threshold velocity for the quenching process. An important conclusion of Ref. [203] is as follows. Since the position of the peak in the

resonance contribution is sensitive to the electron affinity, the experimental and theoretical studies of quenching of Rydberg states with
large quantum defects provide an efficient tool for the investigation
of electron affinities of atoms.

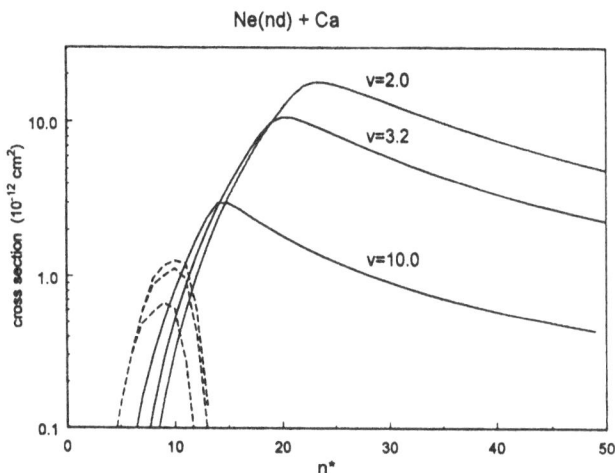

Figure 45. The same as in Fig. 44 for the Ne(*nd*) states [203].

In contrast, two contributions to the quenching of Rydberg Ne(nd)
states, Fig. 45, overlap substantially. Moreover, the impulse contribution is significantly larger in this case and peaks at relatively low
n_* ($n_* = 20$). This makes the resonance contribution hardly de-
tectable, and, therefore, not suitable for the purpose of studies of
small electron affinities.

These conclusions are illustrated further in Fig. 46 by presenting quenching cross sections for Ne(ns)−Yb collisions assuming that
there is only one bound state of Yb$^-$ with the binding energy 2 meV.
The magnitude of the impulse contribution in this case is approximately the same as for Ne(ns)−Ca collisions. However, the resonance
contribution is substantially higher making this case even more favorable for observation of the resonance quenching. The value of n_*
at which the peak is observed, n_*^{max}, varies with collision velocity,
being higher for lower velocities.

In summary, the authors of Ref. [203] have discussed a new tool for the investigation of negative ions with small binding energies. Observation of the previously used process of ion-pair formation (see discussion and references at the beginning of Sect. 5.7.1) is limited by sufficiently high collision velocities and principal quantum numbers n due to the conservation of energy restrictions. The resonance quenching is not limited in this regard. This process can be observed for low collision velocities and low n, where the quenching cross section is very large.

Ne(ns) + Yb

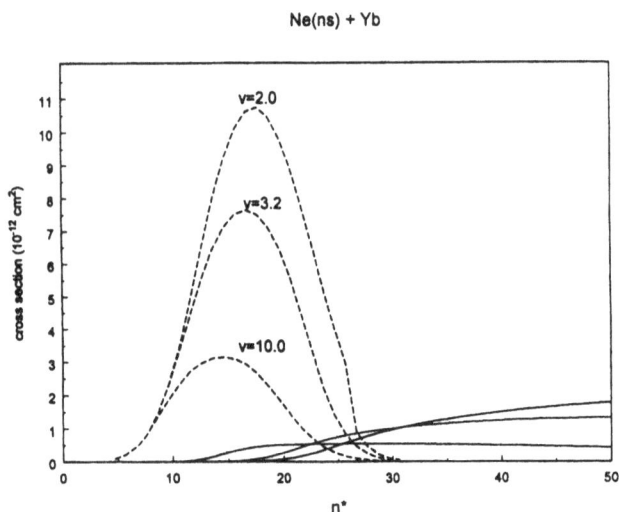

Figure 46. Quenching of Rydberg Ne(ns) states in collisions with the ground-state Yb atoms [203]. Notations are the same as in Fig. 44.

However, the resonance quenching can be masked by the impulse process. For the Rydberg states with small quantum defects the cross section for the impulse mechanism is very large, and it is unlikely that the resonance contribution can be detected experimentally. However, for the Rydberg states with a large non-fractional part of the quantum defect (e.g., Ne(ns) states) the impulse contribution is substantially lower and peaks at much larger n than the impulse cross section for Rydberg states with small quantum defects (e.g., Ne(nd) states). In addition, the resonance process is more efficient for ns states. All these factors make the process of quenching ns-states very favorable for studying properties of negative ions.

All these features were illustrated in Ref. [203] by considering the quenching of Rydberg states by the ground-state Ca and Yb atoms. The electron affinities of Ca are well established [116], and experiments with Ca could serve as a verification of the reliability of the theory. The electron affinity of Yb is unknown. Moreover, it is not clear if a stable Yb$^-$ exists [205]. Therefore calculations [203] can provide a guide for future experiments attempting to detect stable Yb$^-$. An advantage of studies of the quenching process is that they can provide information about the negative ion in both cases: the case of a stable anion and the case of a resonance state. In the former the quenching cross section exhibits a peak, in the latter it exhibits oscillations as a function of the principal quantum number n [110, 187]. Another advantage of the quenching process is that it can be observed not only in a collision experiment but also in a spectroscopic experiment by observation of the pressure broadening of Rydberg states by ground-state atoms (see [3] and references therein).

6.8 Rotationally Inelastic Collisions of a Rydberg Atom with Molecules

Mechanism of Quasi-resonant Energy Transfer. Collisions of Rydberg atoms with molecules have some specific features due to the presence of internal rotational degrees of freedom and the long-range nature of an electron–molecule interaction (see Sect. 2.8.2). Their revalation is especially significant in the n-changing and ionization processes with sufficiently large energy transfer to the Rydberg electron. The behavior of the depopulation cross sections for the rotationally elastic processes (in which the rotational energy of molecule remains unchanged) is qualitatively similar to the case of an atomic projectile. For instance, in quasielastic l- and J- mixing processes the nitrogen molecule as the perturber is similar to the rare gas atoms. However, for strongly polar molecules (such as HCl, HF, NH$_3$ etc), the maximum magnitudes of the quasielastic l-mixing cross sections of Rydberg states turn out to be substantially greater than for neutral atoms, while the position of the maximum is shifted toward a higher principal quantum number $n \sim 50 - 60$. On the whole, the polar molecules in such processes are qualitatively similar to strongly polarizable alkali-metal atoms. This is the result of the major con-

tribution of the second order term from the dipole potential, which
falls off asymptotically as r^{-4}.

On the other hand, the inelastic n-changing transitions with large
energy change ΔE_{f_i} as well as the ionization of Rydberg atom

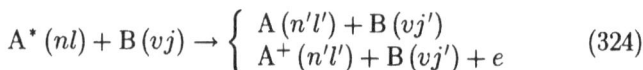

$$A^* (nl) + B (vj) \rightarrow \begin{cases} A (n'l') + B (vj') \\ A^+ (n'l') + B (vj') + e \end{cases} \tag{324}$$

may be the result of rotational $j \rightarrow j'$ (and sometimes vibrational
$v \rightarrow v'$) deexcitation or excitation of a molecule. In spite of the large
value of energy gained or released by the highly excited atom, these
processes may occur with very large cross sections provided a quasi-
resonant energy exchange between the Rydberg electron and internal
rotational degrees of freedom of a molecule takes place. For such
n-changing and ionizing collisions, the kinetic energy transferred to
the translational motion of colliding particles $\Delta \mathcal{E} = \Delta E_{jj'} - \Delta E_{\alpha'\alpha}$ is
small in comparison with both the energy change $|\Delta E_{\alpha'\alpha}|$ of the Ry-
dberg atom A^* and the rotational energy change $|\Delta E_{j'j}|$ of molecule
B. In the opposite case, when the kinetic energy defect $\Delta \mathcal{E}$ is notice-
able, the n-changing and ionization cross sections become small. The
first results in this field were reviewed in [82, 236]. Further detailed
calculations were made in [239, 170, 237].

In the impulse approximation the cross sections $\sigma_{\alpha'\alpha}^{j'j}$ of the inelastic $nl \rightarrow n'$ and $n \rightarrow n'$ transitions induced by slow collisions
$(V \ll v_0/n)$ with molecules may be evaluated by the formulae (196)
(or (221) combined with the binary-encounter expressions for the
form factors. The required expressions for the amplitude squared
$\left| f_{eB}^{j'j} (Q) \right|^2$ of electron scattering by the non-polar and polar molecules
are presented in Sect. 2.8.2. As is evident from these formulae, the
cross sections of the n-changing transitions are determined by the
value of the momentum transferred to the kinetic energy of colliding
particles $A^* + B$. The quasi-resonant nature of these transitions in
its dependence on the value of

$$Q_{\min} \approx \frac{|\Delta \mathcal{E}|}{\hbar V} = \frac{\left| Ry/(n'_*)^2 - Ry/n_*^2 - (E_j - E_{j'}) \right|}{\hbar V} \tag{325}$$

was investigated in [150, 170]. It was shown that for a given mag-
nitude of $|\Delta n|$, the resonant behavior of the cross sections $\sigma_{n',n}^{j\pm1,j}$ of
the $n, j \rightarrow n', j \pm 1$ transitions induced by collisions with dipolar
molecules becomes more pronounced as n grows. For a given value of

n, the form of the normalized function $\sigma_{n',n}^{j\pm1,j}\left(Q_{\min}\right)/\sigma_{n',n}^{j\pm1,j}\left(0\right)$ becomes broader as $|\Delta n|$ increases. In the resonant case the n-changing cross sections for dipole-induced collisions are about several orders of magnitude higher than the those for quasielastic l-mixing process induced by the short-range electron–molecule interaction. Meanwhile, the quadrupole-induced n-changing collisions are much less efficient in comparison with the dipole case because of the shorter range of the Rydberg electron–molecule interaction.

Ionization Cross Sections. The ionization of a Rydberg atom is most efficient, when it is accompanied by rotational deexcitation $j \to j'$ of a molecular projectile $(\Delta E_{jj'} = E_j - E_{j'} > 0)$. Then, the ionization threshold

$$\mathcal{E}_{\min} = \hbar^2 q_{\min}^2/2\mu = |E_{nl}| - \Delta E_{jj'} > 0$$

of reaction (324) may exist only in the range of $n_* < n_*^{(0)}$ (where $n_*^{(0)} = (Ry/|\Delta E_{jj'}|)^{1/2}$). In the opposite case, i.e. for high principal quantum numbers $n_* > n_*^{(0)}$ (when $|E_{nl}| - \Delta E_{jj'} < 0$ and $\mathcal{E}_{\min} = 0$), the resonant energy transfer to the Rydberg electron from the internal rotational motion of molecule may occur for all possible values of the kinetic energy of heavy particles. For this range of $n_* > n_*^{(0)}$, *Matsuzawa* [155] has derived the following approximate expression for the ionization cross section

$$\sigma_{nl}^{\text{ion}}\left(j',j\right) \approx$$

$$\frac{4\pi v_0}{V} \int\limits_{0}^{\infty} \left|f_{eB}^{j'j}\left(Q\right)\right|^2 \left[\frac{dF_{E,nl}(Q)}{d(E/2Ry)}\right]_{E=E_0} \left(Qa_0\right)^2 d\left(Qa_0\right) \tag{326}$$

neglecting the energy exchange between the internal rotational motion of a molecule and the kinetic energy of relative motion of colliding particles. Here the energy of the ejected electron is taken to be equal to

$$E = E_0 = |\Delta E_{jj'}| - |E_{nl}|,$$

and the binding energy of the Rydberg electron is $|E_{nl}| = Ry/n_*^2$. According to the binary-encounter theory the form factor density $dF_{E,nl}\left(Q\right)/dE$ of the bound-free $nl \to E$ transition per unit energy interval is determined by the following expression

$$\frac{dF_{E,nl}(Q)}{dE} = \frac{m}{2\hbar Q} \int\limits_{|p_0|}^{\infty} |g_{nl}(p)|^2 \, p \, dp \,, \tag{327}$$

$$p_0(Q) = \frac{m(E - E_{nl}) - \hbar^2 Q^2/2}{\hbar Q} \,. \tag{328}$$

Using (326) and the Born approximation for the electron–perturber scattering amplitude square (77), the cross section $\sigma_{nl}^{ion}(j-2,j)$ of ionization of Rydberg atom induced by rotational deexcitation $j \to j-2$ by a quadrupolar molecule can be written as [82]

$$\sigma_{nl}^{ion}(j-2,\ j) \approx \frac{8\pi a_0^2 v_0}{15V} \left(\frac{Q}{ea_0^2}\right)^2 \frac{j(j-1)}{(2j+1)(2j-1)} I_q(nl,\ E_0) \,,$$
$$I_q(nl,\ E_0) = \int\limits_0^{\infty} \left[\frac{dF_{E,nl}(Q)}{d(E/2Ry)}\right]_{E=E_0} (Qa_0)^2 \, d(Qa_0) \tag{329}$$

in the range of the principal quantum number $n_* > n_*^{(0)}$. A similar expression for the ionization cross section of the Rydberg atom by a diatomic polar molecule is given by

$$\sigma_{nl}^{ion}(j-1,\ j) \approx \frac{16\pi a_0^2 v_0}{3V} \left(\frac{D}{ea_0}\right)^2 \frac{j}{(2j+1)} I_d(nl,\ E_0),$$
$$I_d(nl,\ E_0) = \int\limits_0^{\infty} \left[\frac{dF_{E,nl}(Q)}{d(E/2Ry)}\right]_{E=E_0} d(Qa_0) \,. \tag{330}$$

A crude estimate, based on the comparison of (329) and (330), shows that the relative efficiency of the quadrupole- and dipole-induced ionization processes for a given velocity V of colliding particles is determined by a factor [36]

$$\frac{(\sigma_n^{ion})_q}{(\sigma_n^{ion})_d} \sim \frac{1}{n^2} \left(\frac{Q/ea_0^2}{D/ea_0}\right)^2 \ll 1, \tag{331}$$

which becomes particularly important at high n. Thus, the ionization cross sections of Rydberg atom by strongly polar molecules are about several order of magnitude greater than the corresponding values for non-polar molecules.

6.9   Collisions with Electron-Attaching Molecules: Free-Electron Model

Thermal collisions of a Rydberg atom with molecules (B = CD) that attach slow free electrons lead to efficient ionization processes of two types:

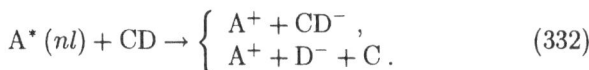

$$A^* (nl) + CD \rightarrow \begin{cases} A^+ + CD^- , \\ A^+ + D^- + C . \end{cases} \qquad (332)$$

The first electron transfer process is accompanied by the formation of positive and negative ions in the final channel, while the second one corresponds to electron transfer with simultaneous dissociative attachment of electron to the molecule. At high n these processes can be described in terms of the quasifree electron model, in which only the interaction between the Rydberg electron and the perturbing molecule is important. Since the electron is in a bound state of the negative ion after collision, the rate coefficients

$$K_{nl}^{(1)} = \left\langle V \sigma_{nl}^{(1)} \right\rangle_T , \qquad K_{nl}^{(2)} = \left\langle V \sigma_{nl}^{(2)} \right\rangle_T \qquad (333)$$

of processes (332) can be expressed through the cross sections of electron attachment $\sigma_a (k)$

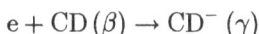

$$e + CD (\beta) \rightarrow CD^- (\gamma)$$

and dissociative attachment $\sigma_{da} (k)$

$$e + CD (\beta) \rightarrow CD^- (\gamma) \rightarrow C + D^-$$

of the free electron to the molecule. The final expressions are given by [82]

$$K_{nl}^{(1)} = \int_0^\infty (\hbar k/m) \, \sigma_a (k) \, W_{nl} (k) \, dk = \langle v \sigma_a (v) \rangle_{nl} , \qquad (334)$$

$$K_{nl}^{(2)} = \int_0^\infty (\hbar k/m) \, \sigma_{da} (k) \, W_{nl} (k) \, dk = \langle v \sigma_{da} (v) \rangle_{nl} , \qquad (335)$$

where $W_{nl} (k) = |g_{nl}(k)|^2 k^2$ is the momentum distribution function of the Rydberg electron in the given state normalized to unity.

6.10   Destruction of Rydberg States by Molecular Targets

Collisions with Polar and Non-polar Molecules. In order to demonstrate typical behavior of the depopulation cross sections of Rydberg states by non-polar, weakly polar and strongly polar molecules we present the experimental [170, 240] and corresponding theoretical results for the $Rb(nS)$ and $Rb(nD)$ atoms perturbed by N_2, CO, and HF in Figs. 47 and 48. As can be seen from Fig. 47, the agreement between theoretical calculations based on the impulse approximation and experimental data for N_2 and CO molecules is quite satisfactory in the studied range of $20 < n < 50$. For thermal collisions ($T = 293$ K) with a non-polar nitrogen molecule the total quenching cross section does not reveal the oscillatory behavior in dependence on n, since the n-changing and ionizing collisions via rotational $j \rightarrow j \pm 2$ transitions induced by the quadrupole part of $e-N_2$ interaction are not efficient. The main contribution to the total quenching cross section σ_{ns}^Q is determined by the $ns \rightarrow n'l'$ transitions to neighboring degenerate sublevels with $n' = n - 3$ and $l' > 2$ ($\delta_s^{Rb} = 3.13$). These transitions are primarily due to the short-range part of the $e - N_2$ interaction.

However, the situation is changed radically even in the case of weakly polar molecule such as CO. One can see from Fig. 47 that in the range of the principal quantum number $20 < n < 50$ the total depopulation cross section of the $Rb(ns)$ states by a CO molecule (at $T = 293$ K) turns out to be one order of magnitude higher than for a N_2 molecule as the perturber. In this case the quenching process is predominantly determined by quasi-resonant n-changing upward and downward transitions accompanied by rotational dipole-deexcitation ($j \rightarrow j - 1$) and excitation ($j \rightarrow j + 1$) of CO [170]. The ionization process via rotational deexcitation $j \rightarrow j - 1$ remains negligible in the studied range of n. The contribution of quasielastic $ns \rightarrow n - 3$, $l' > 2$ transitions without change of the rotational energy of a CO molecule (the upper dashed curve in Fig. 47) is about up 3 to 6 times less than the measured values of σ_{ns}^Q. Thus, the long–range electron–dipole interaction becomes dominant in the quenching of high–Rydberg levels even at small magnitudes of the dipole moment. The oscillatory structure of the quenching cross section of the $Rb(ns)$ by a CO molecule, observed in Ref. [170], reflects the resonant nature of the n-changing collisions via rotationally inelastic $j \rightarrow j \pm 1$

transitions (i.e., strong dependence on the value of the energy defect of the reaction (324).

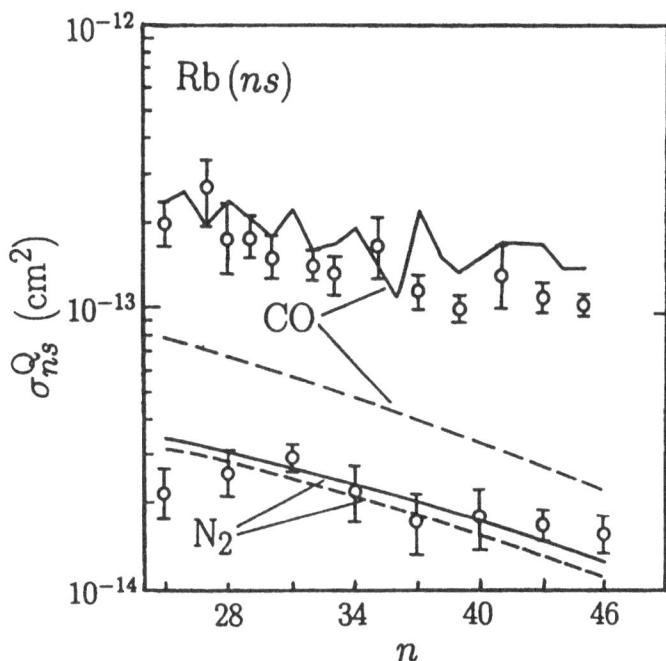

Figure 47. Quenching cross sections of Rb (nS) states by N_2 and CO molecules $(T = 293$ K). Empty circles, experimental data [170]. Full curves, impulse–approximation calculations [170] of the total depopulation cross sections σ_{ns}^Q including the quasielastic $nS \rightarrow n', l' > 2$ transitions to neighboring hydrogen–like levels with $n' = n - 3$ and $n' = n - 4$; inelastic n–changing $nS \rightarrow n', l' > 3$ transitions via rotational deexcitation and excitation of molecule $(j \rightarrow j \pm 2$ and $j \rightarrow j \pm 1$ for N_2 and CO, respectively); and ionization via rotational deexcitation. Dashed curves, the contribution $\sigma_{n-3,ns} + \sigma_{n-4,ns}$ of quasielastic transitions induced by the short–range part of the electron–molecule interaction.

For thermal collisions of high-Rydberg atoms $(n \sim 20 - 50)$ with strongly polar HF molecule $(\mathcal{D} = 0.72 \ ea_0)$ the quenching cross sections become particularly large $\sigma_{nl}^Q \sim (1 - 2) \cdot 10^{-11}$ cm^2 (see

Refs. [170] and [239-241]). The corresponding rate constants $K_{nl}^Q = \left\langle V\sigma_{nl}^Q \right\rangle_T$ may achieve values up to $(1-5)\cdot 10^{-7}$ cm^3·s^{-1} (see Fig. 48). They are about two orders of magnitude greater than for the CO molecule as the perturber. This fact is in reasonable agreement with the impulse approximation (221) with the Born expression (79) for the differential cross section of rotational excitation (deexcitation) of a polar molecule by slow electron impact, according to which σ_{nl}^Q is proportional to the square of projectile dipole moment $\mathcal{D}^2$ (for example, $\mathcal{D}_{HF}^2/\mathcal{D}_{CO}^2 \approx 266$). Nevertheless, the possibility of using the impulse approximation for quantitative calculations of collisions involving the Rydberg atoms and strongly polar molecules is not rigorously justified in the range of moderately high $n \sim 20-50$.

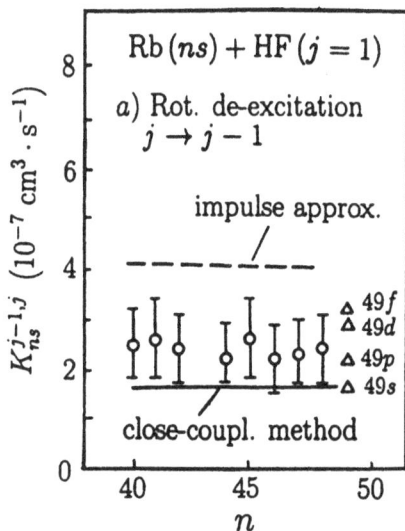

Figure 48. Rate constants $K_{ns}^{j-1,j}$ of the inelastic n-changing upward transitions induced by rotational deexcitation $j \to j-1$ of HF molecule in thermal collisions of Rb(nS) + HF $(j=1)$. Full curves, semiclassical close-coupling calculations [239]. Dashed curves, impulse–approximation calculations. Empty triangles show the dependence of the n-upward rate constants on the initial orbital angular momentum l of the ns, np, nd, and nf states for $n = 49$. Empty circles, experimental data [240].

Some results of semiclassical close-coupling calculations by *Kimura* and *Lane* [239] of quasielastic n, l-changing and inelastic n-changing collisions between the Rydberg Rb(nl) atom and HF molecule are presented in Fig. 48 together with the corresponding experimental data and the impulse-approximation results [240]. One can see that the impulse approximation significantly overestimates the cross sections and rate constants when the energy defect of the reaction is sufficiently small. In particular, the impulse approximation becomes inapplicable for describing the rotationally elastic $ns \to n - 3$, $l' > 2$ transitions (see Fig. 48) in thermal collisions of the Rydberg Rb(ns) atoms with a HF molecule provided the principal quantum number is less than about $50 - 60$. However, the qualitative agreement between the impulse and close coupling calculations at high n is reasonable. On the other hand, the semiclassical close-coupling method [239] provides a satisfactory quantitative description of identified processes in a wide range of n and transition frequencies. Moreover, the semiclassical calculations [239] give a good description of the observed [241] increase of the n-upward rate constants with increasing the initial orbital momentum l of the Rydberg atom $10^7 \cdot K_{nl}^{n-\text{ch}} = 1.6, 2.1, 2.8$ and 3.1 cm^3·s^{-1} for the s-, p-, d-, and f-states of Rb(49l) in collision with HF($j = 1$). The rate constants of the inelastic n-changing downward transitions via rotational excitation $j \to j+1$ of HF($j = 1$) turn out to be one order of magnitude lower than for the n-upward transitions induced by rotational deexcitation process $j \to j - 1$ (for $40 < n < 50$).

The detailed experimental data and corresponding theoretical calculations [121] of the rate constants for destruction of very high-Rydberg states ($90 \leq n \leq 400$) and l-mixing process in collisions of K(nP) + HF are presented in Fig. 49. The rotational close-coupling calculations [121] include the effects associated with the dipole–supported real or virtual states in the $e-$HF system with an energy about $1-1.5$ meV for $j = 0$. The destruction of high-Rydberg states is mainly a result of a collisional ionization process via a rotational deexcitation $j \to j - 1$ of a HF molecule

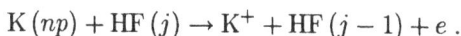

$$K\,(np) + \text{HF}\,(j) \to \text{K}^+ + \text{HF}\,(j - 1) + e\,.$$

As is apparent from Fig. 49, the corresponding rate constant increases with n-growing. *Hill* et al. [121] have also measured the rate constant for the rotationally elastic l-changing collision processes

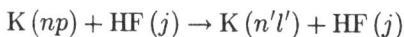

$$K\,(np) + \text{HF}\,(j) \to \text{K}\,(n'l') + \text{HF}\,(j)$$

which populate the Rydberg states with a broad distribution function of the orbital angular momentum l' in the n-manifolds close to
the parent state. The rate constant of this process decreases with
increasing n, in accord with the simple estimate by *Omont's* formula
[128]. Both theoretical dependencies of the rate constants on the
principal quantum number are in good agreement with the measured
values.

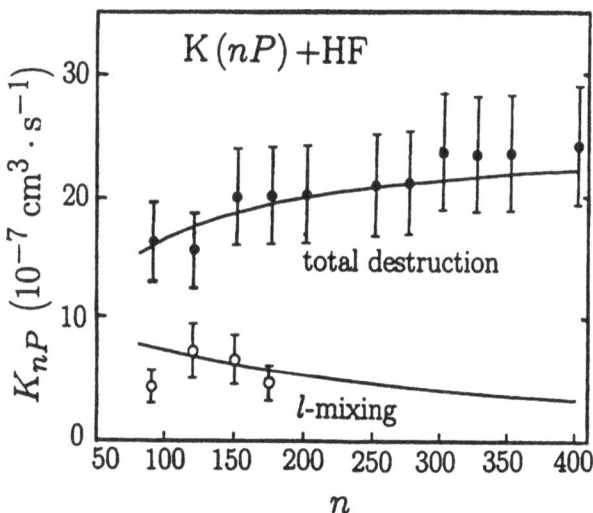

Figure 49. Experimental and theoretical dependencies [121] of the rate
constants for collisional destruction (full circles) and l-mixing process (empty
circles) of high-Rydberg K (nP)-states by HF molecule.

Another interesting example of rotationally inelastic collisions of
Rydberg atoms with polar molecular targets leading to destruction
of highly excited states via the ionization

$$K\,(nl) + CH_3Cl\,(j, K) \rightarrow K^+ + CH_3Cl\,(\dot{j} - 1, K) + e$$

was studied both experimentally and theoretically in the range of
very high principal quantum numbers $100 \leq n \leq 1100$ [120]. Here
j and K are the rotational angular momentum and its component
about the symmetry axis, respectively. According to the results of

Frey et al. [120], thermal collisions ($T = 300$ K) involving a CH_3Cl molecule lead to rapid l-mixing of Rydberg states. As a result, the l-mixed population may be treated as a single reservoir, from which atoms may be removed via the dipole-allowed transfer of rotational energy $j \to j - 1$ of CH_3Cl to the Rydberg electron. Experimental and theoretical results [120] for the rate constants of destruction of $K(nl)$ states in collisions with CH_3Cl are presented in Fig. 50.

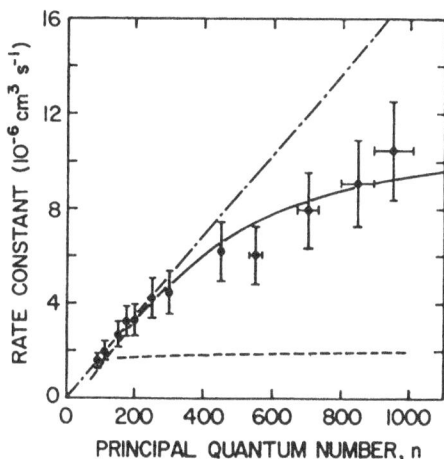

Figure 50. The results of Ref. [120] for the rate constants of collisional destruction of high–Rydberg $K(nl)$–states by CH_3Cl. Theoretical curves present the n-dependencies obtained using the independent–particle approximation and free-electron model assuming that the cross sections for electron–attachment varies as $1/v^{1.05}$ (dashed curve), as $1/v^2$ (dashed–dotted curve), and assuming a virtual (or bound) dipole–supported states with the energy 55 μeV (full curve). Full circles, experimental data.

As follows from the results of Ref. [120], theoretical calculations are in agreement with experimental data if the existence of dipole supported (virtual or real) states with very small energy ~ 55 μeV is assumed. Recent calculations of *Fabrikant* and *Wilde* [242] have confirmed this conclusion. Whereas a previously detected dipole–supported state of negative ion CH_3Cl^- was identified as a virtual state with the energy $\epsilon \approx 30~\mu eV$ at $j = 0$ and $K = 0$.

The detailed comparison of available experimental data for destruction of Rydberg atomic states via rotationally inelastic energy transfer from different molecular targets to the highly excited electron was performed by *Frey* et al. in Ref. [119]. The corresponding results for the rate constant of Rydberg potassium atom at very high magnitudes of principal quantum number $100 \leq n \leq 400$ are shown in Fig. 51 for several molecules such as H_2S, NH_3, HF, CH_2Br_2, and $C_6H_5NO_2$.

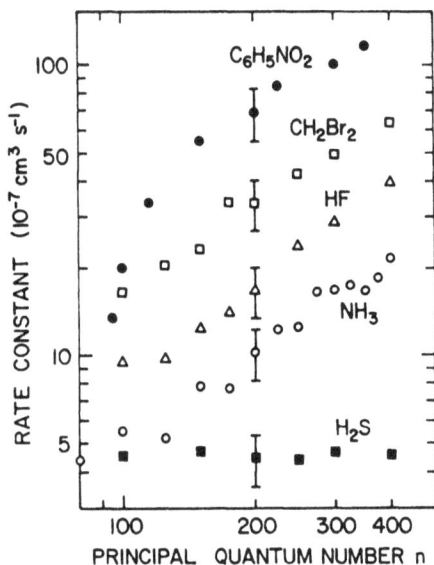

Figure 51. The data presented in Ref. [119] for the comparison of rate constants for collisional destruction of high–Rydberg atomic states by different molecular targets. The experimental results for H_2S and HF correspond to destruction of $K(np)$ states; the results for NH_3, CH_2Br_2, and $C_6H_5NO_2$ refer to the case of an l-mixed population.

As noted in the paper [119], in this region of n the rotational energy transfer in thermal collisions with HF and NH_3 targets leads directly to ionization of a Rydberg atom. For collisions with CH_2Br_2, the large majority of the possible rotationally inelastic transitions yields the ionization of Rydberg atoms. The experimental data for a HF molecule correspond to the destruction of parent K(np) atoms, while the results for NH$_3$ and CH$_2$Br$_2$, and C$_6$H$_5$NO$_2$ refer to the case of an l-mixed population of Rydberg states.

Ion-Pair Formation: Rydberg Atom–Molecule Collisions.
One of the most important example of the ion-pair formation processes in thermal collisions of highly excited atoms with electron–attaching molecules is a collision of the Rydberg atom with a SF$_6$ molecule. The main contribution to collisional ionization results predominantly from the reaction

$$A^*\,(nl) + SF_6 \rightarrow A^+ + SF_6^- \,, \qquad (336)$$

the rate constants of which are presented in Figs. 52 and 53. This reaction was intensively studied experimentally for the Rydberg rare gas and Rydberg alkali-metal atoms (see reviews [82, 236, 90]).

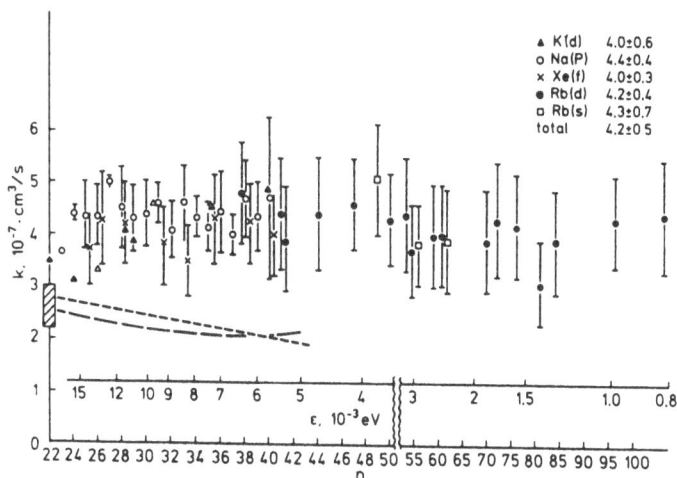

Figure 52. Rate constants K_{nl}^{ion} for the collisional ionization (336) of Rydberg atoms A(nl) by SF$_6$ molecule at high magnitudes of n. Experimental data (Na(np) – empty circles, K(nd) – full triangles, Rb(ns) – empty squares, and Rb(nd) – full circles) are those from [243, 244, 245].

It can be seen that at high $n > 25$ (see Fig. 52) they are practically independent of n and do not reveal a significant dependence on the orbital angular momentum l of the Rydberg atom. Thus, the cross section for free electron attachment to SF_6 is inversely proportional to the electron velocity v $(\sigma_a \propto \epsilon^{-1/2})$ at low energies $\epsilon \sim 1-10$ meV. Averaging over the results of different experimental works yields the following magnitude $K^{ion} = (4.3 \pm 0.4) \cdot 10^{-7}$ cm^3·s^{-1} [243, 245] for the ionization rate constant of high-Rydberg atom by SF_6.

At low and intermediate n the quasifree electron model does not hold since the molecular projectile interacts with the Rydberg electron and ionic core simultaneously. A significant reduction of the ionization rate constant in the range of $n < 25$ with decreasing n (see Fig. 53) occurs because many of the negative and positive ion-pairs initially formed remain electrostatically bound due to their Coulomb interaction in the final channel (see [243, 244]).

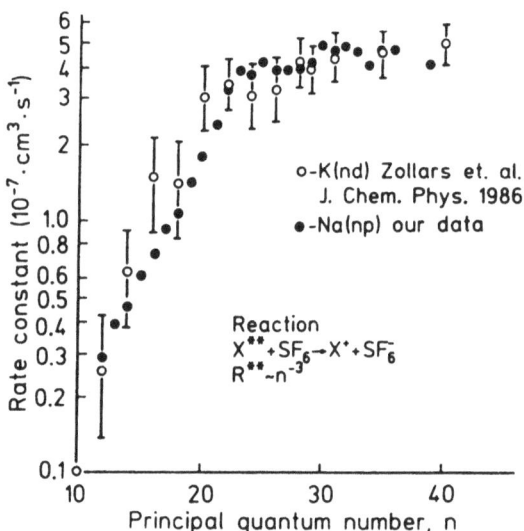

Figure 53. The same as in Fig. 52 for the intermediate magnitudes of the principal quantum number n.

According to [245], the ionization rate constants of reaction (336) averaged over the Maxwellian distribution of velocities in atomic beam is described by the expression [245]

$$K^{\text{ion}} = K_0 \left\{ \frac{\sqrt{3}}{2} \exp \left[\frac{1.19}{n^2} \left(\frac{cv_0}{V_T} \right)^{2/3} \right] + 1 - \frac{\sqrt{3}}{2} \right\}^{-1}, \qquad (337)$$

which can be used both at high and intermediate n. Here $V_T = (2kT/\mu)^{1/2}$ is the relative velocity of colliding particles (T is the temperature of beam source); and c is the constant coefficient of the order of unity. Processing of experimental data for the ionization rate constants of reaction $\text{Na}(nP) + \text{SF}_6 \rightarrow \text{Na}^+ + \text{SF}_6^-$ at intermediate $n \sim 10 - 40$ leads to the following result $K^{\text{ion}} = K_0 \exp \left(-n_0^2/n^2 \right)$ with $K_0 = \left(5.5^{+1.4}_{-1.1} \right) \cdot 10^{-7} \text{ cm}^3 \cdot \text{s}^{-1}$ and $n_0 = 20.1 \pm 0.3$ [245]. Figure 54 illustrates a comparison of theoretical results and experimental data obtained in the work by *Beterov* et al. [245] in the range of intermediate magnitudes of the principal quantum number n.

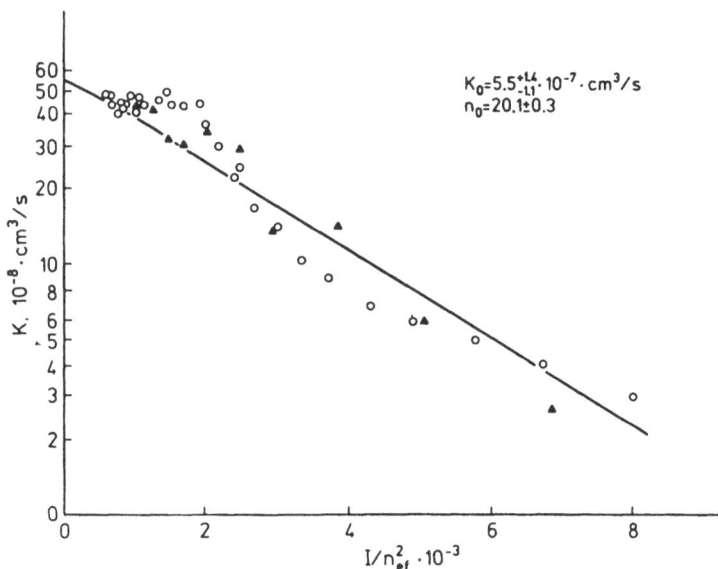

Figure 54. Comparison of theoretical and experimental results [245] for the destruction of Rydberg $\text{Na}(nP)$-states by SF_6 via the ionization reaction (336). The rate constant K^{ion}_{np} is plotted against the binding energy Ry/n_*^2 ($n_* = n - \delta_p$, $\delta_p = 0.86$) of excited electron in units of 10^{-3} eV.

Another vivid example of thermal collisions between a Rydberg atom and an electron-attaching molecule is the destruction process

$$K\left(nl\right) + CCl_4 \rightarrow K^+ + \left(CCl_4^-\right)^* \rightarrow K^+ + CCl_3 + Cl^-$$

via an ionization reaction accompanied by the dissociative attachment to a CCl_4 molecule. The experimental data [120, 246] for the rate constant of this process are shown in Fig. 55.

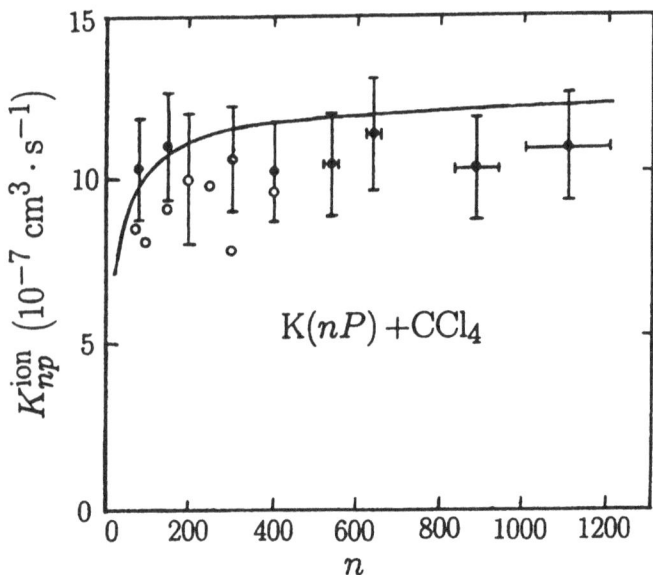

Figure 55. Rate constants K_{np}^{ion} for destruction of Rydberg K(nP)-states in collisions with a CCl_4 molecule. Full circles, experimental data [120] at very high principal quantum numbers up to $n = 1100$. Empty circles, measurements [246]. Full curve, calculations [120] in a quasifree electron model.

The measurements were carried out at very high principal quantum numbers n up to 1100. The full curve presents the rate constant calculated in the quasifree electron model (335). The experimental data indicate that the rate constant of this process is practically independent of n and, hence, is independent of the specific form of the distribution function $W_{nl}(k) = |g_{nl}(k)|^2 k^2$ of the Rydberg electron momenta. This means, that the cross section of dissociative attachment of the quasifree electron to CCl_4 molecule is inversely proportional to the electron momentum, i.e. $\sigma_{da}(k) \sim 1/k$. This fact is in full correspondence with the Wigner threshold law $\sigma_{da}(k) \sim 1/k^{2\ell-1}$, where $\ell = 0$ for an inelastic scattering process with the incident partial s-wave [37]. It should be noted that the detailed experimental studies of collisions between the high–Rydberg atoms and electron–attaching molecules SF_6 and CCl_4 have been performed by the Dunning's group in a series of papers [119, 120, 247, 248].

7   Processes Induced by Core–Projectile Interaction

One of the most important aims of this section is to study the associative and direct ionization processes and inelastic transitions between highly excited states with large energy transfer in collisions of Rydberg atoms with neutral atomic particles. In contrast to the previous theoretical consideration, our attention here will be, primarily, focused on the analysis of the alternative physical mechanisms of the identified processes accounted for by the scattering of the perturbing particle B from the parent core A^+ of the Rydberg atom A^*. We shall discuss the most efficient mechanism of resonant energy exchange between a highly excited electron and the inner electrons of quasimolecule $BA^+ + e$ temporarily formed during the scattering of atom B from the ion core A^+. This mechanism, associated with the potential energy curves crossing, turns out to be especially important for the processes of associative and direct ionization as well as for the inelastic $n \rightarrow n'$ transitions between Rydberg states with a change of the principal quantum number. We shall also consider the dipole-induced n-changing transitions and ionization processes due to the mechanism of energy exchange between the Rydberg electron and translational motion of colliding atoms. A theoretical description of such processes will be given within the framework of the stationary quasimolecular approach based on the semiclassical impact-parameter method.

In this section we shall also give a self-consistent quantal description of the shake-induced transitions between highly excited states and the direct ionization of a Rydberg atom associated with a sudden perturbation of its parent ion core in collision with a neutral perturbing particle. The analysis of this mechanism will be based on two physical approaches: time-dependent sudden perturbation theory and the impulse approximation in the momentum representation.

Another aim of this section is to apply the theory of resonance quenching of highly excited atomic states by neutral particles to the analysis of an efficient mechanism of the three-body electron–ion recombination. We shall present the kinetic model of recombination and relaxation processes taking into account both the collisions with the plasma free electrons and neutral atoms of the buffer gas. A theoretical description of the recombination processes will be given within the framework of the modified diffusion approximation. At the end of this section we shall discuss the basic aspects of the method of the classical distribution and partition functions of a diatomic system since it is widely used in the theory of Rydberg-atom–neutral-particle collisions.

7.1   Overview of the Ideas

As follows from the theory of spectral-line broadening [249], the asymptotic behavior of the impact width for high-Rydberg levels is determined by the integral cross section for elastic scattering of a perturbing atom B by the ionic core A^+. For quasielastic and inelastic collisions of Rydberg atoms with atomic particles several authors investigated the non-inertial mechanism (proposed by *V.Smirnov* [250]) of the l-mixing process (*Flannery* [159, 251], *Hickman* [252], *Matsuzawa* [253, 254], *Kaulakys* [255], *Gounand* and *Petitjean* [228]) and n-changing transitions (*Lebedev* and coworkers [230, 231]). Such transitions are due to the inertial force acting on the Rydberg electron as a result of acceleration of the Coulomb center A^+ upon collision with a neutral perturber B. The identified mechanism does not make, as a rule, any substantial contribution to the quasielastic $nl \rightarrow nl'$ transitions in the experimentally studied range of n. Such transitions are primarily due to the quasifree-electron–perturber scattering.

The influence of the perturber–core scattering on the behavior of the l-mixing process becomes important only in the range of very high n [228, 231] or for low and intermediate n, where the quasifree electron model does not hold. As was shown by *Hahn* [168] and de *Prunelé* [229], at low enough values of n the distortion corrections to the impulse approximation due to the perturber–core interaction are large, so that the $B-e$ and $B-A^+$ collisions cannot be considered independently. However, for the inelastic n-changing transitions the predominant contribution to the cross section $\sigma_{n'n}$ in a wide range of n can be a result of the perturber–core scattering alone [231], while the interaction between the neutral particle B and the outer electron of the Rydberg atom A* can be neglected.

The mechanisms of the perturber–core scattering become particularly important for the ionization processes [257–263] and n-changing transitions (see [230, 231] and [256, 264, 265]) with large energy transfer ΔE_{fi} to the Rydberg electron. That is why the cross sections of inelastic processes induced by scattering of a quasifree electron by a perturber reveal a strong reduction with an increase of ΔE_{fi}. During the scattering of the perturbing atom B from the ionic core A$^+$ of the Rydberg atom A* the quasimolecular BA$^+ + e$ system is temporarily formed with a large orbital radius $r_n \sim n_*^2 a_0$ of the external electron as compared to the internuclear distance R. As a result, the different types of the inelastic transitions and ionization in such a system can occur because of the interaction between the Rydberg electron and the quasimolecular ion BA$^+$.

The excitation and ionization processes induced by the perturber–core scattering are particularly efficient when there is a resonant (or quasiresonant) exchange of the Rydberg electron energy $\Delta E_{fi} = \hbar\omega$ with the energy of the electronic shell of a quasimolecular ion BA$^+$. Such processes occur near the crossing point R_ω of the electronic terms of the quasimolecule BA$^+ + e$, in which the energy defect ΔE_{fi} of the $|i\rangle \rightarrow |f\rangle$ transition becomes equal to the energy splitting $\Delta U_{fi}(R_\omega)$ of electronic terms of quasimolecular ion BA$^+$. This mechanism of excitation and ionization takes place, for example, in thermal collisions of a Rydberg atom A* with the ground-state parent atom A. It is well known that in this case the n-changing transitions with sufficiently large energy transfer as well as direct and associative ionization are due to the dipole transitions between the symmetrical and antisymmetrical terms of the homonuclear A$_2^+$ ion. Such a mechanism (proposed in [264, 265]) has been studied by *Devdariani* et al. [257], *Janev* and *Mihajlov* [259], *Duman* and *Shmatov* [260] for the ionization processes. The detailed experimental studies of associative and direct ionization have been carried out for the Rydberg alkali-metal atoms in symmetrical A(nl)+A and non-symmetrical A(nl)+B thermal collisions with ground-state alkali-metal atoms (see [266] for more details).

The resonant energy exchange of external and inner electrons of a quasimolecular ion BA^+ has been also investigated for the inelastic quenching [267] and ionization [262] of the Rydberg rare-gas atoms $A\left[n_0p^5\left(^2P_{3/2}\right)nl\right]$ in thermal collisions with the ground-state $B\left(^1S_0\right)$ atoms of the buffer rare gas. In this case the excitation (deexcitation) or ionization of the Rydberg electron is accompanied by the

$$A_1\left|j_i=3/2,\Omega_i=3/2\right\rangle \Longleftrightarrow X\left|j_f=3/2,\Omega_f=1/2\right\rangle$$

transition between two split terms of the heteronuclear BA^+ ion with different $\Omega_i=3/2$ and $\Omega_f=1/2$ projections of the electronic angular $j=3/2$ moment on the internuclear $\mathbf{R}$ axis. Due to the selection rules the main role in the mechanism involved is played by the quadrupole and short-range parts of the Coulomb interaction between the Rydberg electron and inner n_0p^5 electrons of the atomic core A^+.

Another mechanism of inelastic $n\to n'$ transitions [231] and ionization [261, 262] associated with the scattering of perturbing B atom on the ionic core A^+ has been proposed by *Lebedev* et al. [230]. In this mechanism, the Rydberg electron transition is accompanied by the translational transition of colliding particles A^+ and B within one electronic $U(R)$ term of a quasimolecular BA^+ ion. The identified processes are due to the interaction of the Rydberg electron of a heteronuclear quasimolecular $BA^+ + e$ system with the dipole moment of the BA^+ ion in the given electronic state. This mechanism can play an essential role for collisions of atoms A^* and B with small reduced mass, provided no resonance transfer of energy from the inner electronic shell of BA^+ system to the outer electron occurs (for example, for $H(n)+He$).

7.2 Shake-Induced Transitions of Highly Excited Atom

First consider the processes of excitation (deexcitation) and ionization of the Rydberg electron due to the transitions within one electronic $U(R)$ term of the quasimolecular BA^+ ion. They are the result of direct exchange of the Rydberg electron energy with the kinetic energy of relative motion of colliding A^+ and B particles. The magnitudes and the behavior of the cross sections for such processes are qualitatively different in the range of small $\omega \ll V/\rho_{cap}$ and large $\omega > V/\rho_{cap}$ transition frequencies. Here ρ_{cap} is the impact parameter for the capture of the perturbing atom B by the ionic core A^+ of the Rydberg atom, which determines the characteristic dimension of the internuclear separation R_{BA^+} , and V is the relative velocity of colliding particles.

7.2.1 Sudden and Impulse Approximations

Basic Expressions of Sudden Perturbation Theory. In the range of low frequencies $\omega \ll V/\rho_{\text{cap}}$ quasielastic and inelastic transitions of Rydberg electron occur in the range of internuclear separations $\rho_{\text{cap}} \underset{\sim}{<} R_\omega \ll n^2 a_0$. Thus, for calculating the transition cross sections it is natural to use the separated-atoms approach based on sudden perturbation theory [159, 231] or on the impulse approximation [228, 254]. The shake-up model of an electron, proposed by *Migdal* for ionization of atoms by neutrons (see [143]), was applied to calculating the cross sections of the n-changing, l-mixing, quenching and ionization processes involving a highly excited atom and neutral projectile in [106, 231, 261]. In its general form the shake-up model of a quantum system was formulated in [269].

Within the framework of sudden perturbation theory the inelastic and ionizing transitions of a highly excited electron induced by the scattering of neutral projectile on the ion core of the Rydberg atom can be described using the two general approaches. The first one corresponds to the shake-up model with a sudden change of the Hamiltonian of the quantum system. The second approach corresponds to the typical shake-up model of the scattering kind. We start our consideration with a brief description of the basic idea of the first approach, and then we present the main equations of the second model. It is important to stress that in the zero-order approximation both approaches lead to completely equivalent results.

Let the velocity of the ion core A^+ be changed by a value $\Delta \mathbf{V}_{A^+}$ as a result of a binary encounter with projectile B. For small enough transition frequencies ω_{fi} the collision time $\tau_{A+B} \sim \rho_{\text{cap}}/V$ of these particles is small compared to the characteristic period $1/\omega_{f_i}$ of the Rydberg electron motion, which can be estimated as $T_n \sim 2\pi n^3 \left(a_0/v_0\right)$. Therefore, the interaction between the ion core and neutral projectile can be regarded as a sudden perturbation of the Rydberg atom Hamiltonian:

$$H\left(t\right) = \left\{ \begin{array}{ll} H_1 \equiv H_A\left(\mathbf{r}\right)\,, & t < 0\,, \\ H_2 \equiv H_A\left(\mathbf{r}'\right)\,, & t \gg \tau_{A+B}\,. \end{array} \right. \tag{338}$$

Here $\mathbf{r}$ and $\mathbf{r}'$ are the radius vectors of a highly excited electron in the frames of reference moving with the ion core before and after collision so that

$$\mathbf{r}' = \mathbf{r} - (\Delta\mathbf{V}_{A^+})\,t\;,$$

and

$$H_A\,(\mathbf{r}) = -\frac{\hbar^2}{2m}\Delta_\mathbf{r} + U\,(r) \qquad (339)$$

is the Hamiltonian of the Rydberg atom. Since the orbital radius of a highly excited electron $r_n \sim n^2 a_0$ is large ($r_n \gg \rho_{cap}$), the displacement

$$\Delta R_{A^+} \sim (\Delta V_{A^+})\,(\tau_{A+B}) \sim (\mu/M_A)\,\rho_{cap} \qquad (340)$$

of the ion core during the time interval τ_{A+B} of the core–projectile interaction is small $\Delta R_{A^+} \ll r_n$ compared to the characteristic size of the Rydberg atom $\Delta R_{A^+} \ll r_n$, and hence can be neglected. As a result, one can assume that the electron coordinates r' and r directly before and after a collision practically coincide.

Further let the perturber–core encounter leads to the $\alpha \to \alpha'$ transition between two eigenstates of the Rydberg atom. It is important to stress that the initial eigenwave function $\psi_\alpha\,(\mathbf{r})$ of the Hamiltonian H_1 is not the eigenfunction of the Hamiltonian H_2. The initial wave function of a Rydberg atom in the reference frame moving with the ion core directly after a collision is

$$\psi_i^{(0)}\,(\mathbf{r}',t=0) = \psi_\alpha\,(\mathbf{r}')\,\exp\left(-i\mathbf{K}\cdot\mathbf{r}'\right)\;, \qquad (341)$$

where $\mathbf{K} = (m\Delta\mathbf{V}_{A^+})/\hbar$ is the electron wave vector corresponding to the change of the parent core velocity $\Delta\mathbf{V}_{A^+} = (\mathbf{q}' - \mathbf{q})/M_A$ in its collision with the perturbing particle B. Hence, using the basic formula of sudden perturbation theory for the transition amplitude

$$a_{fi}\,(t=\infty) = \left\langle \psi_f\,(\mathbf{r}')\,\middle|\,\psi_i^{(0)}\,(\mathbf{r}',t=0)\right\rangle\;, \qquad (342)$$

from the initial α state to the final eigen state $f = \alpha'$ and for the corresponding transition probability W_{fi}, we have

$$a_{\alpha'\alpha}\,(t=\infty) = \langle\psi_{\alpha'}\,(\mathbf{r})|\,\exp\left(-i\mathbf{K}\cdot\mathbf{r}\right)\,|\psi_\alpha\,(\mathbf{r})\rangle\;,$$

$$W_{\alpha'\alpha} = |a_{\alpha'\alpha}\,(t=\infty)|^2\;, \qquad (343)$$

where we additionally put $\mathbf{r}' = \mathbf{r}$ in accordance with the discussion presented above.

Below we shall show that this simple result can be deduced from the general sudden perturbation theory in the time-dependent representation [269]. Consider the total Hamiltonian H of an atom A consisting of the unperturbed Hamiltonian H_A of the Rydberg atom (339) and sudden perturbation $V_{AB}(t)$ associated with its interaction with a perturbing particle B

$$\widehat{H} = \widehat{H}_A + \widehat{V}_{AB}(t) . \tag{344}$$

Suppose that the characteristic time τ of this interaction is sufficiently small compared to the typical period $1/\omega_{fi}$ of the Rydberg electron motion. We also assume that the operators of interaction $\widehat{V}_{AB}(t)$, corresponding to the different time moments, commute with each other. Then, we introduce the wave functions of the total quantum system in the interaction representation $|\Psi\rangle$ and in the Schrödinger representation $|\Phi\rangle$. For the different time moments t and t' these wave functions can be written as

$$|\Psi(t)\rangle = \widehat{S}(t, t') |\Psi(t')\rangle , \tag{345}$$

$$|\Phi(t)\rangle = \widehat{U}(t, t') |\Phi(t')\rangle , \tag{346}$$

where $\widehat{S}(t, t')$ and $\widehat{U}(t, t')$ are the corresponding unitary operators of the time-evolution.

The evolution operator $\widehat{S}(t, t')$ satisfies the following differential equation

$$i\hbar \frac{\partial \widehat{S}(t, t')}{\partial t} = \widehat{W}_{AB}(t) \widehat{S}(t, t') , \quad \widehat{S}(t, t' = t) = I . \tag{347}$$

Here $\widehat{W}_{AB}(t)$ is the operator of the interaction potential written in the ineraction representation

$$\widehat{W}_{AB}(t) = \exp\left(\frac{i}{\hbar}\widehat{H}_A t\right) \widehat{V}_{AB}(t) \exp\left(-\frac{i}{\hbar}\widehat{H}_A t\right) . \tag{348}$$

Let $|i\rangle = |\alpha\rangle$ and $|f\rangle = |\alpha'\rangle$ be the initial and final eigenstates of the unperturbed atomic Hamiltonian $\widehat{H}_A$. Then, the probability W_{fi} of the $|i\rangle \rightarrow |f\rangle$ transition between these eigenstates of an atom A induced by the time-dependent interaction $\widehat{V}_{AB}(t)$ is determined from the following basic formula

$$W_{fi} = \left| \left\langle f \left| \widehat{S} \left(t = +\infty, \, t' = -\infty \right) \right| i \right\rangle \right|^2 . \tag{349}$$

Within the framework of time-dependent sudden perturbation theory a general solution of the equation (347) is usually presented in a series power of the $\omega_{fi}\tau$ parameter (see [269] for more details)

$$\widehat{S}(t, t') = \widehat{S}_0(t, t') + \widehat{S}_1(t, t') + \widehat{S}_2(t, t') + \ldots \quad . \tag{350}$$

In the zero-order approximation over the $\omega_{fi}\tau$ parameter the resulting expression for the time-evolution operator $\widehat{S}(t, t') \approx \widehat{S}_0(t, t')$ takes the form

$$\widehat{S}_0(t, t') = \exp \left(-\tfrac{i}{\hbar} \int_{t'}^{t} \widehat{W}_{AB}(\tau)(\tau) \, d\tau \right)$$

$$= \exp \left(\tfrac{i}{\hbar} \widehat{H}_A t \right) \exp \left(-\tfrac{i}{\hbar} \int_{t'}^{t} \widehat{V}_{AB}(\tau) \, d\tau \right) \exp \left(-\tfrac{i}{\hbar} \widehat{H}_A t \right) \tag{351}$$

in accordance with the general result presented in Ref. [269]. On inserting Eq. (351) into (349), the zero-range approximation for the transition probability can be converted to its final form

$$W_{fi} = \left| \left\langle f \left| \exp \left(-\tfrac{i}{\hbar} \int_{-\infty}^{\infty} \widehat{V}_{AB}(\tau) \, d\tau \right) \right| i \right\rangle \right|^2 . \tag{352}$$

The next step consists in the application of this general expression to the problem of the collision of a neutral particle B with the ion core A^+ of the Rydberg atom A^*. As noted previously, such a collision leads to the change of the weakly-bound electron wave-vector $\mathbf{K} = (m\Delta \mathbf{V}_{A^+})/\hbar$ corresponding to the change $\Delta \mathbf{V}_{A^+}$ of the ion–core velocity. Hence, according to Eq. (352) the transition probability between the two eigenstates $|i\rangle \rightarrow |f\rangle$ of an atom takes the form

$$W_{fi} = |\langle f | \exp(-i\mathbf{K} \cdot \mathbf{r}) | i \rangle|^2 , \tag{353}$$

$$\hbar \mathbf{K} = \int_{-\infty}^{\infty} \mathbf{F}(t) \, dt , \qquad \mathbf{F} = -\nabla_{\mathbf{R}} V_{AB} . \tag{354}$$

Here $\widehat{V}_{AB}(t) = \mathbf{r} \cdot \mathbf{F}(t)$ is the perturbation potential, and $\mathbf{F}$ is the corresponding force acting on the Rydberg electron.

It is apparent that for transitions of highly excited electron induced by the scattering of a neutral particle B on the parent core A^+ of the Rydberg atom A^*, the electron–projectile interaction V_{eB} in the total potential $V_{AB} = V_{eB} + V_{A+B}$ (see Sect. 2.6) can be neglected. This is justified because the characteristic size r_{eB} ($r_{eB} \sim n^2 a_0$) of the electron–projectile distance is much greater than the characteristic internuclear separations R, which are responsible for the shake-induced transitions. At the same time, the core–projectile interaction V_{A+B} is actually reduced to the potential energy $U(R)$ of the B and A^+ particles for transitions within one electronic term of the BA^+ ion.

Thus, such transitions are mainly determined by the second term in Eq. (54) so that the perturbation $V(t)$ and the corresponding force $\mathbf{F}$ are given by

$$V \approx -\frac{m}{M_A} \mathbf{r} \cdot \boldsymbol{\nabla}_{\mathbf{R}} U = \frac{e^2 \mathbf{r} \cdot \mathbf{R}_{A^+}}{r^3} \,, \tag{355}$$

$$\mathbf{F} \approx -\frac{m}{M_A} \boldsymbol{\nabla}_{\mathbf{R}} U = m \frac{d^2 \mathbf{R}_{A^+}}{dt^2} \,. \tag{356}$$

As noted previously (see Sect. 2.6), this interaction corresponds to the inertial force $\mathbf{F} = m \, d^2 \mathbf{R}_{A^-}/dt^2$ acting on the Rydberg electron in the non-inertial frame moving with the Coulomb centre A^+. The second relation on the right–hand side of Eq. (355) proceeds directly from the Ehrenfest theorem for the electron and ion–core radius vector operators

$$m \frac{d^2 \mathbf{r}}{dt^2} = -\boldsymbol{\nabla}_{\mathbf{r}} \left(-\frac{e^2}{r} \right) = \frac{e^2 \mathbf{r}}{r^3} \,, \tag{357}$$

$$\mu \frac{d^2 \mathbf{R}_{A^+}}{dt^2} = -\boldsymbol{\nabla}_{\mathbf{R}} U(R) \,, \tag{358}$$

where $\mathbf{R}_{A^+} = \mu \mathbf{R}/M_A$.

Thus, in the shake-up model the probability of the $nl \to n'l'$ transition between the Rydberg levels with given values of the principal and orbital quantum numbers caused by the perturber–core scattering is given by

$$W_{n'l',nl}(K) =$$

$$\frac{1}{2l+1} \sum_{mm'} |\langle \psi_{n'l'm'}(\mathbf{r})| \exp(i\mathbf{K} \cdot \mathbf{r}) \, |\psi_{nlm}(\mathbf{r})\rangle|^2 . \tag{359}$$

Here $\mathbf{r} = (r, \theta, \varphi)$ is the radius vector of the highly excited electron relative to the parent core A^+, $\psi_{nlm}(\mathbf{r}) = R_{nl}(r) Y_{lm}(\theta, \varphi)$ is the atomic wave function. The electron wave vector $\mathbf{K}$ corresponding to the change of the parent core velocity $\Delta \mathbf{V}_{A+}$ in its collision with the perturbing particle B can be expressed in terms of the momentum transfer vector $\mathbf{Q}$. The values of $\mathbf{K} = (m/M_A)\mathbf{Q}$ and $\mathbf{Q} = \mathbf{q}' - \mathbf{q}$ in the Rydberg-atom–neutral collision are given through their scattering angle $\theta_{\mathbf{q}'\mathbf{q}}$ in the center of mass system by the relation

$$K^2 = \left(\frac{m}{M_A}\right)^2 \left(q^2 + q'^2 - 2qq' \cos \theta_{\mathbf{q}'\mathbf{q}}\right)$$

$$\approx 2a_0^{-2} \left(\frac{\mu V}{v_0 M_A}\right)^2 (1 - \cos \theta_{\mathbf{q}'\mathbf{q}}) . \tag{360}$$

Here $q = \mu V/\hbar$ and $q' = \mu V'/\hbar$ are the relative wave numbers of a Rydberg atom A^* and perturber B (μ is their reduced mass), whereas

$$q' = \left[q^2 + 2\mu \Delta E_{nl,n'l'}/\hbar^2\right]^{1/2} \approx q$$

at $|\Delta E_{n'l',nl}| \ll \hbar^2 q^2/2\mu$.

The summed probability $W_{nl}^{in} = 1 - W_{nl}^{el}$ of all inelastic bound–bound and bound–free transitions of the Rydberg atom (excluding only the contribution of elastic scattering $W_{nl}^{el} \equiv W_{nl,nl}$) can be written as [106]

$$W_{nl}^{in}(K) = 1 - (2l+1) \sum_{\text{œ}=0}^{2l} (2\text{œ}+1) \begin{pmatrix} l & \text{œ} & l \\ 0 & 0 & 0 \end{pmatrix}^2$$

$$\times |\langle nl \, |j_{\text{œ}}(Kr)| \, nl \rangle|^2 . \tag{361}$$

This result directly proceeds from (359) at $|n'l'\rangle = |nl\rangle$ using the expansion in spherical Bessel functions for the $\exp(i\mathbf{K} \cdot \mathbf{r})$ factor.

The cross section of the $nl \rightarrow n'l'$ transition can be expressed in the shake-up model through the probability $W_{n'l',nl}[K(\theta_{\mathbf{q}'\mathbf{q}})]$ and the differential cross section $d\sigma_{A+B}/d\Omega_{q'q} = |f_{A+B}(q, \theta_{q'q})|^2$ for the elastic scattering of perturbing particle B by the ionic core A^+ [159]

$$\sigma_{n'l',nl} = \int W_{n'l',nl} |f_{A+B}|^2 \, d\Omega_{q'q} \,. \tag{362}$$

Here f_{A+B} is the elastic perturber–core scattering amplitude. Then, using (359)–(362) and replacing the integration with respect to the solid angle $d\Omega_{q'q}$ by integration over the momentum transfer dQ, we obtain [106]

$$\sigma_{n'l',nl} = \frac{2\pi\hbar^2}{\mu^2 V^2(2l+1)} \sum_{mm'} \int_{Q_{min}}^{Q_{max}} |f_{A+B}(Q)|^2$$

$$\times \left| \left\langle n'l'm' \left| \exp\left[i\left(\frac{m}{M_A}\right) \mathbf{Q} \cdot \mathbf{r} \right] \right| nlm \right\rangle \right|^2 Q \, dQ \,. \tag{363}$$

Here the lower $Q_{min} = |q' - q|$ and upper $Q_{max} = q' + q$ limits of integration are given by

$$Q_{min} \approx |\Delta E_{n'l',nl}|/\hbar V \,, \qquad Q_{max} \approx 2\mu V/\hbar$$

if the transition energy satisfies the relation $|\Delta E_{n'l',nl}| \ll \hbar^2 q^2/2\mu$.

Impulse Treatment of Core–Projectile Effects. This result for the contribution of the core-perturber scattering mechanism to the transition cross section can also be derived in the impulse approximation similar to the case of transitions induced by the electron–perturber scattering considered previously in Sect. 4.2.2. Indeed, in accord with [254, 228] the basic expression for the matrix elements of the T_{A+B} scattering operator is expressed

$$\langle \mathbf{q}', f \,|\, T_{A+B}(\mathsf{E}) \,|\, \mathbf{q}, i \rangle =$$

$$\int G^*_{\alpha'}(\boldsymbol{\kappa}') \langle \mathbf{q}' \,|\, t_{A+B}(\mathcal{E}) \,|\, \mathbf{q} \rangle \, G_\alpha(\boldsymbol{\kappa}) \, d\boldsymbol{\kappa} \tag{364}$$

in terms of the two-body t_{A+B} operator over the plane waves for the core-perturber relative motion

$$t_{A+B}(\mathbf{q}', \mathbf{q}; \mathcal{E}) =$$

$$\left\langle \mathbf{q}' \left| V_{A+B} + V_{A+B} \left(\frac{1}{\mathcal{E} - K_{A+B} - V_{A+B} + i0} \right) V_{A+B} \right| \mathbf{q} \right\rangle \,, \tag{365}$$

in full correspondence with the analogos result (167). Here we have additionally used the mass disparity relations ($m \ll M_A$) according to which

$$\kappa' \approx \kappa + (m/M_A) \, \mathbf{Q} \, ,$$

where $\mathbf{Q} = \mathbf{q}' - \mathbf{q}$. In the coordinate representation expression (364) can be rewritten as

$$
\langle \mathbf{q}', f \, |T_{A+B} \, (\mathsf{E})| \, \mathbf{q}, i \rangle = \left\langle \psi_f \, (\mathbf{r}) \, \left| \exp \left(-\tfrac{m}{M_A} \mathbf{Q} \cdot \mathbf{r} \right) \right| \, \psi_i \, (\mathbf{r}) \right\rangle
$$
$$
\times \left\langle \mathbf{q}' \, \left| t_{A+B} \left(\mathcal{E} = \hbar^2 \mathbf{q}^2 / 2\mu \right) \right| \mathbf{q} \right\rangle .
\tag{366}
$$

Here the matrix elements of the two-body (core–perturber) operator over the plane waves are related with the amplitude $f_{A+B} \, (Q)$ for elastic scattering of the A^+B-particles

$$
f_{A+B} \, (Q) = -\frac{\mu}{2\pi\hbar^2} \left\langle \mathbf{q}' \, \left| t_{A+B} \left(\mathcal{E} = \hbar^2 \mathbf{q}^2 / 2\mu \right) \right| \mathbf{q} \right\rangle .
\tag{367}
$$

Thus, the final formula of the impulse approximation for the contribution of the core–perturber scattering to the amplitude of the inelastic transition between Rydberg states of an atom takes the following form

$$
f_{fi}^{A^+B} \, (Q) = \left\langle f \, \left| \exp \left(-\frac{m}{M_A} \mathbf{Q} \cdot \mathbf{r} \right) \right| \, i \right\rangle f_{A+B} \, (Q) \, .
\tag{368}
$$

It is evident that integration of the amplitude square $\left| f_{fi}^{A^+B} \, (Q) \right|^2$ over the momentum transfer $Q \, dQ$ leads directly to (363). Therefore, for the perturber–core scattering mechanism, the shake-up model used by *Lebedev* and *Marchenko* [231] for transitions between Rydberg states and the impulse approximation used by *Gounand* and *Petit-jean* [228] should lead to completely equivalent results. Comparison of this expression with the basic formula of the impulse approximation for the electron–perturber scattering shows that usually the core–perturber scattering mechanism yields a small contribution to the cross sections of the inelastic transitions $i \neq f$ due to the presence of a small factor (m/M_A) in the momentum transferred to the Rydberg atom $\mathbf{K} = (m/M_A) \, \mathbf{Q}$. However, we shall show in this section that in some cases it becomes important.

7.2.2   Analytic Expressions of Shake-up Model

High-n Limit for Ionization.   *Flannery* has shown [159] that the
total cross section σ_i^{tot} for all inelastic and elastic transitions caused
by the mechanism under consideration is determined by the integral
cross section for elastic scattering of a neutral projectile B on the
ionic core A^+ of the Rydberg atom, i.e.

$$\sigma_i^{tot} = \sum_f \sigma_{fi} \equiv \sigma_i^{el} + \sigma_i^{in} = \sigma_{A+B}^{el} \, ,$$

$$\sigma_{A+B}^{el}(V) = \int |f_{A+B}|^2 \, d\Omega_{\mathbf{q}'\mathbf{q}} \, . \tag{369}$$

Formula (369) follows from the basic equations (359) and (362) taking
into account the summation rule

$$W_i^{tot} = \sum_f W_{fi} = 1$$

for the total transition probability. Thus in accord with [159], the
cross section σ_{fi} of any bound–bound or bound–free transitions induced by the perturber–core scattering mechanism is limited by the
magnitude of the elastic scattering cross section σ_{A+B}^{el}.

In the asymptotic region of very high principal quantum num-
bers $n \gg v_0 M_A / V\mu$ ($n \gg 10^3 - 10^4$ at thermal velocities) the cross
section σ_{nl}^{in} of all inelastic transitions tends asymptotically to the integral cross section for elastic scattering of a perturbing particle B by
the parent core A^+ of the Rydberg atom A^*. Correspondingly, the
contribution $\sigma_{nl}^{el}(V)$ of elastic scattering of the perturber B on the
Rydberg atom becomes very small and can be neglected, so that

$$\sigma_{nl}^{in}(V) \xrightarrow[n \to \infty]{} \sigma_{A+B}^{el}(V) \, , \qquad \sigma_{nl}^{el}(V) \xrightarrow[n \to \infty]{} 0. \tag{370}$$

As has been shown in [132], in this range of $n \gg v_0 M_A / \mu V$ the main
contribution to the cross section $\sigma_i^{in} = \sum_{f \neq i} \sigma_{fi}$ makes the bound–free
transitions of the highly excited electron. This means that at very
high magnitudes of n the cross section of all inelastic transitions is
primarily determined by the ionization of the Rydberg atom, i.e.,

$$\sigma_{nl}^{in}(V) \approx \sigma_{nl}^{ion}(V) \, .$$

State-Changing Collisions in the Dipole Approximation. Simple expressions for the cross sections of inelastic and quasielastic transitions between the Rydberg states can be derived in the shake-up model for the range $n \ll (v_0 M_A / V\mu)^{1/2}$ of the principal quantum number. At thermal velocities V this condition usually leads to the following restriction $n \underset{\sim}{<} 30 - 100$ on the principal quantum numbers (depending on the kind of colliding particles). Due to the relation $Kn^2 a_0 \ll 1$ the exponential factor in (360) can be expanded in a series of (Kr). Then, in the dipole approximation, the transition probability $W_{n'l',nl}$ of the $nl \to n'l'$ transitions (for which $l' \neq l$ at $n' = n$) is given by

$$W_{n'l',nl}(K) = \frac{K^2}{3(2l+1)} \sum_{mm'} \left| \langle n'l'm' | \mathbf{r} | nlm \rangle^2 \right| . \qquad (371)$$

Substitution the expression (371) into the basic formula (362) yields the following result for the corresponding cross section

$$\sigma_{n'l',nl} = \frac{2}{3} \left(\frac{\mu V}{v_0 M_A} \right)^2 \mathcal{M}_{n'l',nl} \sigma_{A^+B}^{tr} , \qquad (372)$$

$$\mathcal{M}_{n'l',nl} = \frac{l_>}{(2l+1)a_0^2} \left| R_{nl}^{n'l'} \right|^2 . \qquad (373)$$

Here the dimensionless quantity $\mathcal{M}_{n'l',nl}$ is determined by the radial matrix element $R_{nl}^{n'l'} = \langle n'l' | r | nl \rangle$ of the Rydberg coordinate ($l' = l \pm 1$, $l_> = \max\{l, l'\}$), and

$$\sigma_{A^+B}^{tr}(V) = \int |f_{A^+B}|^2 \left(1 - \cos \theta_{\mathbf{q'q}}\right) d\Omega_{\mathbf{q'q}} \qquad (374)$$

is the momentum transfer cross section for the elastic scattering of a neutral particle B by the ionic core A^+.

The formula for the total cross section σ_{nl}^{in} of all inelastic $nl \to n'l'$ transitions (i.e. the quenching cross section of the nl-level) caused by the perturber–core scattering [106]

$$\sigma_{nl}^{in}(V) = \underset{n'l'}{\sum}{}' \sigma_{n'l',nl}$$

$$= \frac{2}{3} \left\langle \left(\frac{r}{a_0} \right)^2 \right\rangle_{nl} \left(\frac{\mu V}{v_0 M_A} \right)^2 \sigma_{A^+B}^{tr}(V) . \qquad (375)$$

directly follows in the dipole approximation from (361). Here $\langle r^2 \rangle_{nl}$ is determined by the simple analytic expression [42] in the case of hydrogen-like degenerate nl-levels, whereas

$$\langle r^2 \rangle_{nl} \approx 5n^4 a_0^2 / 2 , \qquad l \ll n .$$

As is evident from (372) and expressions for the dipole matrix elements $R_{nl}^{n,l\pm1}$, the contribution of pure quasielastic $nl \rightarrow n,l \pm 1$ transitions $\Delta E_{n'l',nl} = 0$ (considered analytically in [231, 255]) to the total quenching cross section is predominant. The contribution of all other inelastic $nl \rightarrow n',l \pm 1$ transitions with $n' \neq n$ (n-changing processes) is appreciably smaller for quenching of degenerate hydrogen-like nl-levels. The corresponding $\mathcal{M}_{nl}^{l-\text{mix}}$ and $\mathcal{M}_{nl}^{n,l-\text{ch}}$ terms in (372) determining the l-mixing and n-changing ($n' \neq n$) cross sections $\sigma_{nl}^{l-\text{mix}}$ and $\sigma_{nl}^{n,l-\text{ch}}$ are given by (at $l \ll n$)

$$\mathcal{M}_{nl}^{l\text{-mix}} = \sum_{l'=l\pm1} \mathcal{M}_{nl',nl} \approx \frac{9n^4}{4} , \tag{376}$$

$$\mathcal{M}_{nl}^{n,l\text{-ch}} = \sum_{n'\neq n} \sum_{l'=l\pm1} \mathcal{M}_{n'l',nl} \approx \frac{n^4}{4} . \tag{377}$$

Provided equally populated lm-sublevels within a given principal quantum number n, the shake-up model leads to the simple analytic expression [231]

$$\sigma_{n'n}(V) = \sum_{ll'} \left(\frac{2l+1}{n^2} \right) \sigma_{n'l',nl}$$

$$= \frac{4\mathcal{G}(\Delta n)}{3^{3/2}\pi} \left(\frac{n}{\Delta n} \right)^4 \left(\frac{\mu V}{v_0 M_A} \right)^2 \sigma_{A+B}^{\text{tr}}(V) \tag{378}$$

for the cross section of the inelastic $n \rightarrow n'$ transitions in the dipole range of $n \ll (v_0 M_A / \mu V)^{1/2}$. Here $\mathcal{G}(\Delta n)$ is the Gaunt factor (see, for example, [36])

$$\mathcal{G}(\Delta n) = \pi\sqrt{3} |\Delta n| J_{\Delta n}(\Delta n) J'_{\Delta n}(\Delta n) , \tag{379}$$

where $J_\nu(x)$ and $J'_\nu(x)$ are the Bessel function and its derivative, respectively. Note that $\mathcal{G} = 1$ in the Kramers approximation. The expression (378) is valid for sufficiently small transition frequencies

$$\omega = 2Ry |\Delta n| / \hbar n^3 \ll V / \rho_{\text{cap}} .$$

Now we will discuss the behavior of the elastic scattering cross section $\sigma_{nl}^{\text{el}} \equiv \sigma_{nl,nl}$ of a neutral particle B by the Rydberg atom $A(nl)$ for the perturber–core scattering mechanism. In the dipole range of $n \ll (v_0 M_A/\mu V)^{1/2}$ the probability of elastic scattering $W_{nl}^{\text{el}} = 1 - W_{nl}^{\text{in}}$ in the shake-up model is approximately equal to

$$W_{nl}^{\text{el}}(K) = 1 - O\left[\left(Kn^2 a_0\right)^2\right],$$

i.e. it is equal to unity in the zeroth approximation over the $\left(Kn^2 a_0\right)$ parameter (see (361)). Thus, in this approximation $(\sigma_{nl}^{\text{el}} \gg \sigma_{nl}^{\text{in}})$ the elastic scattering cross section is independent of the principal quantum number n and is determined by the integral cross section σ_{A+B}^{el} for perturber–core scattering, so that

$$\sigma_{nl}^{\text{el}}(V) = \sigma_{A+B}^{\text{el}}(V) - \sigma_{nl}^{\text{in}}(V)$$

$$\approx \sigma_{A+B}^{\text{el}}(V), \qquad n \ll \left(\frac{v_0 M_A}{\mu V}\right)^{1/2}. \tag{380}$$

At thermal energies, a major contribution to the integral σ_{A+B}^{el} and the momentum transfer σ_{A+B}^{tr} cross sections is determined by the long-range part of the perturber–core interaction $U(R)$. For the power approximation $U(R) = -C_\nu/R^\nu$ the semiclassical expression for the integral cross section (381) is given by [37]

$$\sigma_{A+B}^{\text{el}}(V) = \gamma_\nu \left(\frac{C_\nu}{\hbar V}\right)^{2/(\nu-1)},$$

$$\gamma_\nu = 2\pi^{\nu/(\nu-1)} \sin\left[\frac{\pi}{2}\left(\frac{\nu-3}{\nu-1}\right)\right] \Gamma\left(\frac{\nu-3}{\nu-1}\right) \left[\frac{\Gamma\left(\frac{\nu-1}{2}\right)}{\Gamma\left(\frac{\nu}{2}\right)}\right]^{2/(\nu-1)}. \tag{381}$$

For a given relative velocity V of colliding particles, the magnitudes of the momentum transfer cross section $\sigma_{A+B}^{\text{tr}}(V)$ (374) are significantly smaller than $\sigma_{A+B}^{\text{el}}(V)$, since the contribution of small scattering angles $\theta_{\mathbf{q'q}}$ turns out to be partially compensated by a factor $(1 - \cos\theta_{\mathbf{q'q}})$. As a result, the value of σ_{A+B}^{tr} is closer to that of the cross section $\sigma_{A+B}^{\text{cap}}$ for the capture of a perturbing atom B by the ionic core A^+. The capture cross section is given by the simple expression [38]

$$\sigma_{A+B}^{\text{cap}}(V) = \pi\rho_{\text{cap}}^2(V) = \pi\left(\frac{\nu}{\nu-2}\right)^{\nu/(\nu-2)} \left(\frac{\nu C_\nu}{2\mathcal{E}}\right)^{2/\nu}, \tag{382}$$

where $\mathcal{E} = \mu V^2/2$, and ρ_{cap} is the capture impact parameter. In the special case of polarization interaction

$$U_{\text{pol}} = -\alpha e^2/2R^4$$

(when $\nu = 4$ and $C_4 = \alpha e^2/2$) between the B and A^+ particles expressions, the capture cross section as well as the momentum transfer and integral elastic scattering cross sections can be written as

$$\sigma^{\text{el}}_{A+B} = \tfrac{\pi^{5/3}}{2^{4/3}} \, \Gamma\left(\tfrac{1}{3}\right) \left(\tfrac{\alpha e^2}{\hbar V}\right)^{2/3} \approx 7.16 \left(\tfrac{\alpha v_0}{V}\right)^{2/3},$$

$$\sigma^{\text{tr}}_{A+B} \approx 1.12 \sigma^{\text{cap}}_{A+B}, \qquad \sigma^{\text{cap}}_{A+B} = \pi\left(\tfrac{2\alpha e^2}{\mathcal{E}}\right)^{1/2}, \tag{383}$$

where $\mathcal{E} = \mu V^2/2$. It is seen that the $\sigma^{\text{el}}_{A+B}/\sigma^{\text{tr}}_{A+B}$ ratio is approximately determined by the parameter

$$\sigma^{\text{el}}_{A+B}/\sigma^{\text{tr}}_{A+B} \sim \left[(\mu/m)^2 \, (\mathcal{E}/2Ry) \, (\alpha/a_0^3)\right]^{1/6} \tag{384}$$

whose value turns out to be of the order of $5 - 50$ at thermal energies depending on the polarizability α of the neutral perturber B and reduced mass μ of the colliding partners.

Thus, as follows from the results of the shake-up model of the Rydberg atom A^* by a neutral particle B, the quenching cross section $\sigma^{\text{Q}}_{nl} \equiv \sigma^{\text{in}}_{nl}$ of a highly excited nl-level increases rapidly with the principal quantum number ($\sigma^{\text{Q}}_{nl} \propto n^4$) in the dipole range of $n \ll (v_0 M_A/\mu V)^{1/2}$. It becomes of the order of the momentum transfer cross section σ^{tr}_{A+B} for the perturber-core scattering at $n \sim (v_0 M_A/\mu V)^{1/2}$. A further increase of n leads to a slower growing of the total inelastic cross section. The limiting value σ^{el}_{A+B} of the total inelastic cross section $\sigma^{\text{in}}_{nl} \approx \sigma^{\text{ion}}_{nl}$ for the perturber–core scattering mechanism is achieved only in the range of very high principal quantum numbers $n \gg v_0 M_A/\mu V$, which of no importance for experimental applications at thermal energies. Therefore, the ratio of the total inelastic cross section magnitudes at very high n and at $n \sim (v_0 M_A/\mu V)^{1/2}$ may be estimated by the relation (384).

Symmetrical Atomic Collisions. All results presented above describe the case of collisions between the Rydberg atom A^* and the perturbing atom or molecule B of the buffer gas. Some comments should be made on symmetrical collisions of the highly excited atom A^* with the ground-state parent atom A. In this case the scattering of atom A on the ionic core A^+ leads to charge exchange of the valence electron of the projectile. Hence, the velocity of the Coulomb center, in which the outer Rydberg electron e is localized, periodically changes from V_1 to V_2, where V_1 and V_2 are velocities of the heavy particles A^+ and A. Therefore, as in the case of non-symmetrical $B \rightarrow A^*$ collisions, transitions of the Rydberg electron may be a result of the inertial force acting on this electron due to the change of the coordinate frame of reference connecting one or another Coulomb center. However, for the symmetrical Rydberg atom-neutral $A \rightarrow A^*$ thermal collisions, such transitions are primarily accounted for by charge exchange of the inner electron of the composite $A_2^+ + e$ system (i.e. of the valence electron of the ground-state atom A) but not the momentum transfer to the Coulomb center A^+ in its collision with the projectile A. This is due to the small contribution of the perturber–core scattering $A \rightarrow A^+$ to the momentum transfer cross section σ_{A+A}^{tr} as compared to the charge exchange effect at thermal energies, so that

$$\sigma_{A+A}^{tr} \approx 2\sigma_{A+A}^{ex} . \qquad (385)$$

The cross section σ_{A+A}^{ex} for resonance charge exchange of an atomic ion A^+ by the parent atom A is determined by the well known expression [37]

$$\sigma_{A+A}^{ex}(V) = \int_0^\infty \sin^2\left[\eta\left(\rho, V\right)\right] 2\pi\rho \, d\rho, \qquad (386)$$

$$\eta(\rho, V) = \frac{1}{2\hbar} \int_{-\infty}^\infty \Delta U_{gu}\left[R\left(t\right)\right] dt, \qquad (387)$$

where $\Delta U_{gu}(R) = |U_g(R) - U_u(R)|$ is the exchange splitting of even and odd lower terms of a quasimolecular ion A_2^+, and $V = |V_1 - V_2|$ is the relative velocity of the colliding A^+ and A particles.

As has been shown in [256, 255], analytic formulae of the dipole approximation (372) and (378) presented above for non-symmetrical collisions with sufficiently small transition frequencies also remain the same in the symmetrical case. However, it should be taken into account that the reduced mass of the colliding atoms is $\mu = M_A/2$ and the momentum transfer cross section $\sigma_{A+A}^{tr} \approx 2\sigma_{A+A}^{ex}$ is given by expression the (386). In the range of $n \ll (v_0/V)^{1/2}$, this leads to the following final expressions for the cross sections of quasielastic l-mixing $(nl \rightarrow n, l \pm 1)$

$$
\begin{aligned}
\sigma_{nl}^{l\text{-mix}} &= \tfrac{3}{4}\, n^2 \left(n^2 - l^2 - l - 1\right) \left(\tfrac{V}{v_0}\right)^2 \sigma_{A+A}^{ex}(V) \\
&= \tfrac{4\mathcal{G}(\Delta n)}{3^{3/2}\pi} \left(\tfrac{n}{\Delta n}\right)^4 \left(\tfrac{\mu V}{v_0 M_A}\right)^2 \sigma_{A+B}^{tr}(V) ,
\end{aligned}
\tag{388}
$$

and inelastic n-changing transitions

$$
\sigma_{n'n} = \frac{2\mathcal{G}(\Delta n)}{3^{3/2}\pi} \left(\frac{n}{\Delta n}\right)^4 \left(\frac{V}{v_0}\right)^2 \sigma_{A+A}^{ex}(V) .
\tag{389}
$$

The last expression is valid if the transition frequency is small enough $\omega_{n'n} \ll V/R_{ex}$, where the characteristic size $R_{ex} \sim \left(\sigma_{A+A}^{ex}/\pi\right)^{1/2}$ of internuclear separation R is determined by the value of the charge exchange cross section.

7.3   Stationary Quasimolecular Approach

For large transition frequencies $\omega > V/\rho_{cap}$ the shake model of the Rydberg electron based on the sudden perturbation approximation becomes inapplicable. For symmetrical collisions this occurs at $\omega > V/R_{ex}$, where the characteristic internuclear distance R_{ex} is determined by the cross section σ_{A+A}^{ex} of the resonance exchange process (see (386)). In this range of ω the inelastic transitions of highly excited electron, induced by the perturber–core scattering, occur at sufficiently small internuclear distances $R_\omega < \rho_{cap}$ $(R_\omega < R_{ex})$ as compared to the orbital radius $r_n \sim n^2 a_0$ of a Rydberg electron. Thus, it is natural to use the quasimolecular approach in the theoretical analysis of both transitions within one and between different electronic terms of the BA^+ ion. In this approach, a simple description of excitation (deexcitation) and ionization processes may be given on the basis of stationary perturbation theory over the interaction V between the Rydberg electron e and quasimolecular BA^+ ion.

The total Hamiltonian H of the quasimolecular $BA^+ + e$ system can be presented as

$$
H = -\frac{\hbar^2}{2\mu}\Delta_R + H_{BA^+}(r_\kappa, R) + H_e(r) + V ,
\tag{390}
$$

$$H_e = -\frac{\hbar^2}{2m}\Delta_\mathbf{r} - \frac{e^2}{r} . \tag{391}$$

Here H_e is the Hamiltonian of the outer electron in the Coulomb field of the atomic core A^+

$$H_e\psi_{nlm}(\mathbf{r}) = E_{nl}\psi_{nlm}(\mathbf{r}) , \tag{392}$$
$$H_e\psi_\mathbf{k}^\pm(\mathbf{r}) = E\psi_\mathbf{k}^\pm(\mathbf{r}) , \tag{393}$$

and $\psi_{nlm}(\mathbf{r})$ is the atomic wave function with the given quantum numbers n, l, and m and with energy $E_{nl} = -Ry/n_*^2$, while $\psi_\mathbf{k}^\pm(\mathbf{r})$ are the Coulomb wave functions of a continuous spectrum with wavevector $\mathbf{k} = (k,\theta_\mathbf{k},\varphi_\mathbf{k})$ and energy $E = \hbar^2 k^2/2m$ of the ejected electron; H_{BA^+} is the Hamiltonian of the electronic shell of the BA^+ ion

$$H_{BA^+}\varphi_i(\mathbf{r}_\kappa,\mathbf{R}) = U_i(R)\varphi_i(\mathbf{r}_\kappa,\mathbf{R}) , \tag{394}$$
$$H_{BA^+}\varphi_f(\mathbf{r}_\kappa,\mathbf{R}) = U_f(R)\varphi_f(\mathbf{r}_\kappa,\mathbf{R}) , \tag{395}$$

and $\varphi_i(\mathbf{r}_\kappa,\mathbf{R})$ and $\varphi_f(\mathbf{r}_\kappa,\mathbf{R})$ and $U_i(R)$ and $U_f(R)$ are its wave eigenfunctions and the corresponding electronic terms in the initial $|i\rangle$ and final $|f\rangle$ states ($\mathbf{R} = \mathbf{R}_B - \mathbf{R}_A$ is the radius vector joining the nuclei A and B); V is the operator of the Coulomb interaction of the outer electron ($\mathbf{r}$) with all the inner electrons ($\mathbf{r}_\kappa$, $\kappa = 1,2,\ldots,N$) and nuclei (Z_A and Z_B) of the $BA^+ + e$ system

$$V = \sum_{\kappa=1}^N \frac{e^2}{|\mathbf{r} - \mathbf{r}_\kappa|} - \frac{Z_A e^2}{|\mathbf{r} - \mathbf{R}_A|} - \frac{Z_B e^2}{|\mathbf{r} - \mathbf{R}_B|} + \frac{e^2}{r} . \tag{396}$$

It is important to stress that the Coulomb term $\left(e^2/r\right)$ is to be included in (396) to compensate the corresponding $\left(-e^2/r\right)$ term in the Hamiltonian of the zeroth approximation

$$H_0 = -\hbar^2\Delta_\mathbf{R}/2\mu + H_{BA^+} + H_e . \tag{397}$$

Within the framework of the first-order of stationary perturbation theory and impact-parameter approach, the cross section $\sigma_{n'l',nl}^{f,i}$ of the inelastic $|i, nl\rangle \rightarrow |f, n'l'\rangle$ transition can be presented as

$$\sigma_{n'l',nl}^{f,i}(\mathcal{E}) = \frac{4\pi^3 g_f}{(2l+1)g_i q^2}$$

$$\times \sum_{mm'} \int_0^\infty \left| \left\langle \chi_{\mathcal{E}'J}^{(f)}(R) \right| V_{i,nlm}^{f,n'l'm'}(R) \left| \chi_{\mathcal{E}J}^{(i)}(R) \right\rangle \right|^2 2J \, dJ \ . \tag{398}$$

Here $V_{i,nlm}^{f,n'l'm'}$ is the electronic matrix element of perturbation interaction (396)

$$V_{i,nlm}^{f,n'l'm'}(R) = $$

$$\left\langle \psi_{n'l'm'}(\mathbf{r}) \left| \langle \varphi_f(\mathbf{r}_\kappa, R) | V | \varphi_i(\mathbf{r}_\kappa, R) \rangle \right| \psi_{nlm}(\mathbf{r}) \right\rangle \tag{399}$$

for transitions between different potential energy curves of the quasimolecule $BA^+ + e$

$$U_{i,nl}(R) = U_i(R) + E_{nl} \ , \tag{400}$$

$$U_{f,n'l'}(R) = U_f(R) + E_{n'l'} \ , \tag{401}$$

while g_i and g_f are the statistical weights of the BA^+ ion. Radial nuclear wave functions $\chi_{\mathcal{E}J}^{(i)}(R)$ and $\chi_{\mathcal{E}'J}^{(f)}(R)$ of the continuous spectrum with definite magnitudes of the kinetic energy ($\mathcal{E} = \hbar^2 q^2/2\mu$ and $\mathcal{E}' = \hbar^2 q'^2/2\mu$) and orbital angular momentum $J = q\rho$ for the perturber–core relative motion correspond to the initial $|i\rangle$ and final $|f\rangle$ electronic terms of the quasimolecular BA^+ ion, respectively; ρ is the impact-parameter of the colliding particles B and A^+.

For the cross section $\sigma_{nl}^{\mathrm{d.i.}}(\mathcal{E})$ of the direct ionization process

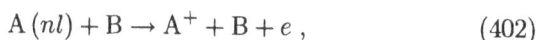

$$A(nl) + B \rightarrow A^+ + B + e \ , \tag{402}$$

caused by the perturber–core scattering mechanism, a similar result can be written as [262]

$$\sigma_{nl}^{\mathrm{d.i.}} = \frac{4\pi^3 g_f}{(2l+1)g_i q^2} \sum_{l'} \sum_{mm'} \int_0^{E_{\max}^{\mathrm{d.i.}}} dE \int_0^\infty 2J \, dJ$$

$$\times \left| \left\langle \chi_{\mathcal{E}'J}^{(f)}(R) \right| V_{i,nlm}^{f,El'm'}(R) \left| \chi_{\mathcal{E}J}^{(i)}(R) \right\rangle \right|^2 \ , \tag{403}$$

where integration over dE is taken over all possible values of the ejected electron energy. Here the electronic matrix element

$$V_{i,nlm}^{f,El'm'}(R) = \langle \psi_{El'm'} | \langle \varphi_f | V | \varphi_i \rangle | \psi_{nlm} \rangle$$

corresponds to the bound–free transition of the outer electron, so that its final wave function in (403) should be replaced by the Coulomb wave function

$$\psi_{El'm'}(\mathbf{r}) = \mathcal{R}_{El'}(r) Y_{l'm'}(\theta, \varphi)$$

of the continuous spectrum normalized to the δ-function of energy.

The semiclassical expression for the cross section

$$\sigma_{nl}^{\text{a.i.}}(\mathcal{E}) = \sum_{vJ} \sigma_{nl}^{fi}(\mathcal{E} \to vJ) \tag{404}$$

of associative ionization

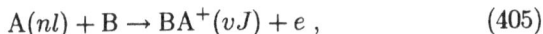

$$A(nl) + B \to BA^+(vJ) + e\ , \tag{405}$$

summed over all possible values of vibrational v and rotational J quantum numbers of the molecular $BA^+(vJ)$ ion in the final electronic term $U_f(R)$, is given by [262]

$$\sigma_{nl}^{\text{a.i.}} = \frac{4\pi^3 g_f}{(2l+1)g_i q^2} \sum_{l'} \sum_{mm'} \int_0^{v_{\text{max}}} dv \int_0^\infty 2J\, dJ \tag{406}$$

$$\times \left| \left\langle \chi_{vJ}^{(f)}(R) \left| V_{i,nlm}^{f,El'm'}(R) \right| \chi_{\mathcal{E}J}^{(i)}(R) \right\rangle \right|^2 .$$

Here summation over all possible values of v and J is replaced by integration over dv and dJ. This procedure corresponds to the approximation of the quasidiscrete spectrum for energy levels $\mathcal{E}_{vJ}$ of the molecular BA^+ ion.

In the JWKB-approximation the radial nuclear wave functions $\chi_{\mathcal{E}J}^{(i)}(R)$ and $\chi_{\mathcal{E}'J}^{(f)}(R)$ in the continuous or discrete $\chi_{vJ}^{(f)}(R)$ spectra have the form

$$\chi(R) = \frac{C}{R\sqrt{V(R)}} \cos\left(\int_b^R q(R)\, dR - \frac{\pi}{4} \right), \tag{407}$$

$$V(R) = \sqrt{\left(\frac{2}{\mu}\right) \left[\mathcal{E} - U(R) - \frac{\hbar^2(J+1/2)^2}{2\mu R^2} \right]}. \tag{408}$$

Here $q_i(R) = \mu V_i(R)/\hbar$ and $q_f(R) = \mu V_f(R)/\hbar$ are the wave numbers, $V_i(R)$ and $V_f(R)$ are the relative velocities of nuclei, b_i and b_f are the left turning points for the motion of A^+ and B particles in the initial and final terms of the quasimolecular (or molecular) BA^+ ion, respectively; C is the normalizing constant equal to $C_{\mathcal{E}} = (2/\pi\hbar)^{1/2}$ in the continuous ($\mathcal{E} > 0$) and $C_{vJ} = 2/\left(T_{vJ}^{(f)}\right)^{1/2}$ in the discrete ($\mathcal{E}_{vJ} < 0$) spectra, and $T_{vJ}^{(f)}$ is the classical period of vibrational–rotational motion of nuclei in the final $U_f(R)$ term with energy $\mathcal{E}_{vJ}$. wave functions $\chi_{\mathcal{E}J}^{(i)}(R)$ and $\chi_{\mathcal{E}'J}^{(f)}(R)$ are normalized to the delta function of energy $\delta(\mathcal{E} - \mathcal{E}')$, while $\chi_{vJ}^{(f)}(R)$ is normalized to unity.

7.4   Electron–Translational Energy Exchange

One of the possible mechanism of dipole-induced transitions associated with the scattering of perturbing atom B on the ion core A^+ corresponds to excitation and ionization processes of the Rydberg atom A^*, which do not affect the variation of electronic states of the quasimolecular ion BA^+ (i.e. when $U_i(R) = U_f(R)$ and $\varphi_i = \varphi_f$ in the basic equations of Sect. 7.3). In contrast to the case of small transition frequencies (see Sect. 7.2), our attention here will be focused on the inelastic n-changing transitions [231] and ionization [262] of a highly excited atom with large enough values of ω ($\omega > V/\rho_{\text{cap}}$ and $R_\omega \leq \rho_{\text{cap}} \ll n^2 a_0$). A simple description of such processes may be given on the basis of the general semiclassical formulae of stationary perturbation theory presented above.

Transition Matrix Elements.   Due to a sufficiently large electron orbital radius $r_n \sim n^2 a_0$ compared with internuclear distances R_ω of the quasimolecular BA^+ ion the main contribution to the excitation and ionization cross section is determined by the interaction

$$\langle \varphi_i(\mathbf{r}_\kappa, \mathbf{R})| \, V \, |\varphi_i(\mathbf{r}_\kappa, \mathbf{R})\rangle = -\frac{e\mathbf{D}(R)\,\mathbf{r}}{r^3} + O(r^{-3})$$
$$= -\frac{e\mathbf{D}_1(R)\,\mathbf{r}}{r^3} - \frac{e\mathbf{D}_2(R)\,\mathbf{r}}{r^3} + O\left(r^{-3}\right) \tag{409}$$

of the Rydberg electron with the total dipole moment

$$\mathbf{D}(R) = \langle \varphi_i(\mathbf{r}_\kappa, \mathbf{R})| \, \mathbf{D}(\mathbf{r}_\kappa, \mathbf{R}) \, |\varphi_i(\mathbf{r}_\kappa, \mathbf{R})\rangle \tag{410}$$

of the BA^+ ion (relative to its center of mass, $\mu = M_B M_A / (M_B + M_A)$ is the reduced mass) in the electronic state $|\varphi_i\rangle$ under consideration. The dipole moment involves two terms

$$\mathbf{D}(R) = \mathbf{D}_1(R) + \mathbf{D}_2(R) .$$

The first one determines the contribution of the positive Coulomb A^+ center

$$\mathbf{D}_1 = e\mathbf{R}_A = (\mu/M_A) e\mathbf{R} . \tag{411}$$

As has been shown in [231], its contribution to the transition cross section corresponds to the non-inertial effect. The second term $\mathbf{D}_2(R)$ (which tends to zero at $R \to \infty$) is due to the displacement (polarization) of the inner electrons of the BA^+ ion with respect to its nuclei induced by the scattering of perturbing atom B on the ionic core A^+.

The transition matrix elements of the dipole interaction (409) can be written as (see review [36])

$$\frac{1}{2l+1} \sum_{l'} \sum_{mm'} \left| \left\langle \chi_{\mathcal{E}'J}^{(i)} \middle| V_{i,nlm}^{f,n'l'm'} \middle| \chi_{\mathcal{E}J}^{(i)} \right\rangle \right|^2 = \frac{m^2\omega^4}{3e^2} |D_{\mathcal{E}'\mathcal{E}}|^2$$

$$\times \left(\frac{l}{2l+1} \left| R_{nl}^{n',l-1} \right|^2 + \frac{l+1}{2l+1} \left| R_{nl}^{n',l+1} \right|^2 \right) . \tag{412}$$

Here

$$D_{\mathcal{E}'\mathcal{E}}(J) \equiv \left\langle \chi_{\mathcal{E}'J}^{(i)} \middle| D(R) \middle| \chi_{\mathcal{E}J}^{(i)} \right\rangle \tag{413}$$

is the radial matrix element of dipole moment of the quasimolecular ion for the free-free $(\mathcal{E} \to \mathcal{E}')$ transition of nuclei within one electronic term $U(R)$; and

$$R_{nl}^{n',l\pm1} = \langle n', l \pm 1 | r | nl \rangle \tag{414}$$

is the radial matrix element of the Rydberg electron coordinate. Expressions for the free–bound $(\mathcal{E} \to v)$ transition of nuclei B and A^+ and bound–free $(n \to E)$ electron transitions have the same form.

In thermal collisions of a Rydberg atom A^* with ground-state B atoms of the buffer gas, the processes considered here prove to be most efficient, provided the electronic term $U(R)$ has a deep well $\mathcal{E}_0 \equiv |U(R_e)|$ and a large value ω_e of the lower vibrational quantum. For this case the dipole matrix elements $D_{\mathcal{E}'\mathcal{E}}(J)$ and $D_{v\mathcal{E}}(J)$ for translational transitions of nuclei within one electronic term of heteronuclear BA^+ ion near its dissociation limit $\mathcal{E}, \mathcal{E}' \ll \mathcal{E}_0$ (or $|\mathcal{E}_{vJ}|$,

$|\mathcal{E}_{v'J}| \ll \mathcal{E}_0$) have been calculated in [231]. These semiclassical calculations are based on the JWKB-approximation for the wave functions (407) of the perturber–core relative motion and the method of the Fourier components (the correspondence principle) for the transition matrix elements (at $\Delta E = \hbar\omega \ll \mathcal{E}_0$). The final result for the free–free $\mathcal{E} \rightarrow \mathcal{E}'$ transitions can be presented as

$$D_{\mathcal{E}',\mathcal{E}}(J) =$$

$$\begin{cases} D_{\mathcal{E}',\mathcal{E}}(0) = \frac{\pi}{2} C_{\mathcal{E}'} C_{\mathcal{E}} \frac{d(\omega)}{\omega} \exp\left(-\omega\tau\right), & J \leq J_{\text{cap}}, \\ \propto D_{\mathcal{E}',\mathcal{E}}(0) \exp(-\omega\rho/V), & J \geq J_{\text{cap}}. \end{cases} \qquad (415)$$

The matrix elements $D_{v,\mathcal{E}}(J)$ of the free–bound $\mathcal{E} \rightarrow v$ are determined by the same expression, but with the other value of the normalizing constant C_v of the semiclassical wave function in the final channel (407). Here $J_{\text{cap}} = q\rho_{\text{cap}}$ is the angular momentum corresponding to the capture of the perturbing particle B by the ionic core A^+; and τ is the collision time of B and A^+ particles in the left turning point b_1 on the repulsive branch of the term $U(R)$. The value $d(\omega) \sim (dD/dR)|_{R_\omega} \Delta R_\omega$ is determined by the increment of the dipole moment of the BA^+ ion over the characteristic length $\Delta R_\omega = |U'(R_\omega)/U''(R_\omega)|$ of the $U(R)$ term in the vicinity of the distance R_ω, which makes a major contribution to the transition with the frequency ω. The particular form of the $d(\omega)$ function for various approximations of the $U(R)$ term is presented in [231]. For instance, for the Morse potential

$$U(R) = \mathcal{E}_0 \left[\exp\left(-2\alpha\frac{R - R_e}{R_e}\right) - 2\exp\left(-\alpha\frac{R - R_e}{R_e}\right) \right], \quad (416)$$

with linear approximation of the dipole moment in the vicinity of the well bottom $R = R_e$

$$D(R) = D_e + \left(\frac{dD}{dR}\bigg|_{R_e}\right)(R - R_e) \qquad (417)$$

calculation of $d(\omega)$ function and the collision time τ of B and A^+ particles in the zeroth approximation with respect to the parameter $|\mathcal{E}|/\mathcal{E}_0 \ll 1$ yields

$$d(\omega) = \left(\frac{dD}{dR}\bigg|_{R_e}\right)\frac{R_e}{\alpha}, \qquad \tau = \omega_e^{-1} = \frac{R_e}{\alpha(2\mathcal{E}_0/\mu)^{1/2}}. \qquad (418)$$

Thus, in this special case the quantity $d(\omega)$ is independent of the transition frequency, i.e. $d(\omega) = const$. Moreover, the quantity reciprocal to τ coincides with the value of the lowest vibrational quantum ω_e of the molecular BA^+ ion, which separates the regions of adiabatic $\omega \gg \tau^{-1}$ and non-adiabatic $\omega \ll \tau^{-1}$ transition frequencies. Hence the use of the approximations (416) and (417) allows to express the quantities d and τ in terms of several parameters of the potential $U(R)$ and the dipole moment $D(R)$ alone.

Excitation and Ionization. As is evident from expression (415), in the most interesting case of small kinetic energy $\mathcal{E} = \mu V^2/2$ of colliding atoms as compared to the depth of the potential well $\mathcal{E}_0$, the matrix elements $D_{\mathcal{E}'\mathcal{E}}(J)$ (or $D_{\nu\mathcal{E}}(J)$) with impact parameters $\rho \leq \rho_{\mathbf{cap}}$ are practically independent of ρ and are determined only by the transition frequency ω. This range of ρ corresponds to the capture of a perturbing atom B by the ionic core A^+ in the region of small internuclear distances $R_\omega < \rho_{\mathbf{cap}}$. In the opposite case $\rho \gg \rho_{\mathbf{cap}}$, they reveal a sharp exponential reduction with ρ. Hence the contribution of large values of ρ to the excitation and ionization cross sections can be neglected. Substituting (412) and (415) into the general formula (398) and performing an integration over the $2\pi\rho\,d\rho$, we obtain the final analytic expressions for the transition cross section [231]

$$\sigma_{n'n}(\mathcal{E}) = \frac{8\pi\mathcal{G}(\Delta n)}{3^{3/2}n^5(n')^3}\left(\frac{2Ry}{\hbar\omega_{n'n}}\right)^2\left(\frac{d(\omega_{n'n})}{ea_0}\right)^2$$

$$\times \exp\left(-2\omega_{n'n}\tau\right)\sigma_{\mathrm{A+B}}^{\mathbf{cap}}(\mathcal{E})\ . \tag{419}$$

The final result for the rate constant $K_{n'n} = \langle V\sigma_{n'n}\rangle_T$ of deexcitation $n \to n'\ (n > n')$ of a Rydberg atom takes the form

$$K_{n'n}(T) = \Gamma\left(2 - \frac{2}{\nu}\right)\langle V\rangle_T\,\sigma_{n'n}(V_T)\ . \tag{420}$$

Here $\langle V\rangle_T = (8kT/\mu)^{1/2}$ and $\sigma_{nn'}(V_T)$ is the cross section at $\mathcal{E} = kT$ (and $\mathcal{G} = 1$ in the Kramers approximation). These expressions are valid provided the following restrictions on the transition frequency and principal quantum number are obeyed

$$V/\rho_{\mathbf{cap}} \ll \omega \ll \mathcal{E}_0/\hbar, \qquad n^2a_0 \gg \rho_{\mathbf{cap}}\ . \tag{421}$$

The excitation cross section and rate constant can be determined from the detailed balance relations.

To explain the physical meaning of the results presented above we take into account only the major non-inertial effect (corresponding to the contribution $\mathbf{D}_1 = (\mu/M_{\mathrm{A}})\,e\mathbf{R}$ of the Coulomb center A^+ to the dipole moment) and completely neglect the induced dipole interaction. This allows rewriting the result (419) in a simple form which clarifies the dependence of the deexcitation cross section $\sigma_{n'n}$ on the principal quantum number n and transition frequency $\omega_{n'n}$, on the mass M_{A}, M_{B} and relative velocity $V\,(R_\omega)$ of colliding atoms. It is interesting to compare this expression with the result (378) of the shake-up model. One can see that the magnitudes of the transition cross section $\sigma_{n'n}$ turns out to be proportional to a factor $n^4 V^2\,(R_\omega)$ for the entire considered region of $n \lesssim (v_0 M_{\mathrm{A}}/\mu V)^{1/2}$. However, in the shake-up region of small frequencies $\left(\omega_{n'n} \lesssim V/\rho_{\mathrm{cap}}\right)$ they fall with decreasing n like n^4, since a major contribution to the inelastic or quasielastic transition is made by large distances $R_\omega \gtrsim \rho_{\mathrm{cap}}$ (where the relative velocity of the colliding A^+ and B particles is determined by $V\,(R_\omega) \sim V = (2\mathcal{E}/\mu)^{1/2}$). In the opposite limiting case $\omega_{n'n} \gg V/\rho_{\mathrm{cap}}$, the behavior of inelastic transitions is changed radically. Such transitions take place in the region of small internuclear distances $R_\omega \ll \rho_{\mathrm{cap}}$, in which the relative velocity $V\,(R_\omega) \approx \left(2\,|U\,(R_\omega)|\,/\mu\right)^{1/2}$ of a perturbing atom B and ionic core A^+ increases significantly with increasing $\omega_{n'n}$, owing to their acceleration in the potential well. In particular, for transition frequencies $\omega_{n'n} \sim \tau^{-1} \approx \omega_{\mathrm{e}}$ (when R_ω corresponds to the region in the vicinity of the potential well R_{e}) the relative velocity reaches the value of $V_0 = (2\mathcal{E}_0/\mu)^{1/2}$ determined by dissociation energy $\mathcal{E}_0 = |U\,(R_{\mathrm{e}})|$ of the BA^+ ion, so that $V_0 \gg V$ at thermal energies. This increase in the velocity $V\,(R_\omega)$ causes a slight change of the cross section $\sigma_{n'n}$ with decreasing n in the frequency region $V/\rho_{\mathrm{cap}} \lesssim \omega_{n'n} \lesssim \tau^{-1}$. A strong reduction of the inelastic cross sections only appears in the adiabatic region of frequencies

$$\sigma_{n'n} \propto \exp\left(-2\omega_{n'n}\tau\right), \qquad \omega_{n'n} \gg \tau^{-1} \approx \omega_{\mathrm{e}},$$

i.e., for low enough principal quantum numbers $n \ll (2Ry\tau/\hbar)^{1/3}$ at $\Delta n \sim 1$.

All these effects were also taken into account in calculations [262] of the ionization processes of a Rydberg atom A^* induced by the scattering of atom B on the ionic core A^+. A particularly simple formula has been obtained for the total cross section

$$\sigma_{nl}^{\text{ion}} = \sigma_{nl}^{\text{d.i.}} + \sigma_{nl}^{\text{a.i.}}$$

of direct and associative ionization

$$\sigma_{nl}^{\text{ion}} = \frac{1}{\alpha} \left(\frac{m}{e\hbar}\right)^2 \sigma_{\text{A}^+\text{B}}^{\text{cap}} (\mathcal{E})$$

$$\times \int_{\omega_{nl}}^{\omega_{\max}} d^2(\omega)\, \sigma_{nl}^{\text{ph}}(\omega) \exp\left(-2\omega\tau\right)\, \omega\, d\omega\;.$$

(422)

Here $\alpha = e^2/\hbar c = 1/137$, and the lower and upper limits of integration are given by

$$\omega_{nl} = |E_{nl}|/\hbar = Ry/\hbar n_*^2\,, \qquad \omega_{\max} = (\mathcal{E}_0 + \mathcal{E})/\hbar\;.$$

The use of a imple analytic expressions for the photoionization cross section

$$\sigma_n^{\text{ph}}(\omega) = \frac{1}{n^2} \sum_l (2l+1)\,\sigma_{nl}^{\text{ph}}(\omega)$$

averaged over all degenerate lm-sublevels leads to the following final results for the ionization rate constant $K_n^{\text{ion}} = \left\langle V\sigma_n^{\text{ion}} \right\rangle_T$ [262]

$$K_n^{\text{ion}}(T) =$$

$$\frac{32\pi^{1/2}\Gamma(2-2/\nu)}{3^{3/2}n^3} \left(\frac{d(\omega_n)}{ea_0}\right)^2 V_T \sigma_{\text{A}^+\text{B}}^{\text{cap}}(T)\, E_2\left(2\omega_n\tau\right),$$

(423)

where $V_T = (2kT/\mu)^{1/2}$, and

$$E_2(z) = \int_1^\infty t^{-2} e^{-zt}\, dt$$

is the exponential integral of the second order. This formula is valid for the Rydberg levels $n \gg (Ry/\mathcal{E}_0)^{1/2}$ with small binding energy $|E_n|$ compared with the depth $\mathcal{E}_0 = |U(R_e)|$ of the potential well. Another restriction for the permissible values of the principal quantum number in (423) follows from the condition (421), which yields $n \ll (\rho_{\text{cap}} v_0 / a_0 V)^{1/2}$.

Hydrogen–Helium Thermal Collisions. The efficiency of the considered mechanism of excitation, deexcitation and ionization of Rydberg states by neutral atomic particles can be illustrated for thermal collisions between the highly excited hydrogen $H(n)$ atom and the ground-state helium atom $He(1s^2)$. In this case the first excited $A^1\Sigma$ electronic term of the molecular $(HeH)^+$ ion, which correlates with the state $He^+(1s)+H(1s)$, lies much higher (by more than 10 eV) than the electronic ground state of $He(1s^2)+H^+$. Therefore, excitation (deexcitation) and ionization processes are not accompanied by intersection of the bound and repulsive potential energy curves corresponding to different configurations of the quasimolecular $\left[(HeH)^+ + e\right]$ system. Hence, the resonance energy transfer from the inner electronic shell of this system to the Rydberg electron is certainly not realized in thermal $H(n)+He(1s^2)$ collisions.

Moreover, the HeH^+ ion in its ground electronic $X^1\Sigma$ state has small a reduced mass and large enough depth of the potential well $\left(|U(R_e)| = 1.85 \text{ eV}\right)$. As a result, the non-adiabatic regime $(\omega \lesssim \tau^{-1})$ of identified processes occurs even in the case of an appreciable energy transfer from the kinetic energy of colliding $H(n)+He$ atoms to the Rydberg electron, i.e., at $\omega \leq \omega_e$ (where $\hbar\omega_e = 0.363$ eV is the lowest vibrational quantum of a HeH^+ ion).

The behavior of cross sections for deexcitation n-changing and l-mixing processes

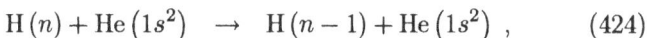

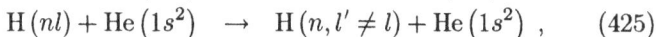

$$H(n) + He(1s^2) \quad \rightarrow \quad H(n-1) + He(1s^2) , \qquad (424)$$
$$H(nl) + He(1s^2) \quad \rightarrow \quad H(n,l' \neq l) + He(1s^2) , \qquad (425)$$

induced by scattering of perturbing He atom on the ion core (proton H^+) of a hydrogen atom, is shown in Fig. 56. This figure illustrates the n-dependencies of cross sections for several values of the kinetic energy of colliding atoms.

We now discuss the n-dependence of cross section at thermal energies $\mathcal{E} = 0.026$ eV. One can see, for the inelastic deexcitation of $H(n)$ atoms with a change in the principal quantum number $n \rightarrow n-1$ the shake-up region with $\omega_{n-1,n} < V/\rho_{\text{cap}}$ (where $\sigma_{n-1,n} \propto n^4$) is realized only for $n > 15$. In the opposite case, with $n < 15$ the behavior of the inelastic cross section $\sigma_{n-1,n}(He - H^+)$ with large enough transition frequencies $\omega_{n-1,n} > V_T/\rho_{\text{cap}}$ is described within the framework of the electron–dipole mechanism. Formula (419) leads to practically n independent magnitudes of the deexcitation cross section $\sigma_{n-1,n}$ for $6 < n < 15$ and to its sharp exponential decrease $\sigma_{n-1,n} \propto \exp(-2\omega\tau)$ only for low principal quantum numbers $n < 6$.

As a result, for the inelastic $n \to n-1$ transitions with a change in the principal quantum number the core–projectile scattering mechanism plays the main role at intermediate and sufficiently low $n < 10$. As follows from calculations by formula (378), the scattering of the perturbing atom He on the ionic core H^+ also becomes predominant in the range of sufficiently high $n > 30 - 35$. The corresponding magnitude of the quenching cross section turns out to be of the order of $\sigma_{tr}^{H^+,He}$ at $n \sim (v_0/V)^{1/2} \sim 40$ ($\sigma_{tr}^{H^+,He} = 6 \cdot 10^{-15}$ cm^2 is the momentum transfer cross section for $\mathcal{E} = 0.026$ eV).

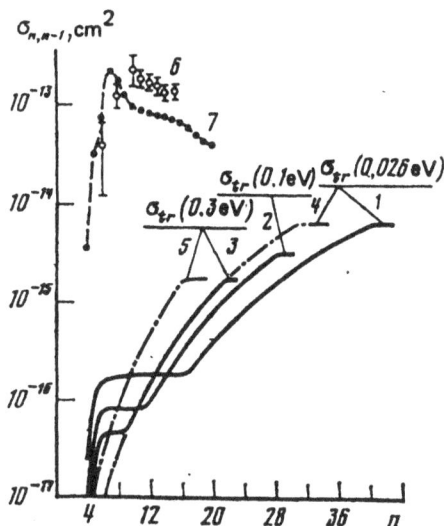

Figure 56. The n-dependencies of inelastic deexcitation cross sections $\sigma_{n-1,n}(\mathcal{E})$ of collision process (424) induced by the core–projectile scattering mechanism [231] for the energies $\mathcal{E} = 0.026$, 0.1, and 0.3 eV (full curves 1, 2, and 3, respectively). Dashed curves 4 and 5 are the l-mixing (425) cross sections for $\mathcal{E} = 0.026$ and 0.3 eV, respectively. Open circles (curve 6) and and full circles (curve 7) represent experimental data [219] and close–coupling calculations [222] for the electron–projectile scattering mechanism of the l-mixing Na(nD)+He collisions.

Of basic interest is the comparison of two competitive mechanisms of inelastic deexcitation process (424) associated with core–projectile and electron–projectile scattering. The results of the calculations [231] of the rate constants for deexcitation $n \rightarrow n - 1$ induced by the scattering of He atom on the ionic core H^+ of the highly excited hydrogen atom are presented in Fig. 57 by full curves for the gas temperatures $T = 250$, 1000, and 4000 K. The dashed-dotted curves indicate the contribution of the traditional perturber–quasifree electron scattering mechanism, obtained by *Bates* and *Khare* [148] for the same values of the gas temperature. The dashed curve represent the analogous quasifree-electron model results of *Flannery* [149] for $T = 300$ K. This comparison points to a predominance of the core–projectile scattering effects for the inelastic transitions with a change in the principal quantum number in the region of $n < 10 - 15$ and $n > 20 - 30$ depending on the specific value of the kinetic energy of colliding atoms.

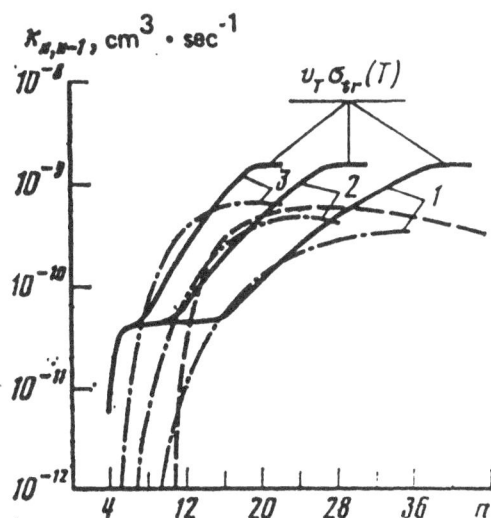

Figure 57. Rate constants $K_{n-1,n}$ of deexcitation process (424). Full curves 1, 2, and 3, calculations [231] of the core–projectile scattering effects at $T = 250, 1000$, and 4000 K, respectively. Dashed–dotted curves, corresponding rate constants obtained [148] in the free electron model. Dashed curve, free electron model calculations [149] at $T = 300$ K.

Figure 58 illustrates the comparison of total rates $\mathcal{K}_n = \sum\limits_{n'<n}$
$K_{n'n}N_{He}$ (in s^{-1}) for collision quenching of H(n) levels in the inelastic transitions $n \to n'$ with their radiative decay $A_n = \sum\limits_{n'<n} A_{n'n}$
averaged over the lm-sublevels. It can be seen that for $n \geq 6$ the collision quenching rates $\mathcal{K}_n$ substantially exceeds the corresponding rates of radiative decay of excited n-levels even at the helium concentrations $N_{He} \geq 10^{17}$ cm^{-3}. This fact makes it possible to measure the rate constants of inelastic n-changing collision transitions in a dense weakly ionized plasma of H_2 and He. These rate constants can be determined, for example, by measuring the dependencies of the emission intensity of highly excited levels of H(n) atoms, produced in such a plasma, on the pressure of the buffer gas (helium).

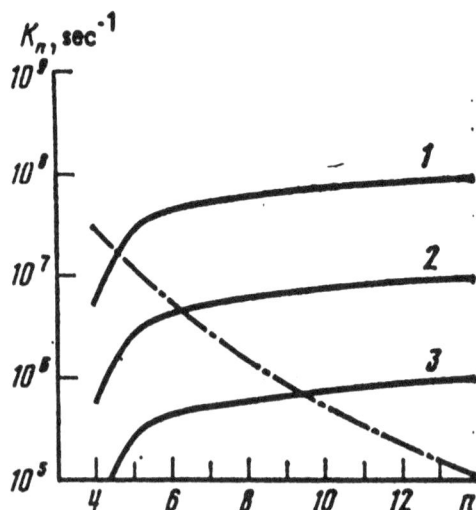

Figure 58. Total rate $\mathcal{K}_n = \sum\limits_{n'<n} K_{n'n}N_{He}$ of collision quenching ($T = $
300 K) of highly excited H(n)-levels by He$(1s^2)$ calculated in Ref. [231]. Full curves 1, 2, and 3 correspond to the helium concentrations $N_{He} = 10^{18}$, 10^{17}, and 10^{16} cm^{-3}, respectively. Dashed curve indicates total rate $A_n = \sum\limits_{n'<n} A_{n'n}$ of radiative decay of the n-level [149].

We now present the most important results of Refs. [262, 263] for the direct and associative ionization of highly excited Hydrogen atoms in thermal collisions with the ground–state helium atoms

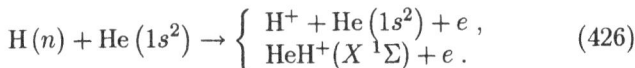

$$H(n) + He\left(1s^2\right) \rightarrow \begin{cases} H^+ + He\left(1s^2\right) + e \,, \\ HeH^+(X \ ^1\Sigma) + e \,. \end{cases} \tag{426}$$

The total ionization rate constant $K_n(T)$ including the integral contribution of both channels of ionization is presented in Fig. 59. One can see that the values of $K_n(T)$ in the core–projectile scattering mechanism (full curve) are practically independent of the gas temperature T and have the maximum $\sim 10^{-12}$ $cm^3 \cdot s^{-1}$ at $n \sim 10$. Thus, for the ionization reactions involving the Rydberg H(n) and the ground-state He atoms, the core–projectile scattering mechanism is more effective than the traditional Fermi mechanism (dashed curves) in the range of $n < 20 - 30$ at $T = 300$ K.

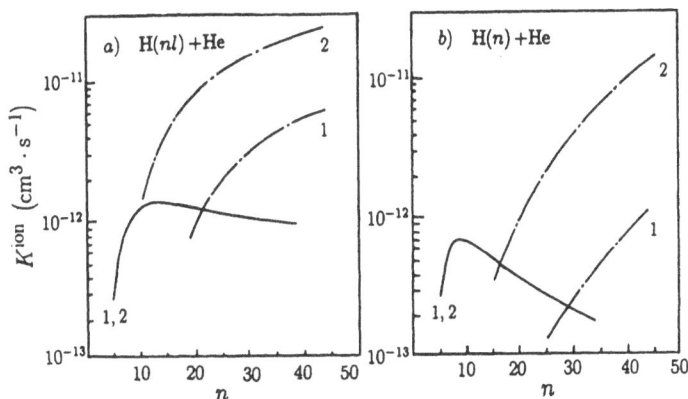

Figure 59. The results of Refs. [231, 263] for total rate constants $K^{ion} = K^{d.i.} + K^{a.i.}$ of direct and associative ionization in thermal collisions of of H(n) +He at $T = 300$ K (curves 1) and $T = 1000$ K (curves 2). Full curves 1 and 2 (coinciding for $T = 300$ K and $T = 1000$ K) are the contributions of the core–projectile scattering. Dashed curves 1 and 2 are the contributions of the quasifree-electron–projectile scattering. Panels a and b are calculations for selectively excited nl-levels ($l \ll n$) and for n-levels (with equally populated sublevels), respectively.

7.5 Potential Curves Crossing Mechanism

The processes of resonant excitation and ionization of a Rydberg atom induced by the scattering of a perturbing atom by the ionic core take place in the vicinity of the crossing point R_ω of two different electronic terms of the quasimolecular $BA^+ + e$ system. Hence, the nuclear matrix elements for such transitions in the general formulae of Sect. 7.3 may be calculated within the framework of the semiclassical theory of Landau–Zener [37]

$$\left|\left\langle \chi_{\mathcal{E}'J}^{(f)}(R) \left| V_{i,nlm}^{f,n'l'm'}(R) \right| \chi_{\mathcal{E}J}^{(i)}(R) \right\rangle\right|^2$$

$$= \frac{\pi\, C_{\mathcal{E}}^2 C_{\mathcal{E}'}^2 \left| V_{i,nlm}^{f,n'l'm'}(R_\omega) \right|^2 \cos^2 S_0(J)}{2V\,\Delta F_{fi}(R_\omega)\left[1 - U_i(R_\omega)/\mathcal{E} - (J+1/2)^2/q^2 R_\omega^2\right]^{1/2}}\,, \tag{427}$$

Here $S_0\,(J)$ is the phase difference of the quasiclassical wave functions

$$S_0\,(J) = \int_{b_i}^{R_\omega} q_i\,(R)\,dR - \int_{b_f}^{R_\omega} q_f\,(R)\,dR + \pi/4\,, \tag{428}$$

and

$$\Delta F_{fi}\,(R_\omega) = |dU_f/dR - dU_i/dR|_{R=R_\omega}$$

is the difference of the derivatives of the potential energy curves of the quasimolecule $BA^+ + e$ at the crossing point R_ω defined by the relation

$$U_{i,nl}\,(R_\omega) = U_{f,n'l'}\,(R_\omega)$$

for the $nl \to n'l'$ transitions (or $U_{i,nl}\,(R_\omega) = U_{f,E}\,(R_\omega)$ for the ionization processes).

Thus, at $R = R_\omega$, the energy $\Delta E_{fi} = \hbar\omega$ transferred to the outer electron in the bound–bound ($\Delta E_{fi} = E_{n'l'} - E_{nl}$) or bound–free ($\Delta E_{fi} = E + |E_{nl}|$) transition becomes equal to the energy splitting of the electronic terms of the quasimolecular BA^+ ion, i.e.

$$\Delta E_{fi} \equiv \hbar\omega = U_i\,(R_\omega) - U_f\,(R_\omega) \equiv \Delta U_{if}\,(R_\omega)\,, \tag{429}$$

where $\hbar\omega = \mathcal{E} - \mathcal{E}'$ for the excitation (deexcitation) and direct ionization, and $\hbar\omega = \mathcal{E} + |\mathcal{E}_{vJ}|$ for the associative ionization.

Figure 60 illustrates the schematic diagram of the potential energy curves of a heteronuclear quasimolecule $YX^+ + e$ and quasimolecular ion YX^+ temporarily formed during the scattering of a projectile atom Y by the ionic core X^+ of highly excited atom X^*. This figure clarifies the main idea of a resonant excitation (deexcitation) mechanism of Rydberg states $n \rightarrow n'$.

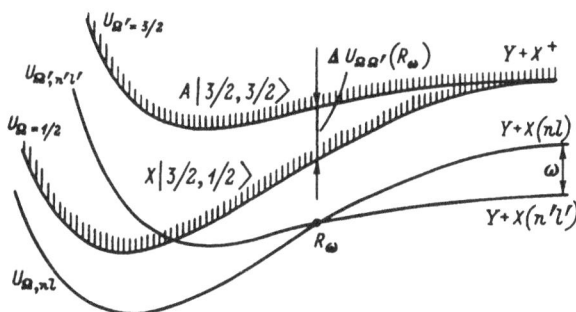

Figure 60. Schematic diagram of the potential energy curves $U_{\Omega=1/2}(R)$ and $U_{\Omega'=3/2}(R)$ for the lowest electronic states $X|j_f = 3/2, \Omega_f = 1/2\rangle$ and $A_1|j_i = 3/2, \Omega_i = 3/2\rangle$ of a heteronuclear rare gas YX^+ ion. $U_{\Omega,nl}(R) = U_\Omega(R) + E_{nl}$ and $U_{\Omega',n'l'}(R) + E_{n'l'}$ are the potrential energy curves of the composite $YX^+ + e$ system correlating with the initial $X[n_0 p^5 (^2P_{3/2}) nl] +$ Y and final $X[n_0 p^5 (^2P_{3/2}) n'l'] + Y$ states of separated particles. j and Ω are the total angular momentum of the molecular ion and its z-projection on the internuclear axis.

The rapidly oscillating $\cos^2 S_0(J)$ function in (427) can be replaced by its average value equal to $1/2$. Further, one should note that integration over dJ in the general semiclassical formula (398) of first-order perturbation theory should actually be performed within the limits of $0 \leq J \leq J_{\max}(R_\omega)$, where

$$J_{\max}(R_\omega) = qR_\omega \left[1 - U_i(R_\omega)/\mathcal{E}\right]^{1/2}. \tag{430}$$

This leads to the following expression for the cross section of the inelastic $nl \rightarrow n'l'$ transition

$$\sigma_{n'l',nl}^{f,i}(\mathcal{E}) = \frac{8\pi^2 g_f R_\omega^2 [1 - U_i(R_\omega)/\mathcal{E}]^{1/2}}{(2l+1) g_i \hbar V \Delta F_{fi}(R_\omega)}$$

$$\times \sum_{mm'} \left| V_{i,nlm}^{f,n'l'm'}(R_\omega) \right|^2 , \tag{431}$$

where $V = \hbar q/\mu = (2\mathcal{E}/\mu)^{1/2}$ is the relative velocity of the colliding $A(nl)$ and B atoms at $R \rightarrow \infty$. Due to the presence of the term

$$V(R_\omega)/V = [1 - U_i(R_\omega)/\mathcal{E}]^{1/2}$$

this formula takes into account the change in the relative velocity of a perturbing atom B and ionic core A^+ in the vicinity of the transition point R_ω in comparison with its value $V \equiv V_\infty = (2\mathcal{E}/\mu)^{1/2}$ at $R \rightarrow \infty$.

A similar semiclassical expression for the total cross section $\sigma_{nl}^{\text{ion}} = \sigma_{nl}^{\text{d.i.}} + \sigma_{nl}^{\text{a.i.}}$ of direct and associative ionization can be written as [262]

$$\sigma_{nl}^{\text{ion}}(\mathcal{E}) = \frac{8\pi^2 g_f}{(2l+1)g_i \hbar V} \int_{R_{\min}}^{R_{\max}} \sum_{l'} \sum_{mm'} \left| V_{i,nlm}^{f,El'm'}(R_\omega) \right|^2$$

$$\times \left[1 - \frac{U_i(R_\omega)}{\mathcal{E}} \right]^{1/2} R_\omega^2 \, dR_\omega . \tag{432}$$

Thus, the final result for the total ionization cross section can be expressed as an integral over the transition frequencies

$$\hbar \, d\omega = d[\Delta U_{fi}(R_\omega)] = \Delta F_{fi}(R_\omega) \, dR_\omega$$

or over the internuclear distances. The lower $R_{\min}$ and upper $R_{\max}$ limits of integration in (432) correspond to the maximum $\omega_{\max}$ and minimum $\omega_{\min}$ possible values of the energy splitting terms of the BA^+ ion. Averaging (432) over the Maxwellian distribution of the relative velocities of colliding atoms, yields

$$K_{nl}^{\text{ion}}(T) = \frac{8\pi^2 g_f}{(2l+1)g_i \hbar} \int_0^{R_{nl}} \sum_{l'} \sum_{mm'} \left| V_{i,nlm}^{f,El'm'}(R_\omega) \right|^2$$

$$\times \exp\left(-\frac{U_i(R)}{kT}\right) \Phi_T^{(c)}(R) R^2 \, dR . \tag{433}$$

Here R_{nl} is defined by the condition

$$\Delta U_{fi}(R_{nl}) = |E_{nl}| = Ry/n_*^2 \,, \qquad (434)$$

and the function $\Phi_T^{(c)}(R)$ assumes a different form in the repulsive $(R < R_0^{(i)}$ and $U_i(R) > 0)$ and attractive $(R \geq R_0^{(i)}$ and $U_i(R) < 0)$ regions of the initial electronic term of quasimolecular BA^+ ion, i.e.

$$\Phi_T^{(c)}(R) = \begin{cases} 1 \,, & R < R_0^{(i)}, \\ \frac{\Gamma(3/2,\,|U_i(R)|/kT)}{\Gamma(3/2)} \,, & R \geq R_0^{(i)}, \end{cases} \qquad (435)$$

where $\Gamma(3/2, z)$ is the incomplete gamma function, $\Gamma(3/2) = \pi^{1/2}/2$, and $U_i\left(R_0^{(i)}\right) = 0$ (see Sect. 7.9 for more details).

The semiclassical formulae (431)–(433) of first-order perturbation theory are valid to an arbitrary form and symmetry of the initial U_i and final U_f terms of a heteronuclear BA^+ or homonuclear A_2^+ quasimolecular ion converging to one dissociation limit (i.e. $U_i(\infty) = U_f(\infty)$). However, in some cases perturbation theory becomes inapplicable. Thus, it is necessary to have simple estimates for the maximum possible values of the excitation and ionization rate constants. They can be obtained assuming the probability of the bound–bound or bound–free transition to be equal to $W_{n'l',nl} = 1$ (or $W_{nl}^{ion} = 1$) for all impact parameters $0 \leq \rho \leq \rho_{max}(\mathcal{E})$. Hence, for the resonant mechanism associated with the crossing of various electronic terms, the maximum value of the cross section can be written as

$$\sigma_{max} = \pi\rho_{max}^2, \qquad (436)$$

$$\rho_{max} = R_\omega \left[1 - U_i(R_\omega)/\mathcal{E}\right]^{1/2} \,. \qquad (437)$$

For the bound–bound transitions of a Rydberg electron the R_ω point is to be found from the relation

$$\Delta U_{if}(R_\omega) = |\Delta E_{n'l',nl}| = \frac{2Ry\left|n_*' - n_*\right|}{n_*^3} \,.$$

Similarly, for the bound–free transitions of the Rydberg electron the R_ω point can be obtained from Eq. (434) so that we have $R_\omega = R_{nl}$.

A simple estimate [262] for the maximum value of the ionization rate constant $K_{max}^{ion}(T) = \langle V\sigma_{max}\rangle$ proceeds directly from (436) on integrating over the Maxwellian distribution of velocities $V = (2\mathcal{E}/\mu)^{1/2}$. Whereas the lower limit of the kinetic energy range $\mathcal{E}_{\min} \leq \mathcal{E} < \infty$ is given by $\mathcal{E}_{\min} = \max\{0, U_i\,(R_{nl})\}$. The final result can be written as

$$K_{\max}^{\text{ion}}\,(T) = \pi R_{nl}^2 \,\langle V\rangle_T \, P_{nl}\,(T) \ , \qquad (438)$$

where the $P_{nl}\,(T)$ quantity is defined by the relation

$$P_{nl}\,(T) = \begin{cases} \exp\left[U_i\,(R_{nl})\,/kT\right] \ , & R_{nl} < R_0^{(\imath)}, \\ 1 + |U_i\,(R_{nl})|\,/kT \ , & R_{nl} \geq R_0^{(\imath)}. \end{cases} \qquad (439)$$

7.6 Associative and Direct Ionization: Symmetric Atomic Collisions

The formulae presented above can be applied to the analysis of the excitation and ionization processes in thermal collisions of Rydberg atoms $A^*\,(nl)$ with the ground-state parent atoms A. In this case they are due to the dipole transition between the symmetrical $U_g(R)$ and antisymmetrical $U_u(R)$ terms of the quasimolecular A_2^+ ion (see Fig. 7.1). Due to the large value of the exchange splitting $\Delta\,(R) \equiv |\Delta U_{gu}\,(R)|$ of these terms the inelastic transitions $n \to n'$ and ionization $n \to E$ of the Rydberg electron occur in the asymptotic range of internuclear distances $R_\omega \gg a_0$, in which the $U_g(R)$ and $U_u(R)$ magnitudes are given by the simple relations [181]

$$U_g\,(R) = -\frac{\Delta\,(R)}{2} - \frac{\alpha_A e^2}{2R^4}, \qquad (440)$$

$$U_u\,(R) = \frac{\Delta\,(R)}{2} - \frac{\alpha_A e^2}{2R^4} \ , \qquad (441)$$

where α_A is the polarizability;

$$\Delta\,(R) = C R^\nu \exp\,(-\gamma R) \ ,$$

and C, ν, γ are the constant coefficients determined by parameters of the ground-state atom A.

The electronic matrix element for the dipole part

$$\mathsf{V} = e\mathbf{r} \cdot \mathbf{D}\,(\mathbf{r}_\kappa)\,/r^3$$

of the Coulomb interaction (396) between the outer electron ($\mathbf{r}$) and the inner electrons ($\mathbf{r}_\kappa$) of the $A_2^+ + e$ system may be written as

$$V_{i,nlm}^{f,n'l'm'}(R) =$$

$$\left\langle \psi_{n'l'm'}(\mathbf{r}) \left| \left\langle \varphi_{A_2^+}^{(f)}(\mathbf{r}_\kappa, \mathbf{R}) \left| \frac{er\,\mathbf{D}(\mathbf{r}_\kappa)}{r^3} \right| \varphi_{A_2^+}^{(i)}(\mathbf{r}_\kappa, \mathbf{R}) \right\rangle \right| \psi_{nlm}(\mathbf{r}) \right\rangle \quad (442)$$

$$= -a_0^{-3} \left(\frac{\hbar\omega_{fi}}{2Ry} \right)^2 \langle n'l'm'| \, e\mathbf{r} \cdot \mathbf{D}_{fi}(R) \, |nlm\rangle .$$

Here

$$\mathbf{D}_{fi}(R) \equiv \left\langle \varphi_{A_2^+}^{(f)}(\mathbf{r}_\kappa, \mathbf{R}) \left| \mathbf{D} \right| \varphi_{A_2^+}^{(i)}(\mathbf{r}_\kappa, \mathbf{R}) \right\rangle$$

is the matrix element of the dipole moment $\mathbf{D} = -e \sum_\kappa \mathbf{r}_\kappa$ of the inner electrons over the initial and final electronic wave functions of the A_2^+ ion (the radius vectors of $\mathbf{r}$ and $\mathbf{r}_\kappa$ are measured from its center of mass). Using (431) and (442) the cross section $\sigma_{n'n}$ of an inelastic n-changing process may be described by the expression

$$\sigma_{n'n}(V) = \frac{8\pi g_f \mathcal{G}(\Delta n) R_\omega^2}{3^{3/2} g_i n^5 (n')^3 (V/v_0)} \left[1 - \frac{U_i(R_\omega)}{\mathcal{E}} \right]^{1/2}$$

$$\times \left(\frac{D_{fi}(R_\omega)}{ea_0} \right)^2 \left(\frac{2Ry}{a_0 \Delta F_{fi}(R_\omega)} \right) . \quad (443)$$

For inelastic transitions between the neighboring Rydberg levels n and $n \pm \Delta n$ ($\Delta n \ll n$) the values of dipole matrix elements are usually large compared to the transition energy $|\Delta E_{n'n}| = 2Ry\,|\Delta n|/n^3$. Therefore, perturbation theory becomes, as a rule, inapplicable for the description of resonant excitation (deexcitation) of a Rydberg atom at $\Delta n \ll n$. Hence the simple expression (443) may be used only for qualitative estimate of the cross section at high enough n.

For the total rate constant $K_{nl}^{\text{ion}}(T)$ of direct and associative ionization of a Rydberg atom $A^*(nl)$ by the ground-state parent A atom, the use of first-order perturbation theory and the dipole approximation for the electronic matrix element $V_{i,nlm}^{f,El'm'}(R_\omega)$ of the bound–free transition yields

$$K_{nl}^{\text{ion}}(T) = \frac{2g_f}{g_i \alpha v_0^2} \int_0^{R_{nl}} \left(\frac{D_{fi}(R_\omega)}{ea_0} \right)^2 \omega^3 \sigma_{nl}^{\text{ph}}(\omega)$$

$$\times \exp\left(-\frac{U_i(R)}{kT} \right) \Phi_T^{(c)}(R_\omega) R_\omega^2 \, dR_\omega , \quad (444)$$

where $\hbar\omega = |\Delta U_{fi}(R_\omega)|$, and the R_{nl} value is determined by the relation

$$|\Delta U_{fi}(R_{nl})| = |E_{nl}| = Ry/n_*^2$$

and $\alpha = e^2/\hbar c \approx 1/137$. It should be emphasized that the dipole matrix element square of the bound–free electron transition in this formula is expressed in terms of the photoionization cross section $\sigma_{nl}^{\rm ph}(\omega)$. The formula (444) involves contributions of both the repulsive $(U_i(R) > 0$ at $R < R_0^{(i)})$ and attractive $(U_i(R) \leq 0$ at $R \geq R_0^{(i)})$ regions of the upper $U_i(R)$ term (see (435)). For the case of a pure repulsive upper term $U_1(R)$ of quasimolecular A_2^+ ion (when $\mathcal{H}(R) = 1$ for all internuclear distances $0 < R < \infty$) (444) is equivalent to the expressions obtained by *Janev* and *Mihajlov* [259] and *Duman* and *Shmatov* [260]. It should be noted that for the matrix element of a dipole moment one can use the asymptotic approximation

$$D_{fi}(R_\omega) = -eR_\omega/2 \,,$$

since the major contribution to the excitation (deexcitation) and ionization cross sections is determined, as a rule, by large enough internuclear distances $R_\omega \gg a_0$.

The formulae presented above are used in several papers for the calculations of ionization [259, 260] and quenching [256] of Rydberg states of alkali-metal atoms in thermal collisions with the ground-state parent atoms Li, Na, K, Rb, and Cs. As follows from the comparison with the available experimental data (see [266] and references therein) they are, on the whole, in reasonable agreement with the measured rate constants of ionization. It should be noted that at thermal energies the major contribution to the total ionization rate constant $K_{nl}^{\rm ion} = K_{nl}^{\rm d.i.} + K_{nl}^{\rm a.i}$ is determined by the process of associative ionization $A^*(nl) + A \rightarrow A_2^+ + e$, while the role of direct ionization $A^*(nl) + A \rightarrow A^+ + A + e$ becomes important at high energies of the colliding particles. As follows from the detailed analysis of available theoretical and experimental data, perturbation theory provides reasonable results in the range of high enough principal quantum numbers $n \gg n_{\max}$ (where $n_{\max} \sim 8 - 15$ at thermal energies $T \sim 300 - 1000$ K), but it does not hold near the maximum $n \sim n_{\max}$ and at low principal quantum numbers $n \ll n_{\max}$. In this range $n \lesssim n_{\max}$ some more realistic magnitudes may be obtained by the simple estimate (438). Note also that the values of the ionization rate constants, measured for the case of non-symmetrical collisions between the Rydberg alkali atoms A^* (nl) and the ground-state alkali B atoms of the buffer gas, turn out to be of the same order of magnitude as those for symmetrical A^* (nl) +A collisions.

We illustrate the main features of the resonant ionization processes induced by the dipole interaction by presenting the results of calculations [262, 263] for collisions of the Rydberg heavy rare gas atoms $X[n_0 p^5 (^2P_{3/2}) nl]$ with the ground-state parent $X(^1S_0)$ atoms. At thermal energies $(kT \sim 0.03 - 0.06$ eV$)$ the value of kT is much less than the spin–orbit splitting $\Delta_{3/2,1/2}$ for the $^2P_{3/2}$ and $^2P_{1/2}$ states of the ground–electronic p^5-shell of the X^+ $[n_0 p^5]$ ion. Thus there are two channels of the ionization processes $($see Fig. 61$)$.

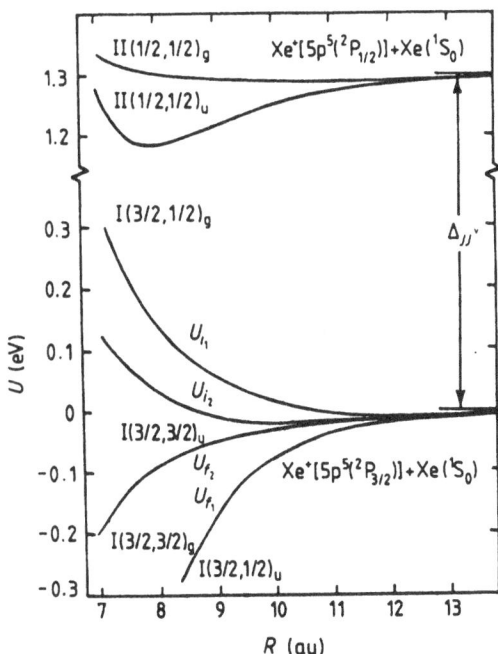

Figure 61. Potential energy curves (U in eV) for the lower electronic terms of homonuclear Xe_2^+ ion at large internuclear distances (R in atomic units).

They correspond to the dipole transitions $|i_1\rangle \to |f_1\rangle$ and $|i_2\rangle \to |f_2\rangle$ between the symmetrical and antisymmetrical four lower terms of the homonuclear X_2^+ ion ($|i_1\rangle = \left|I\left(3/2, 1/2\right)_g\right\rangle$, $|f_1\rangle = \left|I\left(3/2, 1/2\right)_u\right\rangle$, $|i_2\rangle$ $\left|I\left(3/2, 3/2\right)_u\right\rangle$, $|f_2\rangle = \left|I\left(3/2, 3/2\right)_g\right\rangle$.

Due to large $\Delta U_{i_1 f_1}(R)$ and $\Delta U_{i_2 f_2}(R)$ of exchange splitting of the latter terms, the ionization processes (402) and (405) would have to occur in the asymptotic region of internuclear distances R (where Hund's type C-binding takes place). For such transitions, the dipole matrix elements over the wave functions of the $j\Omega$-representation can be described in the approximation of "$R/2$" (see [270]), i.e.

$$\mathbf{D}_{i_1 f_1} = \left\langle I\left(3/2, 1/2\right)_g \left| e \sum_{\kappa=1}^{N} \mathbf{r}_\kappa \right| I\left(3/2, 1/2\right)_u \right\rangle = -\frac{e\mathbf{R}}{2}, \quad (445)$$

$$\mathbf{D}_{i_2 f_2} = \left\langle I\left(3/2, 3/2\right)_u \left| e \sum_{\kappa=1}^{N} \mathbf{r}_\kappa \right| I\left(3/2, 1/2\right)_g \right\rangle = -\frac{e\mathbf{R}}{2}. \quad (446)$$

Thus, the total ionization rate constant

$$K_{nl}(T) = K_{nl}^{i_1 f_1}(T) + K_{nl}^{i_2 f_2}(T) \quad (447)$$

should be calculated by Eq. (444) taking into account the latter two types of dipole transitions. The statistical weights for the lower terms $U_{i_p}(R)$ and $U_{f_p}(R)$ of the rare gas X_2^+ ion (correlated with the $X^+\left[n_0 p^5\left(^2P_{3/2}\right)\right]+X(^1S_0)$ states of separated atom and atomic ion) are equal to $g_{i_p} = g(j\Omega_{i_p}) = 2$; $g_{f_p} = g(j\Omega_{f_p}) = 2$ and $g_{tot} = 2(2j+1) = 8$ (where $p = 1, 2$).

The ionization rate constants $K_{nl}(T)$ (at $l \ll n$) for thermal collisions of $Xe\left[5p^5\left(^2P_{3/2}\right)nl\right]+Xe(^1S_0)$ atoms, calculated [262, 263] within the framework of perturbation theory according to formulae (444)–(447), are shown in Fig. 62 (panel a) by full curves. For comparison, we also present here the corresponding results for the $K_{nl}^{max}(T)$ values (dashed curves) obtained from Eqs. (438) and (439). It can be seen that the use of perturbation theory leads to overestimated $K_{nl}(T)$ values at $n \leq 20$. Thus, the $K_{nl}(T)$ values of dipole ionization rate constants for selectively excited nl levels ($l \ll n$) should be equal to $10^{-10} - 10^{-9}$ cm$^3 \cdot$s^{-1} (see Fig. 62 a). Similar results [262, 263] of numerical calculations for the ionization rate constants $K_n(T)$ for the case of equally populated nlm-sublevels within a given n level are presented in Fig. 62 (panel b). It should be noted that because of the inapplicability of perturbation theory in the range of $n \lesssim 20$, these results also overestimate the values of the ionization $K_n\left(T\right)$ rate constants.

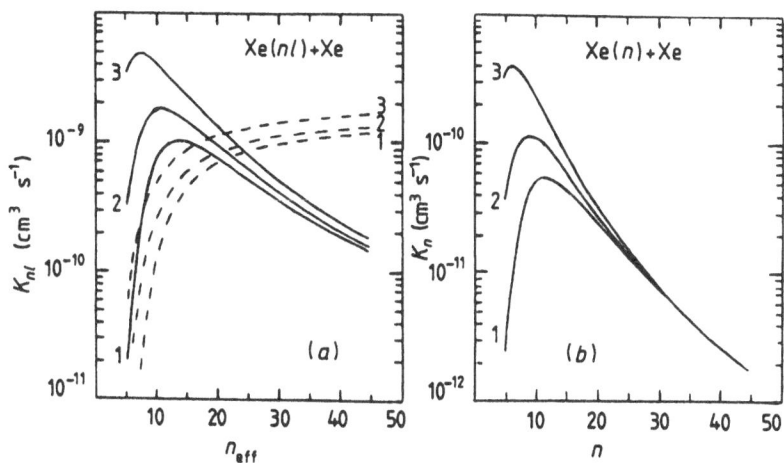

Figure 62. Total rate constants $K_n^{\text{ion}} = K_{nl}^{\text{d.i.}} + K_{nl}^{\text{a.i.}}$ of direct and associative ionization of the Rydberg $5p^5\left({}^2P_{3/2}\right)nl$ (panel a) and $5p^5\left({}^2P_{3/2}\right)n$ (panel b) levels of xenon atoms in thermal collisions with ground–state parent $\text{Xe}\left({}^1S_0\right)$ atoms [262, 263]. Curves 1, 2, and 3 correspond to gas temperatures $T = 300$, 500, and 1000 K, respectively. Full curves are results obtained using the first–order of perturbation theory. Dashed curves are the corresponding results for the maximum values $K_{nl}^{\max}\left(T\right)$ of the ionization rate constants.

7.7 Resonant Deexcitation of Rydberg States in a Buffer Rare Gas

Another example of resonant processes of inelastic $n \rightarrow n'$ transitions [267] and ionization [262] is thermal collisions of the Rydberg rare gas $X^*\left(nl\right)$ atoms with the ground-state $Y\left({}^1S_0\right)$ atoms of the buffer rare gas. In this case the excitation and ionization of the highly excited electron induced by scattering of atom Y on the atomic core $X^+\left[n_0p^5\left({}^2P_{3/2}\right)\right]$ are due to the transition

$$A_1 \left|j = 3/2, \Omega = 3/2\right\rangle \to X \left|j = 3/2, \Omega = 1/2\right\rangle$$

between the first excited $U_i(R)$ and lower $U_f(R)$ electronic terms of the heteronuclear quasimolecular YX^+ ion (while the deexcitation process is accompanied by an inverse transition). These terms, split by electrostatic interaction, correlate with the ground-states of the separated $Y(^1S_0)$ atom and atomic core $X^+ \left[n_0 p^5 \left(^2P_{3/2}\right)\right]$. They correspond to the different $\Omega_i = 3/2$ and $\Omega_f = 1/2$ projections of the total angular momentum $\mathbf{j} = \mathbf{L} + \mathbf{S}$ $(j_i = j_f = 3/2)$ of the quasimolecular YX^+ ion on the internuclear $\mathbf{R}$ axis (see Fig. 61). The second excited $A_2 \left|j' = 1/2, \Omega' = 1/2\right\rangle$ term does not take part in the excitation and ionization processes involved (for the case of heavy rare gas Rydberg Ar^*, Kr^*, and Xe^* atoms under consideration). This is due to a large value $\Delta_{j'j}$ of the spin–orbit splitting for the $^2P_{3/2}$ and $^2P_{1/2}$ states of the ground-electronic p^5-shell of the X^+ ion in comparison with thermal energy $kT \sim 0.03 - 0.1$ eV of colliding atoms and energy splitting $\Delta U_{if}(R)$ of the lower electronic terms. In the zero-approximation the final result of the calculation for the sum of matrix elements determining the values of the quenching and ionization rate constants can be written as

$$\sum_{mm'} \left|V_{i,nlm}^{f,n'l'm'}\right|^2 = (2Ry)^2 \frac{2\gamma_{l'l}}{25 n_*^3 \left(n_*'\right)^3}, \tag{448}$$

$$V_{i,nlm}^{f,El'm'} = \left(\frac{\left(n_*'\right)^3}{2Ry}\right)^{1/2} V_{i,nlm}^{f,n'l'm'}. \tag{449}$$

Here the n-independent value $\gamma_{l'l}$ characterizes the coupling constant of different Rydberg states with different magnitudes l and l' of the orbital momentum. It is determined by the short range $(r < r_\kappa)$ and the long-range $(r > r_\kappa)$ parts of the Coulomb interaction (396) between the Rydberg electron $(\mathbf{r})$ and the electrons $(\mathbf{r}_\kappa)$ of the ionic core X^+ from its ground $n_0 p^5$-shell (interaction with the electrons of the filled $^1S_0$ shell of the rare gas Y atom can be disregarded). As has been shown in [267], the magnitude of $\gamma_{l'l}$ can be expressed in terms of the angular coefficient and the radial matrix elements of the quadrupole type $r_<^2/r_>^3$ over the Rydberg electron wave function $\mathcal{R}_{nl}(r)$ and the Hartree–Fock wave functions $\mathcal{R}_{n_0p}(r_\kappa)$ of the rare gas ion $X^+ \left(n_0 p^5\right)$ with the principal quantum number n_0 and orbital momentum $L = 1$. Here $r_< = \min\{r, r_\kappa\}$ and $r_> = \max\{r, r_\kappa\}$, while $n_0 = 3$, 4 and 5 for Ar^+, Kr^+ and Xe^+, respectively. The dipole part of the interaction does not make any contribution to the transition matrix elements due to the selection rules $\Delta l = 0, \pm 2$, $\Delta m = -1$ for transitions with a change $\Delta\Omega = 1$ of the total angular momentum projections on the internuclear axis.

With the use of general expression of first-order perturbation theory (431), the cross section of the inelastic $nl \to n'l$ transition in the Rydberg rare gas atom $\text{X}\left[n_0 p^5 \left(^2 P_{3/2}\right) nl\right]$ induced by the scattering of the rare gas atom $\text{Y}(^1 S_0)$ on its ionic core $\text{X}^+ \left[n_0 p^5 \left(^2 P_{3/2}\right)\right]$ can be described by the simple analytic formula [267]

$$\sigma_{nl',nl} = \frac{8\pi^2 a_0^2 \gamma_{l'l} \left(A/2Ry\right)^{3/\nu} \left[1 - U_i\left(R_\omega\right)/\mathcal{E}\right]^{1/2}}{25\left(2l + 1\right)\nu\left(V/v_0\right) n_*^{3(1-3/\nu)} \left|n_*' - n_*\right|^{1+3/\nu}} \, . \qquad (450)$$

This formula pertains to the case of a power-law approximation of the energy splitting

$$\left|\Delta U_{fi}\left(R\right)\right| = A\left(a_0/R\right)^\nu \, , \qquad (451)$$

when the crossing point R_ω of the Rydberg $U_{i,nl}\left(R_\omega\right) = U_i\left(R_\omega\right) + E_{nl}$ and $U_{f,n'l'}\left(R_\omega\right) = U_f\left(R_\omega\right) + E_{n'l'}$ terms of the quasimolecule $\left(\text{YX}^+ + e\right)$ is given by the simple relation

$$R_\omega = a_0 \left(\frac{A}{\hbar\omega}\right)^{1/\nu} \, , \quad \hbar\omega = \left|\Delta E_{n'l',nl}\right| = \left|\Delta U_{fi}\left(R_\omega\right)\right| . \qquad (452)$$

A detailed analysis shows, that the most effective transitions in this mechanism are $np \to n'p$ with $l' = l = 1$. The cross sections $\sigma_{n'l',nl}$ of the inelastic $nl \to n'l'$ transitions with $l' \neq 1$ and $l \neq 1$ decreases rapidly with an increase of the orbital quantum number l and l'. The $ns \to n's$ transitions accompanied by the change of the angular momentum projection $\Delta\Omega_{3/2,1/2} = 1$ of the quasimolecular YX^+ ion are forbidden in first-order perturbation theory.

Thus, the total rate constant of collisional deexcitation

$$K_n(T) = \frac{1}{n^2} \sum_{n'<n} \sum_{l'} \left(2l + 1\right) \langle V\sigma_{nl',nl}\rangle$$

of a given n level of the Rydberg rare gas $\text{X}\left[n_0 p^5 \left(^2 P_{3/2}\right) n\right]$ atom by the ground-state rare gas $\text{Y}(^1 S_0)$ atoms of the buffer gas may be approximately given as [267]

$$K_n(T) \approx$$
$$\left(v_0 a_0^2\right) \frac{8\pi^2 \gamma (A/2Ry)^{3/\nu}}{25\,\nu\,n^{5-9/\nu}} \,\zeta\left(1+3/\nu\right) \mathcal{A}\left(\overline{U_i\left(R_\omega\right)/kT}\right). \tag{453}$$

Here the coupling parameter γ in Eq. (453) is primarily determined by the $np \rightarrow n'p$ transitions $(l' = l = 1)$ so that

$$\gamma = \sum_{ll'} \gamma_{l'l} \approx \gamma_{11}\,,$$

and $\zeta(z) = \sum\limits_{k=1}^{\infty} k^{-z}$ is the Riemann zeta function. The $\mathcal{A}$ quantity in (453) is expressed through the $\Phi_T^{(c)}(R_\omega)$ function (435) and the potential energy curve $U_i(R)$ of the lower $X\,|j_i = 3/2, \Omega_i = 1/2\rangle$ electronic term of the quasimolecular YX^+ ion

$$\mathcal{A}\left(|U_i\left(R_\omega\right)|/kT\right) = \exp\left[-U_i\left(R_\omega\right)/kT\right]\,\Phi_T^{(c)}\left(R_\omega\right)\,.$$

A simple estimate shows that for $Xe\left[5p^5\left(^2P_{3/2}\right)n\right] + He\left(^1S_0\right)$ collisions, total rate constant of inelastic deexcitation of the Rydberg n-level is weakly dependent on the gas temperature in the range of $T \sim 300 - 600$ K. Its value can be approximately estimated as $K_n \approx 3 \cdot 10^{-7}/n_*^{3.85}$ cm$^3 \cdot$s^{-1} if the principal quantum number is not too small (when perturbation theory does not hold and the limiting magnitude of K_n is determined by (436)).

The n-dependence and the characteristic values of the total deexcitation rate constant K_n, calculated in the [267] by Eq. (453) for the case of $Xe\left[5p^5\left(^2P_{3/2}\right)n\right] + He\left(^1S_0\right)$ collisions, are shown in Fig. 63 by the full curve. As can be seen from the figure, the deexcitation of highly excited n-levels of xenon in the inelastic n-changing collisions with He via the resonant mechanism proposed in [267] is substantially more effective than that via the traditional Fermi mechanism (dashed–dotted curve) in the region of $n < 15 - 20$. Moreover, the probabilities $W_n^e = \langle v\sigma_n^e\rangle N_e$ for inelastic deexcitation of Rydberg levels of $Xe\left[5p^5\left(^2P_{3/2}\right)n\right]$ atoms by electron impact also turn out to be lower than the probabilities $W_n^{He} = \langle V\sigma_n^{He}\rangle N_{He}$ of deexcitation of these levels in collisions with He atoms for $n < 8$, 12, and 16 at plasma ionization degrees $\alpha = N_e/N_{He} = 10^{-6}$, 10^{-7} and 10^{-8}, respectively.

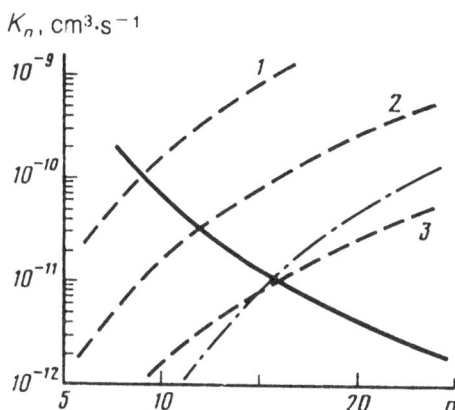

Figure 63. Comparison of the probabilities for inelastic quenching of Rydberg Xe(n) levels by the He buffer-gas atoms and by free electrons in a plasma of the He/Xe mixture $(N_{Xe} \ll N_{He})$. Full curve, total quenching rate constant $W_n^{He} = \langle V\sigma_n^{He} \rangle N_{He}$ for the resonant mechanism [267] of inelastic n-changing transitions induced by the core–projectile scattering. Dashed–dotted curve, rate constant for quenching by the competitive Fermi mechanism. Dashed curves, relative probabilities $W_n^e/N_{He} = \langle v\sigma_n^e \rangle /\alpha$ for inelastic deexcitation of Rydberg n-levels by electron impact $(T_e = 0.2$ eV) for different degrees of plasma ionization $\alpha = N_e/N_{He} = 10^{-6}$, 10^{-7}, and 10^{-8} (curves 1, 2, and 3, respectively).

Hence, the mechanism considered here of inelastic quenching of highly excited atomic Xe(n) states by atoms of the buffer rare gas He play a significant role in the process of three-particle electron–ion recombination. The influence of the inelastic deexcitation of Rydberg Xe(n) states by helium atoms on the rates of the three-body electron–ion recombination will be discussed in details in the following section. Here we note that the ionization rate constants in thermal collisions of the Rydberg rare gas $X^* \left[n_0 p^5 \left(^2 P_{3/2}\right) nl \right]$ atoms with lighter rare gas $Y\left(^1 S_0\right)$ atoms in the ground state have been calculated in Ref. [262]. As follows from the results of this work the characteristic values of the rate constants are equal to $K_{np}^{ion} \sim 10^{-11} - 10^{-10}$ cm^3·s^{-1} for the case of selectively excited np-levels in the range of $n \sim 20 - 60$. In this range of n the mechanism of ionization, associated with the core–projectile scattering and resonance transitions between the potential energy curves near the crossing points, turns out to be predominant. The competitive Fermi mechanism becomes more effective only in the range of high enough principal quantum numbers.

7.8 Three-Body Electron–Ion Recombination

7.8.1 Recombination in Rare-Gas He/Xe Mixture

It is well known that in a low-temperature plasma the three-body recombination of electrons with atomic ions X^+ proceeds as a rule via the formation of X^* atoms in excited and highly excited n-states (see, for example, the books [39, 271, 272] and references therein). This process has the character of diffusion over the highly excited energy $n \to n'$ levels in collisions with free electrons

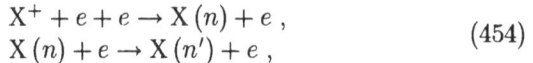

$$
\begin{aligned}
X^+ + e + e &\to X(n) + e \, , \\
X(n) + e &\to X(n') + e \, ,
\end{aligned}
\tag{454}
$$

or neutral atomic particles Y of a buffer gas

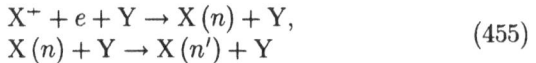

$$
\begin{aligned}
X^+ + e + Y &\to X(n) + Y, \\
X(n) + Y &\to X(n') + Y
\end{aligned}
\tag{455}
$$

(or the parent ground-state atoms X of its own gas). The rates of electron–ion recombination processes via collisions (454) with the plasma free electrons have been calculated in the diffusion approximation by *Belayev* and *Budker* [273] and *Gurevich* and *Pitaevskii* [274, 275]. The other approach, based on numerical solutions of a system of balance equations for the levels populations, has been developed by *Bates, Kingston* and *McWhirter* [276, 277]).

Calculations of the three-body electron–ion recombination coefficients via collisions (455) with buffer-gas atoms have been performed in elegant work by *Pitaevskii* [151] in the diffusion approximation. *Bates* and *Khare* [148] have proposed a classical approach for evaluation of the rate constants $K_{n'n}(T)$ for transitions between highly excited states in collisions with atomic particles and have made detailed calculations of recombination coefficients using a system of balance equations. In these papers [151, 148] (and in most further studies of recombination in collisions with neutral particles, see [272] and references therein) the inelastic transitions between the highly excited $n \to n'$ levels of the Rydberg $X(n)$ atom were considered within the framework of the Fermi mechanism due to scattering of the weakly bound electron by a perturbing particle Y. The role of the ionic core X^+ in this mechanism reduces only to the production of a distribution function of the electron momenta of Rydberg atom $X(n)$ in a Coulomb field. As follows from a great number of numerical calculations, at sufficiently high degrees α of plasma ionization the three-body electron–ion recombination takes place, as a rule, in collisions (454) with free electrons. At the same time, collisions with the buffer-gas atoms Y, via the Fermi mechanism, are substantial only at the very low values of $\alpha \lesssim 10^{-8} - 10^{-7}$.

The authors of Refs. [267, 268] have investigated experimentally and theoretically the recombination of electrons with atomic X^+ ions in a plasma of rare gas mixture X/Y (with concentrations $N_X \ll N_Y$). In this case collisions with neutral atomic particles Y of the buffer gas play a decisive role even at rather high degrees of ionization $\alpha = N_e/N_Y \sim 10^{-8} - 10^{-4}$, which are typical for low-temperature plasmas produced by an electron beam or by a pulse discharge. The experimental studies [267, 268] were performed for afterglow plasma of a pulse discharge in a helium–xenon mixture ($N_{He} = 2.6 \cdot 10^{17} - 4.4 \cdot 10^{18}$ cm^{-3}, $N_{Xe} = 10^{14} - 10^{16}$ cm^{-3}, and $T = 300 - 600$ K). Under these conditions the dissociative recombination of molecular Xe_2^+ ions can be neglected, since $\left[Xe_2^+ \right] \ll \left[Xe^+ \right] \approx N_e$.

The experimental method used in [267, 268] (see also [278]) yielded the dependencies of the coefficients $\beta \left(T_e, N_e \right)$ $\left[\text{cm}^6 \cdot \text{s}^{-1} \right]$ of three-body recombination of electrons with Xe^+ ions and the emission intensities $J_\lambda \left(T_e, N_e \right)$ for a number of xenon atom spectral lines on the temperature and on the electron density in the ranges of $T_e = 400 - 2500$ K and $N_e = 2 \cdot 10^{10} - 3 \cdot 10^{12}$ cm^{-3}. A characteristic feature of the obtained dependencies (see Fig. 64) is a strong decrease (close to $\propto T_e^{-9/2}$) in the region of low electron temperatures $T_e \lesssim 800$ K, and a substantially slower decrease and anomalously high recombination rates at $T_e \gtrsim 800$ K. It was also established [267, 268] that the recombination flux

$$\Gamma \left[\text{cm}^{-3} \cdot \text{s}^{-1} \right] = \beta N_e^3$$

depends considerably on the ionization degree and the He density.

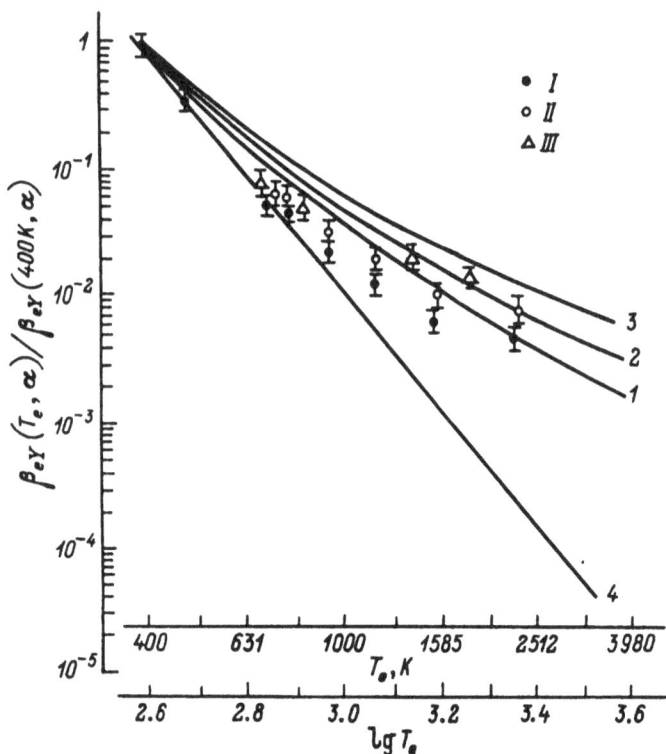

Figure 64. Dependencies of relative coefficients of the three–body recombination $\beta_{e,Y}(T_e,\alpha)/\beta_e(T_e = 400$ K$)$ of electrons with Xe^+ ions on the electron temperature T_e at different degrees of ionization $\alpha = N_e/N_{He}$ of a Xe–He plasma ($P_{He} = 50$ Torr, $P_{Xe} < 0.01$ Torr, and $T_{gas} = 400$ K). Experimental data are those from Refs. [267, 268]: I – $N_e = 3\cdot10^{11}$ cm^{-3}, II – $N_e = 1.5\cdot10^{11}$ cm^{-3}, III – $N_e = 7.5\cdot10^{10}$ cm^{-3}. Full curves 1, 2, and 3 represent corresponding calculations [267, 268] by Eqs. (471), (476), and (478) taking into account both collisions with free electrons and He atoms. Curve 4, calculation by Eq. (477) in the model [274].

This anomalous character of the recombination process cannot be explained by using the known models of recombination with atomic ions in three-body collisions with electrons or with neutral atoms of the buffer gas He. Electron recombination of Xe^+ ions in the three-body collisions with the ground-state Xe atoms [256], via dipole transitions between symmetric and antisymmetric terms $(\Sigma_g \rightarrow \Sigma_u)$ of the quasimolecular Xe_2^+ ion, is also ineffective because of the low xenon concentration.

To explain the experimental results a new recombination model has been proposed in [267, 268]. In this model the capture of the recombining electron by the rare–gas atomic Xe^+ ion and its relaxation over the highly excited levels of the $Xe(nl)$ atom is initially performed, first of all, by diffusion through collisions with electrons (just as in the model of *Gurevich* and *Pitaevskii* [274]). However, starting from certain levels $|E_{nl}| \gtrsim E_Y$, the relaxation of electron energy is primarily determined by an efficient resonant deexcitation mechanism (see Sect. 7.7), associated with scattering of buffer rare-gas He atom on the parent core Xe^+ of the Rydberg $Xe(nl)$ atom. The traditional Fermi mechanism (associated with scattering of the highly excited electron by the perturbing He atom) is not effective in the region of principal quantum numbers n having practical importance for the recombination process.

7.8.2   Kinetic Model of Recombination

Consider the kinetic model of three-body recombination processes developed in [267, 268]. As noted above, in this model the recombining electron relaxes over the highly excited levels of the rare-gas $X(n)$ atom both by collisions with free electrons (454) and by the resonant quenching by collisions with buffer rare-gas Y atoms (455). We assume that the most interesting conditions for practical applications correspond to sufficiently high degrees of ionization $\alpha \gtrsim 10^{-8}$. Then, the diffusion flux over the energy levels is primarily determined by collisions with the free electrons. It is important to stress that substantial deviations from the well known $T_e^{-9/2}$ behavior of the recombination coefficient appear (see [267, 268]), when the electron temperature T_e of a plasma is much greater than the gas temperature T (i.e., the temperature characterizing the translational motion of the heavy X^+ and Y particles). This means that in the range of binding energies of a Rydberg electron $|E| \sim kT_e$ (which is the most important for recombination induced by electron impact) the inelastic collisions with the buffer rare-gas Y atoms lead mainly to the deexcitation of Rydberg electron energy. This fact is in agreement with the well known relation for the rates of excitation and deexcitation reactions

$$W^Y_{n+\Delta n,n}/W^Y_{n-\Delta n,n} \propto \exp\left(-\left|\Delta E_{n+\Delta n,n-\Delta n}\right|/kT\right) . \qquad (456)$$

Besides, as follows from the results of Sect. 7.7, the rates $W_{n,n-\Delta n}$ of resonant deexcitation of Rydberg levels, via the n-changing inelastic transitions, reveal strong drop with an increase of the Δn value. Therefore, it is possible to use the modified diffusion approximation [272] for a description of recombination and relaxation processes.

Within the framework of this approximation one can replace the discrete energy–level spectrum of $X(n)$ atom by a quasicontinuum. Then, the kinetic equation for the distribution function $N(E,t)$ of weakly bound electrons in the quasidiscrete spectrum of energy levels can be written as

$$\frac{\partial N(E,t)}{\partial t} = -\frac{\partial \Gamma(E,t)}{\partial E} , \qquad (457)$$

$$N(E,t) = N_n(E,t)\left|\frac{dn}{dE}\right| . \qquad (458)$$

Here the $N(E,t)$ quantity is the concentration of highly excited atoms per unit energy interval of quasicontinuum with the energy $E \equiv E_n = -Ry/n^2 < 0$, and $\Gamma(E,t)$ is the recombination flux which is given by the following expression

$$\Gamma(E,t) = -D_e(E)\left[\frac{\partial N(E,t)}{\partial E} + \left(\frac{1}{kT_e} + \frac{5}{2E}\right)N(E,t)\right]$$

$$+W_Y(E)N(E,t) . \qquad (459)$$

The first term on the right-hand side of this equation is responsible for collisions with the free electrons of a plasma, whereas

$$D_e(E) = \frac{1}{2}\frac{\partial}{\partial t}\overline{\Delta E^2} \qquad (460)$$

is the corresponding diffusion coefficient in the energy space.

The second term in Eq. (459) corresponds to the total probability $\mathsf{W}_Y(E)$ of deexcitation of the Rydberg electron energy in the quasicontinuous spectrum. This quantity $\mathsf{W}_Y(E)$ [erg· s^{-1}] is expressed in terms of the resonant deexcitation rate $\mathsf{W}_n^Y = N_Y K_n^Y$ [s^{-1}] of the Rydberg $X(n)$ atom by the ground-state Y atom of the buffer rare-gas:

$$\mathsf{W}_Y(E) = \mathsf{W}_{n(E)}^Y \left| dE/dn \right| . \tag{461}$$

By using expression (453) of Sect. 7.7, we have

$$\mathsf{W}_Y(E) = \frac{e^2 \hbar}{m} \frac{2^{8-9/2\nu} \pi^2 \gamma}{25\nu} \zeta \left(1 + 3/\nu\right) \langle \mathcal{A} \rangle$$
$$\times \left(\frac{A}{2Ry}\right)^{3/\nu} \left(\frac{|E|}{2Ry}\right)^{(8\nu-9)/2\nu} N_Y . \tag{462}$$

Note that neglecting the resonant deexcitation term ($\mathsf{W}_Y = 0$) in Eq. (459) (associated with atomic particle collisions) we obtain the Fokker–Planck equation directly from (457). This equation describes the energy relaxation process of a highly excited electron, induced by collisions with a plasma free electrons, within the framework of the diffusion approximation.

The coefficient of recombination can be evaluated in the quasistationary regime, when the recombination flux $\Gamma = const < 0$. In this situation it is convenient to introduce a new variable $\varphi(E) = N(E)/N_0(E)$. It corresponds to the ratio of electron concentration in the quasicontinuous spectrum to its equilibrium value, which is determined by the Saha–Boltzmann formula

$$N_0(E) = \frac{\pi^{3/2} e^6 \exp\left(|E|/kT_e\right)}{2\left(kT_e\right)^{3/2} |E|^{5/2}} N_e N_{X^+} , \tag{463}$$

where N_e and N_{X^+} are the concentrations of free electrons and atomic X^+ ions in a plasma, respectively.

As a result, equation (459) can be rewritten as

$$\Gamma = N_0(E) \left[-D_e(E) \frac{d\varphi(E)}{dE} + \mathsf{W}_Y(E) \varphi(E) \right] . \tag{464}$$

Here the diffusion coefficient is given by the well known expression of *Gurevich* and *Pitaevskii* [274, 275]

$$D_e\left(E\right) = \frac{2\left(2\pi\right)^{1/2} e^4 \Lambda_C \left|E\right| N_e}{3\left(mkT_e\right)^{1/2}} \, , \qquad (465)$$

and Λ_C is the Coulomb logarithm (see also [272]).

The formulation of boundary conditions for equation (459) is based, as usual, on the assumptions that the density of electrons N_e in a recombining plasma is much greater than its equilibrium magnitude $N_0\left(E\right)$ for a given value of electron temperature T_e (see [39, 272]). Then, one can fully neglect the ionization of the $X(n)$ atoms and suppose that a decrease of the free electron density in a plasma is determined by the recombination processes alone. It is also necessary to take into account that for the energy levels with extremely small binding energy $\left|E\right| \ll kT_e$, the Rydberg electrons are in equilibrium with the free electrons in the continuous spectrum. This allows us to write one of the boundary conditions as

$$\varphi\left(0\right) = 1 \, . \qquad (466)$$

This means that $N\left(E\right) = N_0\left(E\right)$ at $E = 0$, where the zero of electron energy corresponds to the ionization limit of the $X(n)$ atom.

In this situation the constant magnitude of the recombination flux $\left(\Gamma = const < 0\right)$ directly yields the value of the recombination coefficient

$$\alpha^{\mathrm{rec}} = -\frac{1}{N_e}\frac{dN_e}{dt} = \frac{\left|\Gamma\right|}{N_e N_{X+}} \quad \left[\mathrm{cm}^3 \cdot \mathrm{s}^{-1}\right] . \qquad (467)$$

The solution of the differential equation (464) taking into account the boundary condition (466) can be presented in the following form

$$\varphi\left(E\right) = P\left(E\right)\left[1 - \left|\Gamma\right| \int_0^{\left|E\right|} \frac{dE'}{N_0\left(E'\right) D_e\left(E'\right) P\left(E'\right)}\right] , \qquad (468)$$

where the $P\left(E\right)$ function is given by the relation

$$P\left(E\right) = \exp\left\{\int_0^{\left|E\right|} dE' \frac{W_B\left(E'\right)}{D_e\left(E'\right)}\right\} . \qquad (469)$$

It can be seen that this function is a solution of the uniform equation (464).

On inserting the specific form of expressions (462) and (465) for the resonant deexcitation rate and for the diffusion coefficient into (469), we obtain the following simple formula for the $P(E)$ function

$$P(E) = \exp\{C\,|E|^\eta\} = \exp\left\{0.2\left(\frac{|E|}{E_Y}\right)^\eta\right\}, \qquad (470)$$

where $\eta = (8\nu - 9)/2\nu$. Here the E_Y quantity takes the form

$$E_Y(T_e, \alpha) =$$

$$Ry\left[\frac{5(8\nu-9)\alpha\Lambda_C}{6(2\pi)^{3/2}\gamma(A)\zeta(1+3/\nu)}\left(\frac{2Ry}{A}\right)^{3/\nu}\sqrt{\frac{2Ry}{kT_e}}\right]^{2\nu/(8\nu-9)}. \qquad (471)$$

Note that the η index is related to the ν index characterizing the power approximation of the energy splitting of potential energy curves of the quasimolecular YX^+ ion (see Eq. (451)).

The physical meaning of the the E_Y quantity is as follows. This is the typical value of an electron binding energy, which separates the quasidiscrete energy spectrum of the $X(n)$ atom into two different regions. In the first region ($|E| \equiv |E_n| < E_Y$) the recombination flux Γ is mainly determined by diffusion over the highly excited levels in collisions with the free electrons. In the second ($|E| \equiv |E_n| > E_Y$) it is determined by the resonant n-changing deexcitation by neutral Y atoms of the buffer rare-gas. It should be noted that the value of E_Y is a function of the electron temperature T_e and the ionization degree $\alpha = N_e/N_Y$ of a plasma of the rare-gas X/Y mixture with $N_X \ll N_Y$.

Indeed, for $|E| < E_Y$ we have $P(E) \approx 1$ so that the solution (468) of equation (464) for the normalized distribution function $\varphi(E)$ actually coincides with the result of *Gurevich* and *Pitaevskii* [274, 275]

$$\varphi(E) = 1 - \frac{|\Gamma|}{\beta_e N_e^2 N_{X^+}}\frac{\gamma(5/2, |E|/kT_e)}{\Gamma(5/2)}, \qquad (472)$$

$$\beta_e = \frac{4\pi(2\pi)^{1/2}e^{10}\Lambda_C}{9\sqrt{m}(kT_e)^{9/2}}, \qquad (473)$$

Here $\gamma\left(5/2, x\right)$ is the incomplete gamma function of the $5/2$-order, and $\beta_{\rm e}\left(T_{\rm e}\right)$ is the well known expression for the coefficient of the three-body electron–ion recombination in collisions with the plasma free electrons. This means, that in the range of $|E| < E_{\rm Y}$ the resonant quenching by neutral Y atoms of the buffer gas is negligible. On the contrary, for $|E| > E_{\rm Y}$ we have from (469) $P\left(E\right) \gg 1$, i.e., collisions with free electrons can be neglected. This means that for the levels with sufficiently large electron binding energy their distribution function is not perturbed by the free electrons. The distribution function of these levels is not dependent on the free electron density $N_{\rm e}$ since it is primarily determined by the equilibrium electron density $N_{\rm e}^{(0)}$ (whereas $N_{\rm e}^{(0)} \ll N_{\rm e}$).

It is important to stress that the transition from the "diffusion" region $|E| < E_{\rm Y}$ to the "predominant drain" region $|E| > E_{\rm Y}$ occurs in a narrow vicinity of the point $E_{\rm Y}$ (see Eq. (471)). We can therefore put $P\left(E\right) = 1$ for all energy levels with $|E| < E_{\rm Y}$. Then, imposing a second boundary condition

$$\varphi\left(E_{\rm Y}\right) = 0 \,, \tag{474}$$

we can determine from (468), (470) and (471) the recombination flux Γ and the coefficient $\beta_{\rm eY}/N_{\rm e}^2 N_{\rm X^+}$ $\left(N_{\rm e} \approx N_{\rm X^+}\right)$ of the three-body recombination for the kinetic model considered [267, 268]

$$\Gamma = -\beta_{\rm eY}\left(T_{\rm e}, \alpha\right) N_{\rm e}^2 N_{\rm X^+} \quad [\text{cm}^{-3} \cdot \text{s}^{-1}], \tag{475}$$

$$\beta_{\rm eY} = \beta_{\rm e}\left(T_{\rm e}\right) \xi^{-1} \left(E_{\rm Y}/kT_{\rm e}\right) \quad [\text{cm}^6 \cdot \text{s}^{-1}], \tag{476}$$

As is apparent from (475) and (476), the final results can be expressed in terms of the standard three-body recombination coefficient $\beta_{\rm e}\left(T_{\rm e}\right) = \alpha_{\rm e}^{\rm rec}/N_{\rm e}$ [cm^6·s^{-1}] (see 473) associated with the electron–ion–electron collision alone, i.e.

$$\beta_{\rm e}\left(T_{\rm e}\right) = 8.8 \cdot 10^{-27} \cdot \Lambda_{\rm C}\left(T_{\rm e} \text{ [eV]}\right)^{-9/2} \quad \left[\text{cm}^6 \cdot \text{s}^{-1}\right], \tag{477}$$

and some quantity $\xi^{-1}\left(E_{\rm Y}/kT_{\rm e}\right)$ which is a function of the electron temperature T_e and the ionization degree $\alpha = N_e/N_{\rm Y}$. This quantity has the form

$$\xi\left(x\right) = \frac{4}{3\pi^{1/2}} \int\limits_{0}^{x} t^{3/2}\, e^{-t}\, dt = \frac{4}{3\pi^{1/2}}\, \gamma\left(5/2, x\right) \,, \tag{478}$$

whereas $0 \leq \xi \leq 1$ at $(0 \leq x < \infty)$. The quantity $\xi^{-1} \left(E_Y / kT_e \right)$ has the physical meaning of the growth of the recombination coefficient due to the resonant deexcitation of Rydberg levels of $X(n)$ atoms in thermal collisions with the buffer rare-gas Y atoms.

As is apparent from the analysis of Eqs. (475), (476) and (478), in the range of $kT_e \sim E_Y$ the recombination regime changes from a strong drop with increase of T_e at low temperatures $kT_e \ll E_Y$

$$\xi \left(E_Y / kT_e \right) \approx 1,$$

$$|\Gamma| \propto N_e^3 T_e^{-9/2} \tag{479}$$

to a substantially slower decrease with temperature and anomalously large values of $\beta_{eY} \gg \beta_e$ at high temperatures $kT_e \gg E_Y$, i.e.

$$\xi \left(E_Y / kT_e \right) \approx \tfrac{8}{15\pi^{1/2}} \left(\tfrac{E_Y}{kT_e} \right)^{5/2},$$

$$|\Gamma| \propto N_e^{(19\nu - 27)/(8\nu - 9)} \, N_Y^{5\nu/(8\nu - 9)} \, T_e^{-[(27\nu - 36)/2(8\nu - 9)]}, \tag{480}$$

For the special case of electron–ion recombination with atomic Xe^+ ions in an He/Xe plasma $\left(N_{Xe} \ll N_{He} \right)$, we obtain

$$|\Gamma| \propto N_e^3 \cdot T_e^{-9/2}, \qquad (kT_e \ll E_B),$$

$$|\Gamma| \propto N_e^{2.3} N_{He}^{0.7} \cdot T_e^{-1.6}, \qquad (kT_e \gg E_B). \tag{481}$$

Owing to the efficient resonant deexcitation mechanism of the Rydberg $Xe(n)$ levels in a buffer helium gas, the recombination coefficient $\beta_{e,He}$ then increases as compared to the values of β_e obtained within the framework of the usual "diffusion" recombination with free electrons. In particular, for a temperature $T_e \approx 0.2$ eV at the ionization degrees $\alpha = N_e / N_{He} = 10^{-5}$, 10^{-6}, 10^{-7} and 10^{-8}, we obtain an increase of the recombination rate $\beta_{e,He} / \beta_e$ by 4, 10, 57 and 325 times from Eqs. (476)–(478) . Thus, the works [267, 268] point to the efficiency of the identified resonant deexcitation mechanism of Rydberg atomic levels by neutral particles in the electron–ion recombination of low-temperature plasma of the rare gas He/Xe mixture even at degrees of ionization of about $10^{-5} - 10^{-4}$.

7.9   Classical Distribution and Partition Functions

A broad class of the problems of molecular and atomic physics, spectroscopy and plasma physics is closely related with evaluation of distribution functions of internuclear separations in molecular and quasimolecular systems as well as internal partition functions of neutral molecules and molecular ions. Similar problems appear in calculations of rate coefficients of collision processes involving Rydberg atoms and neutral targets within the framework of the quasimolecular approach. Some typical examples of applications of such distributions in the collision theory of highly excited atoms with neutral atomic particles are the processes of direct and associative ionization and resonant deexcitation considered in Sects. 7.5–7.7. Here we shall concentrate our attention, first of all, on some calculations of internal partition functions of diatomic systems and on the detailed comparison of purely classical and quantal results.

The internal partition function is a fundamental physical quantity which determines thermodynamic functions of molecular gases (plasmas) [279, 280] and equilibrium constants of chemical reactions [281], molecular opacities [282] and densities of atomic and molecular particles in stellar and planet atmospheres (see also [283, 284] and references therein). Available theoretical approaches for the evaluation of partition functions are described in a number of books (e.g. [279, 280, 288]). Extensive material on the partition functions of molecules having astrophysical importance is given in Ref. [289].

When a gas temperature is not too large, an efficient approach for practical evaluation of partition functions is based on the *Dunham* [285] expansions (see also [37, 286]) of rovibrational energy ϵ_{vJ} in power series of $(v + 1/2)$ and $J(J + 1)$, where v and J are the vibrational and rotational quantum numbers. These expansions follow from the known expansions of the effective potential energy curve

$$U_{\text{eff}}(R) = U(R) + \frac{\hbar^2 J(J+1)}{2\mu R^2}$$

of a molecule in a power series of the $(R - R_e)$ value near the bottom of the well. Here R_e is the equilibrium internuclear separation, and μ is the reduced mass of a molecule. The quantal and purely classical expressions for the partition functions of diatomic molecules take particularly simple form in the rigid-rotator, harmonic-oscillator approximation [279]. Another simple model, which has found wide application in practical calculations, was proposed by *Mayer* and *Göppert-Mayer* [287]. It additionally includes the first few correction terms of a series expansion of rovibrational energy

$$\epsilon_{vJ} = \sum_{pq} \alpha_{pq} \left(v + 1/2\right)^p \left[J(J+1)\right]^q$$

associated with anharmonicity of vibrations and rovibrational interaction. There are also a number of efficient semiemperical methods for the calculation of partition functions and thermodynamic properties of molecular gases (see [288]).

However, when the thermal energy kT significantly exceeds the lowest vibrational quantum $\hbar\omega_e$ of a molecule, and, particularly, provided kT becomes of the order of its dissociation energy D_0) standard "low-temperature" approximations do not hold because the partition function is determined by the contribution of a great number of rovibrational levels. For excited states with large rotational J and vibrational v quantum numbers the amplitude of vibrations is substantially increased so that the effects of anharmonicity and rovibrational interaction become particularly important. Apparently, an exact description of these effects can be given by using quantal *ab initio* calculations of all rovibrational energy levels of a molecule and direct summation over the quantum numbers v and J of the basic expressions for the partition function.

An aternative way is to use purely classical and quasiclassical methods, taking into account correspondence principles between quantal and classical results. Classical and quasiclassical methods have become an efficient tool in the physics of highly excited atomic states [3]. Moreover, as follows from recent works on Rydberg atoms (see [138, 140] and [176, 177]), they turn out to be applicable not only for large quantum numbers but also in some other cases, which have been previously described only quantum-mechanically. This has attracted renewed interest in classical and quasiclassical treatments of rovibrational states. Similar methods are particularly suitable for the description of radiative and collisional processes involving highly excited molecular states for which application of the exact quantum-mechanical formalism is sufficiently difficult due to the highly oscillatory behavior of the wave functions.

Purely classical methods of distribution and partition functions, based on microcanonical distributions (see [279]), were widely used in atomic and molecular physics. For example, there is a series of works by *Bates* and *McKibbin* [290, 291], *Bates* [292] and *Flannery* [293, 294] devoted to different aspects of classical distribution functions and their applications in the theory of the three-body ion–neutral association, ter-molecular recombination, ion–ion and electron–ion recombination processes (see also the references in these papers).

Below we present a simple approach [295] aimed towards the derivation of exact classical result for the internal partition function $Z_{v.r}$ of a diatomic molecule directly from the basic quantal expression. A solution of this problem will be given in a general analytical form within the framework of the quasiclassical approach combined with the approximation of quasicontinuum for rovibrational energy levels. This corresponds to the replacement of summation over the rotational J and vibrational v quantum numbers by integration over an appropriate region of the angular momentum and energy space corresponding to discrete rovibrational states of a molecule.

We proceed directly from the basic quantal expression [279]

$$Z\left(T\right) = \frac{g}{\text{œ}} \sum_{vJ} \left(2J+1\right) \exp\left(-\frac{\epsilon_{vJ}}{kT}\right) . \tag{482}$$

The symmetry factor œ is equal to 1 and 2 for the heteronuclear and homonuclear molecules, respectively; g is the electronic statistical weight. The value of Z is referred to the rotationless ground vibrational state throughout this paper, i.e. $\epsilon_{vJ} = 0$ when $v = 0$, $J = 0$.

By using the quasiclassical approach combined with the approximation of a quasicontinuum for discrete energy levels expression (482) can be rewritten in terms of a double integral over the quantum numbers v and J

$$Z\left(T\right) = \frac{g}{\text{œ}} \int\limits_{0}^{v_{\max}} dv \int\limits_{0}^{J_{\max}(v)} 2J\,dJ\, e^{-(E_{vJ}+D_0)/kT} . \tag{483}$$

Here D_0 is the dissociation energy of a molecule, $E_{vJ} = \epsilon_{vJ}-D_0$ is the rovibrational energy referred to its dissociation limit (i.e. $E_{v=0,J=0} = -D_0$ and the point $E = 0$ separates the discrete spectrum from the continuum).

With the help of the Bohr–Sommerfeld relation, integration over

$$dv = \left(T_{vJ}/2\pi\hbar\right) dE_{vJ} \tag{484}$$

can be replaced by integration over the energy space of bound states $E < 0$

$$Z = \frac{g}{2\pi\hbar c} \int\limits_{-D_0}^{0} \int\limits_{0}^{J_{\text{max}}} e^{-(E_{vJ}+D_0)/kT} T_{vJ} \, dE_{vJ} \, 2J \, dJ \, . \qquad (485)$$

The period T_{vJ} of rovibrational motion over the classically allowed region of internuclear separation R is

$$T_{vJ} = \oint \frac{dR}{\sqrt{(2/\mu)\left[E_{vJ} - U(R) - \hbar^2 J^2/2\mu R^2\right]}} \, , \qquad (486)$$

where $U(R)$ is the potential energy of a molecule.

On inserting Eq. (486) into (485) and changing the order of integration over J and R, the partition function can be rewritten in terms of the classical density $\rho(E)$ of rovibrational states per unit energy interval of the quasicontinuum

$$Z(T) = \int\limits_{-D_0}^{0} \rho(E) \, e^{-(D_0+E)/kT} \, dE \, . \qquad (487)$$

It is given by the relation

$$\rho = \frac{g\sqrt{\mu}}{\pi\hbar c\sqrt{2}} \int\limits_{a_1}^{a_2} dR \int\limits_{0}^{J_{\text{max}}} \frac{2J \, dJ}{\sqrt{E - U(R) - \hbar^2 J^2/2\mu R^2}} \, . \qquad (488)$$

Here the upper limit of integration $J_{\text{max}}(E, R)$ over the rotational quantum numbers is to be found from the relation

$$\hbar^2 J_{\text{max}}^2/2\mu R^2 = E - U(R) \, ,$$

while $a_1(E)$ and $a_2(E)$ are the left and right turning points $U(a_{1,2}) = E$. The resultant expression for the density of states takes the form

$$\rho(E) = \frac{g(2\mu)^{3/2}}{\infty\pi\hbar^3} \int\limits_{a_1(E)}^{a_2(E)} \sqrt{E - U(R)} \, R^2 \, dR \, . \qquad (489)$$

The general classical formula for the partition function can be deduced from Eqs. (487) and (489) on changing the order of integration

over E and R. This allows us to represent $Z(T)$ in terms of the Boltzmann averaged coordinate distribution function $W_T^{(d)}(R)$ taking into
account the integral contribution of all discrete rovibrational states

$$W_T^{(d)}(R) = \frac{g(2\mu)^{3/2}}{\text{œ}\pi\hbar^3} \int\limits_{E_{\min}}^{0} \sqrt{E - U(R)}\, e^{-E/kT}\, dE \,. \qquad (490)$$

Here $E_{\min} = \min\{U(R), 0\}$, whereas $E_{\min} \leq 0$ because the potential energy is referred to the dissociation limit of a molecule, i.e.
$U(R) \to 0$ at $R \to \infty$. The partition function is then the integral of
$W_T^{(d)}(R)$ over an appropriate region of internuclear separation

$$Z(T) = e^{-D_0/kT} \int\limits_{R_0}^{\infty} W_T^{(d)}(R)\, 4\pi R^2\, dR \,, \qquad (491)$$

where R_0 is to be found from the relation $U(R_0) = 0$, i.e. it corresponds to the left turning point $a_1(E)$ at $E \to 0$. The integral
over E in Eq. (490) is reduced to the incomplete gamma function
$\Gamma(3/2, z) = \int_z^{\infty} t^{1/2} e^{-t}\, dt$.

As a result, expression (490) for the distribution function $W_T^{(d)}$ can
be converted to its final analytical form with explicit dependence on
potential energy

$$W_T^{(d)}(R) = \frac{g}{\text{œ}} \left(\frac{\mu kT}{2\pi\hbar^2}\right)^{3/2} \exp\left[-\frac{U(R)}{kT}\right] \Phi_T^{(d)}(R) \,, \qquad (492)$$

$$\Phi_T^{(d)}(R) = \begin{cases} 1 - \frac{2}{\sqrt{\pi}}\Gamma\left(\frac{3}{2}, \frac{|U(R)|}{kT}\right) \,, & R_0 \leq R < \infty, \\ 0 \,, & 0 \leq R < R_0 \,. \end{cases} \qquad (493)$$

According to (492) $W_T^{(d)}$ is a product of the classical Boltzmann
probability and some dimensionless factor (493), the magnitude of
which satisfies the relation $0 \leq \Phi_T^{(d)} < 1$. This factor is the ratio
$\Phi_T^{(d)} = W_T^{(d)}/W_T$ of the distribution function $W_T^{(d)}$, which defines
the integral contribution of the discrete spectrum, to its total value

$$W_T = W_T^{(d)} + W_T^{(c)} = \frac{g}{\text{œ}} \left(\frac{\mu kT}{2\pi\hbar^2}\right)^{3/2} \exp\left[-\frac{U(R)}{kT}\right] \,,$$

including the contribution $W_T^{(c)} \propto \exp\left[-U(R)/kT\right]\, \Phi_T^{(c)}(R)$ of the continuum. Therefore, the incomplete gamma function in (493) removes the integral contribution of the continuous spectrum from the coordinate distribution function of a molecule. Note that a factor similar to $\Phi_T^{(c)}$ (435) appear in the classical theory of radiative transitions between different electronic terms [296] as well as associative ionization [259, 262], and termolecular association [294].

The physical significance of Eqs. (491)–(493) is that they allow us to obtain an exact classical result for the partition function for a given form of the potential energy curve $U(R)$, and, hence, for thermodynamic functions and the equilibrium rate constant [279]

$$K_p\left(T\right) = \frac{g_A\, g_B}{Z_{AB}\left(T\right)} \left(\frac{\mu}{2\pi\hbar^2}\right)^{3/2} (kT)^{5/2}\, e^{-D_0/kT} \qquad (494)$$

of dissociation reaction AB→A+B (g_A, g_B are electronic statistical weights of atoms). The theory presented above is valid when the classical treatment of a rovibrational motion is justified. Its validity criterion can be practically written as $kT \gg \hbar\omega_e/3$ in accordance with comments in Ref. [279].

Consider the limiting cases of the general expressions (491)-(493). For small thermal energies compared with the dissociation energy $kT \ll D_0$, the major contribution to the partition function is determined by the bottom of the well, i.e. $R \sim R_e$ (R_e is the equilibrium point). Then, one can put the lower limit of integration over R to be equal to zero, and $\Phi_T^{(d)}(R) \approx 1$ because of

$$|U|\,/kT \approx D_e/kT \gg 1\ ,$$

where $D_e = |U(R_e)|$ is the depth of the potential well. As a result, we obtain the simple relation

$$Z\left(T\right) \approx \frac{g}{\infty} \left(\frac{\mu kT}{2\pi\hbar^2}\right)^{3/2} \int\limits_0^\infty e^{-[D_0+U(R)]/kT}\, 4\pi R^2\, dR\ . \qquad (495)$$

It corresponds to the low-temperature classical limit for the partition function ($\hbar\omega_e/3 \ll kT \ll D_0$) discussed in Ref. [279]. The use of only the two first terms of expansion of the potential energy

$$U = -D_e + \left(\mu\omega_e^2/2\right)\xi^2 \qquad (496)$$

as a power series in $\xi = R - R_e$, yields

$$Z\left(T\right) \approx \frac{\text{g}}{\text{œ}} \frac{\left(kT\right)^2}{\hbar\omega_e B_e} \, e^{\left(D_e - D_0\right)/kT} \,, \qquad D_e - D_0 \approx \frac{\hbar\omega_e}{2} \,, \qquad (497)$$

where $B_e = \hbar^2/2\mu R_e^2$ is the rotational constant of a molecule. This is the rigid-rotator, classical harmonic-oscillator approximation. The exponential factor appeared in Eq. (497) since the Z value is referred to the ground quantum state $\epsilon_{v=0,J=0} = 0$ in contrast to the standard classical approach [279], when Z is referred to the minimum of potential energy. It improves the correspondence between classical and quantal versions of the rigid-rotator, harmonic-oscillator model in the range of $kT \sim \hbar\omega_e$.

In the opposite limiting case $kT \gg D_0$ the relative contribution $\Phi_T^{(d)}$ of all discrete levels exhibits a strong fall with increasing temperature. This directly follows from Eq. (493) and the known expansion of the incomplete gamma function $\Gamma\left(3/2, z\right)$ at small argument $z \ll 1$ according to which

$$\Phi_T^{(d)} \rightarrow 4z^{3/2}/3\sqrt{\pi}, \qquad z = |U\left(R\right)|/kT \,. \qquad (498)$$

At the same time, one can put $\exp\left[-\left(D_0 + U\right)/kT\right] \rightarrow 1$ in Eqs. (491), (492). Thus, the contribution of all discrete rovibrational levels $\left(E_{vJ} < 0\right)$ into the partition function tends asymptotically to a temperature independent constant

$$\lim_{T \to \infty} Z\left(T\right) = \frac{2\text{g}}{3\pi\text{œ}} \int_{R_0}^{\infty} \left(\frac{2\mu\,|U\left(R\right)|}{\hbar^2}\right)^{3/2} R^2 \, dR \,, \qquad (499)$$

whose value is determined by the reduced mass of a molecule and its potential energy.

We illustrate the main features of the partition function behavior by presenting the results of calculations [295] for $H_2^+(X^2\Sigma_g^+)$ and $Na_2(X^1\Sigma_g^+)$ molecules in Fig. 65. They are based on *ab initio* numerical data of Refs. [297, 298] for the potential energy curves of both systems. The results obtained by Eqs. (491)–(493) are shown by full curves in the whole region of validity of classical treatment $T \geq T_v = \hbar\omega_e/k$ ($T_v = 3340$ K for H_2^+ and $T_v = 229$ K for Na_2). Both curves clearly demonstrate similar qualitative dependencies of partition functions on temperature. But we notice a great difference in magnitudes of $Z_{H_2^+}$ and Z_{Na_2} (about two orders) in all the considered range of T. This is the result of a large differences in masses of molecules and parameters of potential energy curves: $D_0 = 2.651$ eV, $\hbar\omega_e = 0.288$ eV, $B_e = 3.74 \cdot 10^{-3}$ eV for H_2^+ [297] and $D_0 = 0.702$ eV, $\hbar\omega_e = 1.97 \cdot 10^{-2}$ eV, $B_e = 1.92 \cdot 10^{-5}$ eV for Na_2 [298] (see also the data in Refs. [286, 299]).

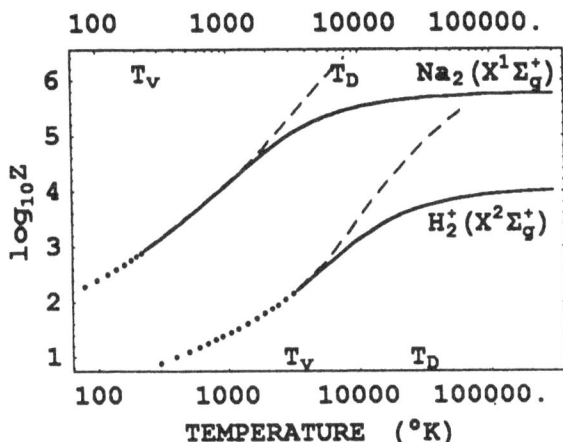

Figure 65. Internal partition functions of $H_2^+(X^2\Sigma_g^+)$ and $Na_2(X^1\Sigma_g^+)$. Full curves: exact classical results [295]. Dashed curves: classical formula (495). Dots represent *ab initio* quantal results for H_2^+ and the quantal model [287] for Na_2, respectively.

In order to demonstrate how classical the low-temperature approximation works in the range of $T \geq T_v$ we also present the results obtained by Eq. (495) (dashed curves) in Fig. 65. Apparently, it is able to reasonably reproduce a rapid increase of $Z_{H_2^+}$ at $3340 \leq T < 7000$ K and Z_{Na_2} at $229 \leq T < 2000$ K. However, a further increase of T leads to incorrect temperature dependencies and to dramatic quantitative deviations from exact classical results. As follows from Eqs. (491)–(493), the partition function behavior at the intermediate $T_v \ll T \ll T_D$ and, particularly, at large values of $T > T_D = D_0/k$, is changed drastically due to a significant decrease of the relative contribution $\Phi_T^{(d)}$ of the discrete spectrum of a molecule into the coordinate distribution function $W_T^{(d)}$. Here $T_D = 30760$ K and $T_D = 8146$ K for H_2^+ and Na_2, respectively.

This effect is illustrated in Fig. 66, where the ratio of $\Phi_T^{(d)} = W_T^{(d)}/W_T$ (493) for an H_2^+ ion (panel a) and a Na_2 molecule (panel b) is plotted against R for several magnitudes of T. It can be seen that the functions $\Phi_T^{(d)}(R)$ exhibit maxima at the equilibrium point R_e and tend to zero at $R \to R_0$ and $R \to \infty$. At low T the $\Phi_T^{(d)}$ value is close to unity excepting very large distances and the nearest vicinity of R_0. (Note that $R_0 = 1.11$ a.u., $R_e = 2.0$ a.u. and $R_0 = 4.3$ a.u., $R_e = 6.0$ a.u. for H_2^+ and Na_2, respectively). However, as T increases this ratio falls monotonically in the whole range of $R_0 \leq R < \infty$. The effect will result in some slower increase of $Z(T)$ at intermediate and high magnitudes of T as compared to the temperature dependence predicted by Eq. (495). Note that asymptotic values of $Z_{H_2^+}$ and Z_{Na_2} obtained by Eqs. (491)–(493) are in full agreement with the simple relation (499).

For a comparison of the forms of the coordinate distribution functions W_T^d corresponding to different gas temperatures it is also convenient to introduce the function

$$w_T^{(d)}(R) = \frac{e^{-D_0/kT}}{Z(T)} \, 4\pi R^2 \, W_T^{(d)}(R) \,, \qquad (500)$$

which is normalized to unity for any arbitrary value of T

$$\int_0^\infty w_T^{(d)}(R) \, dR \to \int_{R_0}^\infty w_T^{(d)}(R) \, dR = 1 \,. \qquad (501)$$

It is important to note that for H_2^+ ion there are *ab initio* calculations [288, 297, 300] of all rovibrational energy levels. This allows us to make a comparison (see Fig. 65) between exact classical calculations of the partition function (Z^{cl}, full curve) and quantal results (Z^q, dotted curve) obtained by direct summation (482) over all discrete states vJ ($E_{vJ} < 0$). A detailed comparison of classical and quantal results for the dissociation reaction

$$H_2^+ (X^2\Sigma_g^+) \to H(1s) + H^+$$

is shown in Fig. 68. It illustrates the ratio $K_p/K_p^{(q)} = Z^{(q)}/Z$ of the rate constant (494) (calculated using different approximations for Z) to its exact quantal value $K_p^{(q)}$ as a function of T.

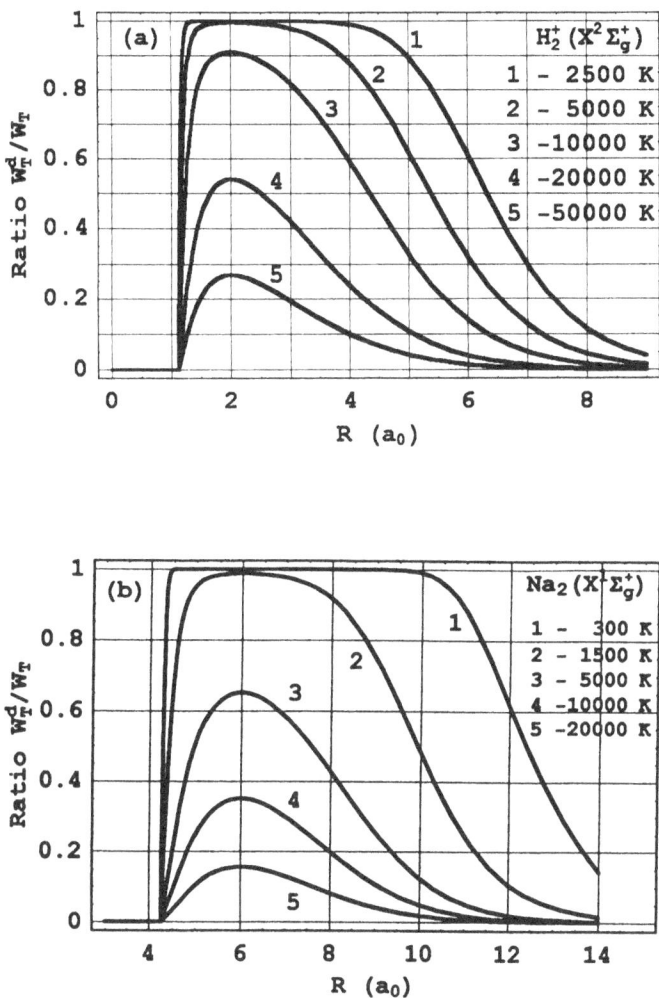

Figure 66. Relative contribution $\Phi_T^{(d)} = W_T^{(d)}/W_T$ of all rovibrational levels of $H_2^+(X^2\Sigma_g^+)$ (panel a) and $Na_2(X^1\Sigma_g^+)$ (panel b) to the coordinate distribution function at different temperatures.

As is evident from Figs. 65 and 68, the agreement between the exact classical and quantal results is excellent in the whole region of $T \geq T_v = 3340$ K. The ratio of $(Z^{(cl)} - Z^{(q)})/Z^{(q)}$ turns out to be equal to 3.1 % at $T = T_v$; it becomes less than 1 % in the range of $T > 5750$ K, and decreases as T grows. Note that the full and dotted curves in Fig. 1 virtually coincide at $T \geq T_v$. According to Fig. 3 substantial deviations from the quantal $Z^{(q)}$ and $K_p^{(q)}$ values appear only in the range of $T < 1000 - 1500$ K.

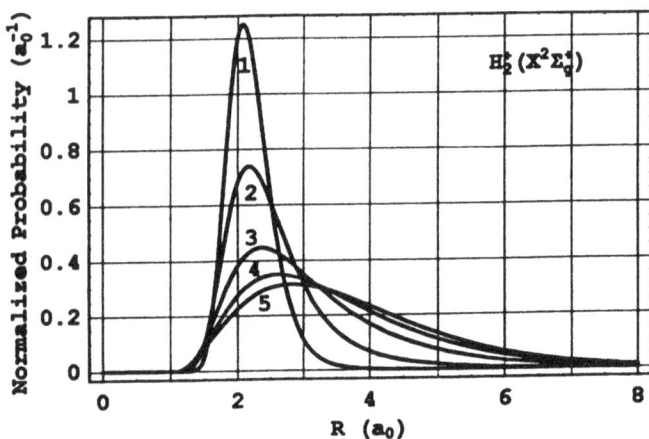

Figure 67. Classical distribution function (500) of $H_2^+ (X^2\Sigma_g^+)$ for $T = 2500$, 5000, 10000, 20000, and 50000 K – curves 1, 2, 3, 4, and 5, respectively.

We also compared purely classical $F^{(cl)}$ and quantal $F^{(q)}$ results for the contribution of the rovibrational motion of H_2^+ ions (N is their concentration) into the Helmholtz thermodynamic potential $F = -NkT \ln Z$ of an equilibrium hydrogen plasma. Here the accuracy of the exact classical approach turns out to be particularly high due to the logarithmic dependence on Z. The ratio of $(F^{(cl)} - F^{(q)})/F^{(q)}$ is only 0.5 % at $T = T_v = 3340$ K. It becomes equal to 4.6 % at $T = 1500$ K and 9.3 % at $T = T_v/3 = 1113$ K, i.e. out of the formal validity condition.

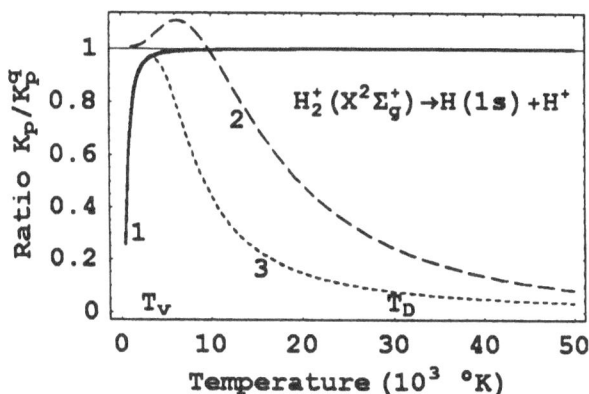

Figure 68. Ratio of dissociation equilibrium rate constant K_p of $H_2^+ (X^2\Sigma_g^+)$ to its exact quantal value $K_p^{(q)}$. Full curve 1: exact classical result. Dashed curves 2 and 3 represent the quantal model [287] and the classical formula (495), respectively.

In summary, the approach, presented above, gives an exact classical description of the internal partition function and integral contribution of all rovibrational states into the coordinate distribution function of a diatomic molecule. It is valid not only at high temperatures but also in the whole region of $kT \geq \hbar\omega_e$ including thermal energies about the lowest vibrational quantum. Therefore, if the ω_e value of a molecule is not too large, the general formulae (491)–(493) turn out to be applicable already at room temperatures (e.g. for Na_2). For the heavy molecules, such as Cs_2 and Rb_2, they become valid starting from $T \approx 60 - 80$ K. The excellent quantitative agreement between classical and quantal results for H_2^+ allows us to conclude that such an approach provides a powerful tool in practical calculations of partition functions, dissociation equilibrium constants and thermodynamic properties of diatomic systems. A key point is that the basic formula (492) for the distribution function $W_T^{(d)}$ contains an explicit analytical dependence on the potential energy $U(R)$. Thus, in contrast to the exact quantum-mechanical technique the classical approach does not require knowledge of all rovibrational energy levels of a molecule. Finally, the present analysis assumes the next extension to polyatomic molecules and clusters where the density of energy levels is particularly large, so that the approximation of a quasicontinuum seems to be especially reasonable.

8   Conclusions and Perspectives

We have considered a broad class of elementary processes and new dynamic phenomena in collisions of Rydberg atoms with neutral targets. Such processes include different types of inelastic state-changing transitions, direct and associative ionization, quenching and ion-pair formation reactions. Studies in this field are required for many fundamental and applied problems of atomic and molecular physics, high-resolution spectroscopy and kinetics of gases and low-temperature plasmas as well as for astrophysics and radioastronomy. Theoretical and experimental data for the rate coefficients of elementary processes involving excited and highly excited atomic states are also important for the physics of gas discharges and atomic beams, and for understanding the physical mechanisms in the active medium of a gas and plasma lasers.

We have presented a number of efficient physical approaches and theoretical techniques for describing the identified processes, whereas our attention was primarily focused on the discussion of new achievements in this field. For example, we have concentrated on the detailed consideration of new resonance phenomena associated with the long-range part of an electron–projectile interaction. Special attention was paid to the analysis of the ion core effects in inelastic and ionizing collisions involving Rydberg atoms with large energy transfer. The theoretical description of the most important reactions was provided by a discussion of the cross section behavior in its dependence on the main physical parameters such as the principal and orbital quantum numbers, transition energy defect, relative velocity of colliding partners etc. As is apparent from the comparison of numerous calculations with available experimental data, modern collision theory involving Rydberg atoms and neutral targets gives a quite adequate explanation and quantitative description of many observed phenomena in the physics of highly excited states.

Nevertheless, a lot of fundamental problems still remain to be solved. This especially concerns further detailed studies of resonant excitation (deexcitation), ionization and recombination processes induced by interaction of a neutral projectile with the ionic core of the Rydberg atom. Studies in this field have attracted renewed interest due to a broad class of applications in the kinetics of relaxation and recombination processes in gases and plasmas. The key point consists in the elaboration of an efficient theoretical approach, taking into account the multi-state character of collisional transitions. There is a number of unsolved problems in the theory of resonance quenching and ion-pair formation reactions in thermal collisions of Rydberg atoms with neutral targets having small electron affinities. Experimental and theoretical investigations in this direction are important for negative-ion spectroscopy.

In our opinion, considerable efforts should be aimed towards a self-consistent extension of collision theory to the case of excited atoms having more than one valence electrons in the open shell. Some close field is connected with the incorporation of autoionizing states and electron correlation effects into a description of new dynamic effects in collisions of Rydberg atoms with neutral particles. This seems to be particularly important since extensive experimental material has been obtained for many complex Rydberg atoms for the past few years. On the other hand, further theoretical studies are certainly required to extend some of the available physical approaches to non-spherical neutral targets. The situation with polyatomic molecules and clusters as perturbing particles remains more complicated compared to the ground-state atoms. Of special interest is to study collision phenomena involving coherent elliptic Rydberg states in electric and magnetic fields.

Acknowledgements

The author is grateful to I.L. Beigman and I.I. Fabrikant for a productive collaboration stimulating part of the present work. Support of part of this work by INTAS (project 99-01326) and RFBR (grants 99-02-16602 and 00-02-17245) is gratefully acknowledged.

References

[1] R.F. Stebbings and F.B. Dunning (eds.) *Rydberg States of Atoms and Molecules* (Cambridge University Press, Cambridge 1983)

[2] T.F. Gallagher: *Rydberg Atoms* (Cambridge University Press, Cambridge 1994)

[3] V.S. Lebedev and I.L. Beigman: *Physics of Highly Excited Atoms and Ions* (Springer, Berlin, Heidelberg 1998)

[4] J-P. Connerade: *Rydberg Atom* (Cambridge University Press, Cambridge 1998)

[5] M.J. Seaton: Rep. Prog. Phys. **46**, 167 (1983)

[6] N.R. Badnell and M.J. Seaton: J. Phys. B**32**, 3955 (1999)

[7] M. Aymar, C.H. Greene and E. Luc-Koenig: Rev. Mod. Phys. **68**, 1015 (1996)

[8] V.S. Lisitsa: *Atoms in Plasmas* (Springer-Verlag, Berlin, 1994)

[9] L.A. Bureeva and V.S. Lisitsa: *Perturbed Atom* (IzdAT, Moscow 1997) (in Russian)

[10] Yong Li and Baiwen Li: J. Phys. B**30**, 547 (1997)

[11] F. Merkt, A. Osterwalder, R. Seiler, R. Signorell, H. Palm, H. Schmutz and R. Gunzinger: J. Phys. B**31**, 1705 (1998)

[12] J.P. Santos, F. Mota-Furtado, M.F. Laranjeira and F. Parente: Phys. Rev. A**59**, 1703 (1999)

[13] V. Averbukh, N. Moiseyev, P. Schmelcher and L.S. Cederbaum: Phys. Rev. A**59**, 3695 (1999)

[14] A. Kips, W. Vassen, W. Hogervorst and P.A. Dando: Phys. Rev. A**58**, 3043 (1999)

[15] P. Sorensen, J.C. Day, B.D. DePaola, T. Ehrenreich, E. Horsdal-Pedersen and L. Kristensen: J. Phys. B**32**, 1125 (1999)

[16] S. Haroche and J. Raimond: Adv. Atom. Molec. Phys. **20**, 347 (1985)

[17] V.V. Apollonov, S.I. Derzhavin, V.I. Kislov, V.V. Kuzminov, D.A. Mashkovsky and A.M. Prohorov: Phys. Rev. A58, 2342 (1998)

[18] J.A.C. Gallas, G. Leuchs, H. Walther and H. Figger: Adv. Atom. Molec. Phys. 20, 413 (1985)

[19] S. Gu, S. Gong, B. Liu, J. Wang, Z. Dai, T. Lei and B. Li: J. Phys. B30, 467 (1997)

[20] J.M. Weber, K. Ueda, D. Klar, J. Kreil, M-W. Ruf and H. Hotop: J. Phys. B32, 2381 (1999)

[21] J. Bömmels, J.M. Weber, A. Gopalan, H. Herschbach, E. Leber, A. Schramm, K. Ueda, M-W. Ruf and H. Hotop: J. Phys. B32, 2395 (1999)

[22] G. Casati, B.V. Chirikov, D.L. Shepelyansky and I. Guarneri: Phys. Rep. 154, 77 (1987)

[23] N.B. Delone and V.P. Krainov: Usp. Fiz. Nauk. 168, 531 (1998)

[24] G. Casati and B. Chirikov: in: *Quantum Chaos: Between Order and Disorder*, ed. by G. Casati and B.V. Chirikov (Cambridge University Press, Cambridge 1995)

[25] J-P. Connerade: J. Phys. B30, L31 (1997)

[26] I.Sh. Averbukh and N.F. Perel'man: Usp. Fiz. Nauk 161, 41 (1991)

[27] M. Strechle, U. Weichmann and G. Gerber: Phys. Rev. A58, 450 (1998)

[28] D. Delande and J. Zakrzewski: Phys. Rev. A 58, 466 (1998)

[29] B.S. Mecking and P. Lambropoulos: J. Phys. B31, 3353 (1998)

[30] H. Carlsen and O. Goscinski: Phys. Rev. A59, 1063 (1999)

[31] Atoms in Astrophysics, ed. by P.G. Burke, W.B. Eissner, D.G. Hummer and I.C. Percival (Plenum Press, New York and London, 1983)

[32] M.A. Gordon and R.L. Sorochenko (eds.) *Radio Recombination Lines: 25 years of Investigations* (Kluwer Academic Publishers, Dordrecht, Boston, London 1990)

[33] R.L. Sorochenko: Astronom. and Astrophys. Trans. 11, 199 (1996)

280   V.S. LEBEDEV

[34] I. Percival and D. Richards: Adv. Atom. Molec. Phys. **11**, 1 (1975)

[35] T.F. Gallagher: Phys. Rep. **210**, 319 (1992)

[36] I.L. Beigman and V.S. Lebedev: Phys. Rep. **250**, 95 (1995)

[37] L.D. Landau and E.M. Lifshitz: *Quantum Mechanics* (Pergamon, Oxford 1977)

[38] L.D. Landau and E.M. Lifshitz: *Mechanics* (Pergamon, Oxford 1969)

[39] E.M. Lifshitz and L.P. Pitaevskii: *Physical Kinetics* (Pergamon Press, Oxford 1981)

[40] V.B. Berestetskii, E.M. Lifshitz and L.P. Pitaevskii: *Quantum Electrodynamics* (Pergamon, Oxford 1982)

[41] K. Pachucki, D. Leibfried, M. Weitz, A. Huber, W. König and T.W. Hänsch: J. Phys. B**29**, 177 (1996)

[42] H.A. Bethe and E.E. Salpeter: *Quantum Mechanics of One- and Two- Electron Atoms*, 2nd edn. (Plenum, New York 1977)

[43] Ya.F. Verolainen and A.Ya. Nikoláich: Usp. Fiz. Nauk **137**, 303 (1982)

[44] K. Bockasten: Phys. Rev. A**9**, 1087 (1974)

[45] C.-J. Lorenzen, K. Niemax and L.R. Pendrill: Opt. Commun. **39**, 370 (1981)

[46] C.-J. Lorenzen and K. Niemax: Phys. Scr. **27**, 300 (1983)

[47] A.I. Ferguson and M.H. Dunn: Opt. Commun. **23**, 227 (1977)

[48] B.P. Stoicheff and E. Weinberger: Can. J. Phys. **57**, 2143 (1979)

[49] C.J. Sansonetti and K.-H. Weber: J. Opt. Soc. Am. B **2**, 1385 (1985)

[50] C.-J. Lorenzen and K. Niemax: Z. Phys. A**311**, 249 (1983)

[51] K.-H. Weber and C.J. Sansonetti: J. Opt. Soc. Am. A**1**, 1233 (1984)

[52] K.-H. Weber and C.J. Sansonetti: Phys. Rev. A**35**, 4650 (1987)

[53] L. Qu, Z. Wang and B. Li: J. Phys. B **31**, 2469 (1998)

[54] M.A. Baig, M. Akram, N.K. Piracha, M.S. Mahmood, S.A. Bhatti and N. Ahmad: J. Phys. B**28**, 1421 (1995)

[55] S. Dyubko, V. Efremov, S. Podnos, X. Sun and K.B. MacAdam: J. Phys. B**30**, 2345 (1997)

[56] S. Kaur and R. Srivastava: J. Phys. B **32**, 2323 (1999)

[57] P. Nosbaum, A. Bleton, L. Cabaret, J. Yu, T.F. Galagher and P. Pillet: J. Phys. B**28**, 1707 (1999)

[58] T.K. Fang and Y.K. Ho: J. Phys. B **32**, 3863 (1999)

[59] R.R. Jones: Phys. Rev. A**58**, 2608 (1998)

[60] C. Laughlin: J. Phys. B**28**, 2787 (1995)

[61] R. van Leewen, W. Ubachs and W. Hogervorst: J. Phys. B**27**, 3891 (1994)

[62] R. van Leewen, M. Aymar, W. Ubachs and W. Hogervorst: J. Phys. B**29**, 1007 (1996)

[63] W. Mende and M. Kock: J. Phys. B**29**, 655 (1996)

[64] C.J. Dai and J. Lu: J. Phys. B**29**, 2473 (1996)

[65] M. Meyer, J. Lacoursiere, P. Morin and F. Combet Farnoux: J. Phys. B**27**, 3875 (1994)

[66] J.J. Groote, M. Masili and J.E. Hornos: J. Phys. B**31**, 4755 (1998)

[67] W.C. Martin: J. Res. Nat. Bur. Std. **64** A, 79 (1960)

[68] W.H. Wing, K.R. Lea, W.E. Lamb, in: *Atomic Physics,* ed. by S.J. Smith and G.K. Walters (Plenum, New York 1973), Vol. 3, p. 119

[69] I.I. Sobel'man: *Atomic Spectra and Radiative Transitions* (Springer-Verlag, Berlin 1992)

[70] D. Klar, K. Ueda, J. Ganz, K. Harth, W. Buβert, S. Baier, J.M. Weber, M-F. Ruf and H. Hotop: J. Phys. B**27** , 4897 (1994)

[71] N.K. Piracha, B. Suleman, S.H. Khan and M.A. Baig: J. Phys. B**28**, 2525 (1995)

[72] N.K. Piracha, M.A. Baig, S.H. Khan and B. Suleman: J. Phys. B30, 1151 (1997)

[73] M. Ahmed, M.A. Zia, M.A. Baig and B. Suleman: J. Phys. B30, 2155 (1997)

[74] L. Avaldi, R. Camilloni, G. Stefani, C. Comicioli, M. Zacchigna, K.C. Prince, M. Zitnik, C. Quaresima, C. Ottaviani, C. Crotti and P. Perfetti: J. Phys. B29, L737 (1996)

[75] S. Yoon and W.L. Glab: J. Phys. B27, 4133 (1994)

[76] A. Ehresmann, G. Mentzel, K-H. Schartner and H. Schmoranzer: J. Phys. B29, 991 (1996)

[77] M. Gisselbrecht, A.Marquette and M.Meyer: J. Phys. B31, L977 (1998)

[78] M.L. Goldberger and K.M. Watson: *Collision Theory* (Wiley, New York 1964)

[79] R.G. Newton: *Scattering Theory of Waves and Particles* (McGraw-Hill, New York 1966)

[80] E. Fermi: Nuovo Cimento 11, 157 (1934)

[81] A.P. Hickman, R.E. Olson and J. Pascale: in: *Rydberg States of Atoms and Molecules*, ed. by R.F. Stebbings and F.B. Dunning (Cambridge University Press, Cambridge 1983) Chap. 6, p. 187

[82] M. Matsuzawa: in *Rydberg States of Atoms and Molecules*, ed. by R.F. Stebbings and F.B. Dunning (Cambridge University Press, Cambridge 1983) Chap. 8, p. 267

[83] M.R. Flannery: in *Rydberg States of Atoms and Molecules*, ed. by R.F. Stebbings and F.B. Dunning (Cambridge University Press, Cambridge 1983) Chap. 11, p. 393

[84] D.W. Norcross and L.A. Collins: Adv. Atom. Molec. Phys. 8, 341 (1982)

[85] Y. Itikawa: Phys. Rep. 46, 117 (1978)

[86] N.F. Lane: Rev. Mod. Phys. 52, 28 (1980)

[87] I. Shimamura: in *Electron–Molecule Scattering*, ed. by I. Shimamura and K. Takayanagi (Plenum Press, New York 1984)

[88] M.A. Morrison: Adv. Atom. Molec. Phys. **24**, 51 (1988)

[89] I.I. Fabrikant: Comments Atom. Molec. Phys. **32**, 267 (1996)

[90] F.B. Dunning: J. Phys. B**28**, 1645 (1995)

[91] L. Spruch, T.F. O'Malley and L. Rosenberg: Phys. Rev. Lett. **5**, 347 (1960); T.F. O'Malley, L. Spruch and L. Rosenberg: Phys. Rev. **125**, 1300 (1962)

[92] T.F. O'Malley: Phys. Rev. **130**, 1020 (1963); Phys. Rev. A **134**, 1188 (1964); T.F. O'Malley and R.W. Crompton: J. Phys. B **13**, 3451 (1980)

[93] K.S. Golovanivskii and A.P. Kabilan: Zh. Eksp. Teor. Fiz. **80**, 2210 (1981)

[94] M. Weyhreter, B. Barzick, A. Mann and F. Linder: Z. Phys. D **7**, 333 (1988)

[95] R.J.Gulley, D.T.Alle, M.J.Brennan, M.J.Brunger and S.J.Buckman: J. Phys. B **27**, 2593 (1994)

[96] J.C. Gibson, R.J. Gulley, J.P. Sulivan, S.J. Buckman, V. Chan and P.D. Burrow: J. Phys. B**29**, 3177 (1996)

[97] K.D. Heber, P.J. West and E. Matthias: Phys. Rev. A **37**, 1438 (1988); J. Phys. D **21**, 63 (1988)

[98] D.C. Thompson, E. Kammermayer, B.P. Stoicheff and E. Weinberger: Phys. Rev. A **36**, 2134 (1987)

[99] A.A. Radtsig and B.M. Smirnov: in *Reference Data on Atoms, Molecules and Ions* Springer Ser. Chem. Phys., Vol. **31**, ed. by V.I. Goldanskii et al (Springer, Berlin 1985)

[100] D.R. Lide (ed.): *Handbook of Chemistry and Physics* 79th edn., (Boca Raton 1999), pp. 10–162; S.H. Patil: Atom. Data Nucl. Data Tables **71**, 41 (1999)

[101] V.P. Shevelko: *Atoms and Their Spectroscopic Properties* (Springer, Berlin 1997)

[102] I.I. Fabrikant, Opt. Spektrosk. **53**, 131 (1982) [Opt. Spectrosc. (USSR) **53**, 223 (1982)]

[103] I.I. Fabrikant: J. Phys. B **19**, 1527 (1986)

[104] A.L. Sinfailam and R.K. Nesbet: Phys. Rev. A **7**, 1987 (1973)

[105] V.S. Lebedev and V.S. Marchenko: Zh. Eksp. Teor. Fiz. **91**, 428 (1986) [Engl. Transl.: Sov. Phys. - JETP **64**, 251 (1986)

[106] V.S. Lebedev and V.S. Marchenko: J. Phys. B **20**, 6041 (1987)

[107] U. Thumm and D.W. Norcross: Phys. Rev. A **45**, 6349 (1992)

[108] E.M. Karule: J. Phys. B **5**, 2051 (1972)

[109] A.R. Johnston and P.D. Burrow: J. Phys. B **15**, L745 (1982)

[110] V.M. Borodin, I.I. Fabrikant and A.K. Kazansky: Phys. Rev. A **44**, 5725 (1991)

[111] I.I. Fabrikant: Phys. Rev. A **45**, 6404 (1992)

[112] H. Heinke, J. Lawrenz, K. Niemax and K.H. Weber: Z. Phys. A **312**, 329 (1983)

[113] D.C. Thompson, E. Weinberger, G.X. Xu and B.P. Stoicheff: Phys. Rev. A **35**, 690 (1987)

[114] M. Hugon, F. Gounand, P.R. Fournier and J. Berlande: J. Phys. B **16**, 2531 (1983)

[115] D. Sundholm: J. Phys. B **28**, L399 (1995)

[116] V.V. Petrunin, H.H. Andersen, P. Balling, and T. Andersen: Phys. Rev. Lett. **76**, 744 (1996)

[117] H.S.W. Massey: Proc. Cambridge Phil. Soc. **28**, 99 (1931)

[118] S. Altshuler: Phys. Rev. **107**, 114 (1957)

[119] M.T. Frey, S.B. Hill, X. Ling, K.A. Smith, F.B. Dunning and I.I. Fabrikant: Phys. Rev. A **50**, 3124 (1994)

[120] M.T. Frey, S.B. Hill, K.A. Smith, F.B. Dunning and I.I. Fabrikant: Phys. Rev. Lett. **75**, 810 (1995)

[121] S.B. Hill, M.T. Frey, F.B. Dunning and I.I. Fabrikant: Phys. Rev. A **53**, 3348 (1996)

[122] J.C. Gibson, L.A. Morgan, R.J. Gulley, M.J. Brunger, C.T. Bundschu and S.J. Buckman: J. Phys. B**29**, 3197 (1996)

[123] C.T. Bundschu, J.C. Gibson, R.J. Gulley, M.J. Brunger, S.J. Buckman, N. Sanna and F.A. Gianturco: J. Phys. B **30**, 2239 (1997)

[124] M.H. Bettega, M.A.P. Lima and L.G. Ferreira: J. Phys. B**31**, 2091 (1998)

[125] M. Allan, K.R. Asmis, D.B. Popovic, M. Stepanovic, N.J. Mason and J.A. Davies: J. Phys. B**29**, 4727 (1996)

[126] I.C. Walker, J.M. Gingell, N.J. Mason and G. Marston: J. Phys. B**29**, 4749 (1996)

[127] J.I. Gersten: Phys. Rev. A **14**, 1354 (1976)

[128] A. Omont: J. de Phys. (Paris) **38** , 1343 (1977)

[129] J. Derouard and M. Lombardi: J. Phys. B **11**, 3875 (1978)

[130] E de Prunelé and J. Pascale: J. Phys. B **12**, 2511 (1979)

[131] B. Kaulakys: J. Phys. B **17**, 4485 (1984)

[132] V.S. Lebedev and V.S. Marchenko: Zh. Eksp. Teor. Fiz. **88**, 754 (1985) [Engl. Transl.: Sov. Phys. - JETP **61**, 443 (1985)]

[133] J.Q. Sun and P.J. West: J. Phys. B **23**, 4119 (1990)

[134] J.Q. Sun, E.Matthias, K.D. Heber, P.J. West and J. Güdde: Phys. Rev. A **43**, 5956 (1991)

[135] L. Sirko and K. Rosinski: J. Phys. B **24**, L75 (1991)

[136] V.S. Lebedev: J. Phys. B **25**, L131 (1992)

[137] V.S. Lebedev: Zh. Eksp. Teor. Fiz. **103**, 50 (1993) [Engl. Transl.: JETP **76**, 27 (1993)]

[138] V.S. Lebedev and I.I. Fabrikant: Phys. Rev. A **54**, 2888 (1996)

[139] V.S. Lebedev and I.I. Fabrikant: J. Phys. B **30**, 2649 (1997)

[140] V.S. Lebedev: J. Phys. B **31**, 1579 (1998)

[141] M.J. Seaton: Proc. Phys. Soc. **79**, 1105 (1962)

[142] D.A. Varshalovich, A.N. Moskalev and V.K. Khersonsky: *Quantum Theory of the Angular Momentum* (World Scientific, Singapore 1988)

[143] A.B. Migdal: *Qualitative Methods in Quantum Theory* (Benjamin, New York 1977)

[144] M. Abramovitz and I.A. Stegun: *Handbook of Mathematical Functions* (Dover, New York 1965)

[145] M. Matsuzawa: J. Phys. B **12**, 3743 (1979)

[146] B. Kaulakys: J. Phys. B **18**, L167 (1985)

[147] F. Gounand and L. Petitjean: Phys. Rev. A **30**, 61 (1984)

[148] D.R. Bates and S.P. Khare: Proc. Phys. Soc. (London) **85**, 2331 (1965)

[149] M.R. Flannery: Ann. Phys. (N.Y.) **61**, 465 (1970); ibid. **79**, 480 (1973)

[150] L. Petitjean and F. Gounand: Phys. Rev. A. **30**, 2946 (1984)

[151] L.P. Pitaevskii: Zh. Eksp. Teor. Fiz. **42**, 1326 (1962)

[152] E. Fermi: Ricerca Sci. **VII-II**, 13 (1936)

[153] G.F. Chew: Phys. Rev. **80**, 196 (1950); G.F. Chew and G.C. Wick: Phys. Rev. **85**, 636 (1952); G.F. Chew and M.L. Goldberger: Phys. Rev. **87**, 778 (1952)

[154] V.A. Alekseev and I.I. Sobel'man: Zh. Eksp. Teor. Fiz. **49**, 1274 (1965) [Sov. Phys. - JETP **22**, 882 (1966)]

[155] M.Matsuzawa: J. Chem. Phys. **55**, 2685 (1971); errata **58**, 2674 (1973); J. Electron. Spectrosc. Relat. Phenon. **4**, 1 (1974)

[156] M. Matsuzawa: Phys. Rev. A **9**, 241 (1974); ibid. **20**, 860 (1979)

[157] G.N. Fowler and T.W. Preist: J. Chem. Phys. **56**, 1601 (1972)

[158] T.W. Preist: J. Chem. Soc. Farad. Trans. Part I, **68**, 661 (1972)

[159] M.R. Flannery: Phys. Rev. A **22**, 2408 (1980)

[160] A.P. Hickman: Phys. Rev. A **18**, 1339 (1978)

[161] A.P. Hickman: Phys. Rev. A **19**, 994 (1979)

[162] A.P. Hickman: Phys. Rev. A **23**, 87 (1981)

[163] M.Hugon, F.Gounand, P.R.Fournier and J.Berlande: J. Phys. B **13**, 1585 (1980); ibid. **12**, 2707 (1979)

[164] M.Hugon, P.R.Fournier and E. de Prunelé: J. Phys. B **14**, 4041 (1981)

[165] M.Hugon, B.Sayer, P.R.Fournier and F.Gounand: J. Phys. B **15**, 2391 (1982)

[166] K.Sasano, Y.Sato and M.Matsuzawa: Phys. Rev. A **27**, 2421 (1983)

[167] Y.Sato and M.Matsuzawa: Phys. Rev. A **31**, 1366 (1985)

[168] Y.Hahn: J. Phys. B **14**, 985 (1981); ibid. **15**, 613 (1982)

[169] M. Inokuti: Rev. Mod. Phys. **43**, 297 (1971)

[170] L. Petitjean, F. Gounand and P.R. Fournier: Phys. Rev. A. **30**, 71 (1984); ibid. **30**, 736 (1984); ibid. **33**, 143 (1986)

[171] V.S. Lebedev: J. Phys. B **24**, 1977 (1991)

[172] B.P.Kaulakys: Zh. Eksp. Teor. Fiz. **91**, 391 (1986) [Engl. Transl.: Sov. Phys. - JETP **64**, 229 (1986)]

[173] B.P.Kaulakys: J. Phys. B **24**, L127 (1991)

[174] H. van Regemorter: J. Phys. B **27**, 3863 (1994)

[175] D. Hoang-Binh and H. van Regemorter: J. Phys. B**28**, 3147 (1995)

[176] D. Vrinceanu and M.R. Flannery: Phys. Rev. Lett. **82**, 3412–3415 (1999)

[177] D. Vrinceanu and M.R. Flannery: Phys. Rev. A **60**, 1053 (1999)

[178] I. Samengo: Phys. Rev. A **58**, 2767 (1998)

[179] Y. Fang, A.N. Vasil'ev and V.V. Mukhailin: Phys. Rev. A **58**, 3683 (1998)

[180] S.T. Butler and R.A. May: Phys. Rev. A **137**, 10 (1965)

[181] B.M. Smirnov: *Ions and Excited Atoms* (Atomizdat, Moscow 1974) (in Russian)

[182] M. Matsuzawa: J. Phys. B **8**, 2114 (1975); corrigendum **9**, 2559 (1976)

[183] D. Hoang Binh and H. van Regemorter: J. Phys. B**30**, 2403 (1997)

[184] V. Fock: Z. Phys. **98**, 145 (1935)

[185] M. Matsuzawa: J. Phys. B **8**, L382 (1975); ibid. **10**, 1543 (1977)

[186] B.P. Kaulakys, L.P. Presnyakov and P.D. Serapinas: Pis'ma Zh. Eksp. Teor. Fiz. **30** 60 (1979) [Engl. Transl.: JETP Lett. **30**, 53 (1980)]

[187] V.M. Borodin and A.K. Kazansky: Zh. Eksp. Teor. Fiz. **97**, 445 (1990) [Engl. Transl.: Sov. Phys. - JETP **70**, 252 (1990)]; J. Phys. B **25**, 971 (1992)

[188] V.M. Borodin, A.K. Kazansky, D.B.Khrebtukov and I.I.Fabrikant: Phys. Rev. A **48**, 479 (1993)

[189] M. Hugon, F. Gounand and P.R. Fournier: J. Phys. B **13**, L109 (1980)

[190] R.K. Janev: Adv. At. Mol. Phys. **12**, 1 (1976)

[191] M.I. Chibisov and R.K. Janev: Phys. Rep. **166**, 1 (1988)

[192] K.W. McLaughlin and Duquette: Phys. Rev. Lett. **72**, 1176 (1974)

[193] C. Desfrançois, H. Abdoul-Carime, C. Adjouri, N. Khelifa and J.P. Shermann: Phys. Rev. Lett. **73**, 2436 (1994)

[194] C. Desfrançois, P. Bailon, J.P. Shermann, S.T. Arnold, J.P. Hendricks and K.H. Bowen: Phys. Rev. Lett. **72**, 48 (1994)

[195] C. Desfrançois, H. Abdoul-Carmine, J.P. Schermann, J.H. Hendricks, S.A. Lyapustina and K.H. Bowen: J. Chem. Phys. **105**, 3472 (1996)

[196] R.N. Compton, H.S. Carman Jr., C. Desfrançois, H. Abdoul-Carmine, J.P. Schermann, J.H. Hendricks, S.A. Lyapustina and K.H. Bowen: J. Chem. Phys. **105**, 3472 (1996)

[197] M. Reicherts, T. Roth, A. Gopalan, M.-W. Ruf, H. Hotop, C. Desfrançois and I.I. Fabrikant: Europhys. Lett. **40**, 129 (1997)

[198] I.I. Fabrikant: J. Phys. B **26** , 2533 (1993)

[199] D.B. Khrebtukov and I.I. Fabrikant: Phys. Rev. A **54**, 2906 (1996)

[200] I.I. Fabrikant: J. Phys. B **31** , 2921 (1998)

[201] I.I. Fabrikant and M.I. Chibisov: Phys. Rev. A **61**, 022718 (2000)

[202] C. Desfrançois: Phys. Rev. A **51**, 3667 (1995)

[203] I.I. Fabrikant and V.S. Lebedev: J. Phys. B **33**, 1521 (2000)

[204] S.J. Buckman and C.W. Clark: Rev. Mod. Phys. **66**, 539 (1994)

[205] H.H. Andersen, T. Andersen, and U.V. Pedersen: J. Phys. B **31**, 2239 (1998)

[206] V.A. Dzuba and G.F. Gribakin: J. Phys. B **30**, L483 (1998)

[207] Yu.N. Demkov and V.I. Osherov: Zh. Eksp. Teor. Fiz. **53**, 1157 (1966) [Sov. Phys. JETP **22**, 804 (1966)]

[208] B.M. Smirnov: Zh. Eksp. Teor. Fiz. **51**, 466 (1966) [Engl. Transl.: Sov. Phys. JETP 24, 314 (1966)]

[209] R.K. Janev and A. Salin: J. Phys. B **5**, 177 (1972)

[210] C. Herring: Rev. Mod. Phys. **34** , 631 (1962)

[211] L.P. Gor'kov and L.P. Pitaevskii: Dokl. Akad. Nauk **151**, 823 (1963) [Engl. Transl.: Sov. Phys.–Dokl. **8**, 788 (1963)]

[212] J.B. Delos: Rev. Mod. Phys. **53**, 287 (1981)

[213] A.A. Radtsig and B.M. Smirnov: Zh. Eksp. Teor. Fiz. **60**, 521 (1971) [Engl. Transl.: Sov. Phys. JETP **33**, 282 (1971)]

[214] L.P. Presnyakov and D.B. Uskov: Zh. Eksp. Teor. Fiz. **86**, 882 (1984) [Engl. Transl.: Sov. Phys. JETP **59**, 515 (1984)]

[215] J. Yuan and L. Fritsche: Phys. Rev. A **55**, 1020 (1997)

[216] J. Yuan and C.D. Lin: Phys. Rev. A **58**, 2824 (1998)

[217] K. Harth, M. Raab, J. Ganz, A. Siegel, M.-W. Ruf, and H. Hotop: Opt. Commun. **54**, 343 (1985)

[218] W.R. Pendleton, Jr., M. Larsson, and B. Mannfors: Phys. Rev. A **28**, 3223 (1983)

[219] T.F. Gallagher, S.A. Edelstein and R.M. Hill: Phys. Rev. A **15**, 1945 (1977)

[220] C.E. Burkhardt and J.J Leventhal: Phys. Rev. A **43**, 110 (1991)

[221] M. Harnafi and B. Dubreil: Phys. Rev. A **31**, 1375 (1985)

[222] R.E. Olson: Phys. Rev. A **15**, 631 (1977)

[223] T.F. Gallagher and W.E. Cooke: Phys. Rev. A **19**, 2161 (1979)

[224] J. Boulmer, J.-F. Delpech, J.-C. Gauthier and K. Safinya: J. Phys. B**14**, 4577 (1981)

[225] M. Chapelet, J. Boulmer, J.C. Gauthier and J.F. Delpech: J. Phys. B **15**, 3455 (1982)

[226] R. Kachru, T.F. Gallagher F. Gounand, K.A. Safinya and W. Sandner: Phys. Rev. A **27** 795 (1983)

[227] E.E. Nikitin and S.Ya. Umansky: *Theory of Slow Atomic Collisions* (Springer, New York, 1984)

[228] F. Gounand and L. Petitjean: Phys. Rev. A **32**, 793 (1985)

[229] E. de Prunelé: Phys. Rev. A **27**, 1831 (1983)

[230] V.S. Lebedev, V.S. Marchenko and S.I. Yakovlenko: Izv. Akad. Nauk SSSR, Ser. Fiz. **45**, 2395 (1981)

[231] V.S. Lebedev and V.S. Marchenko: Zh. Eksp. Teor. Fiz. **84**, 1623 (1983) [Engl. Transl.: Sov. Phys. - JETP. 57, 946 (1983)]

[232] C. Higgs, K.A.Smith, F.B. Dunning and R.F. Stebbings: J. Chem. Phys. **75**, 754 (1981)

[233] J. Supronovicz, J.B. Atkinson and J. Krause: Phys. Rev. A **30**, 112 (1984)

[234] M. Lukaszewski and I. Jackowska: J. Phys. B **21**, L659 (1988); ibid. **24**, 2047 (1991)

[235] H. Liu and B. Li: Phys. Rev. A. **46**, 1291 (1992)

[236] F.B. Dunning and R.F. Stebbings: in *Rydberg States of Atoms and Molecules*, ed. by R.F.Stebbings and F.B.Dunning (Cambridge University Press, Cambridge 1983) Chap. 9, p. 315

[237] S. Preston and N.F. Lane: Phys. Rev. A **33**, 148 (1986)

[238] A. Kumar, N.F. Lane and M. Kimura: Phys. Rev. A **39**, 1020 (1989)

[239] M. Kimura and N.F. Lane: Phys. Rev. A **42**, 1258 (1990)

[240] T. Yoshizawa and M. Matsuzawa: J. Phys. Soc. Jpn. **54**, 918 (1985)

[241] A. Kalamarides, L.N. Goeller, K.A. Smith, F.B. Dunning, M. Kimura and N.F. Lane: Phys. Rev. A **36**, 3108 (1987)

[242] I.I. Fabrikant and R.S. Wilde: J. Phys. B **32**, 235 (1999)

[243] B.G. Zollars, C. Higgs, F. Lu, C.W. Walter, L.G. Gray, K.A. Smith, F.B. Dunning and R.F. Stebbings: Phys. Rev. A **32**, 3330 (1985)

[244] B.G. Zollars, C.W. Walter, F. Lu, C.B. Johnson, K.A. Smith and F.B. Dunning: J. Chem. Phys. **84**, 5589 (1986)

[245] I.M. Beterov, G.L. Vasilenko, B.M. Smirnov and N.V. Fateyev: Sov. Phys. - JETP **93**, 31 (1987)

[246] X. Ling, B.G. Lindsay, K.A. Smith, F.B. Dunning: Phys. Rev. A **45**, 242 (1992)

[247] X. Ling, K.A. Smith and F.B. Dunning: Phys. Rev. A **47**, R1 (1993)

[248] X. Ling, M.T. Frey, K.A. Smith and F.B. Dunning: Phys. Rev. A **48**, 1252 (1994)

[249] I.I. Sobel'man, L.A. Vainshtein and E.A. Yukov: *Excitation of Atoms and Broadening of Spectral Lines* (Springer, Berlin 1995)

[250] V.A. Smirnov: Opt. Spectrosk. **37** , 407 (1974) [Engl. Transl.: Opt. Spectosc. **37**, 231 (1974)]

[251] M.R. Flannery: J. Phys. B **13**, L657 (1980)

[252] A.P. Hickman: J. Phys. B **14**, L419 (1981)

[253] M. Matsuzawa: J. Phys. B **14**, L553 (1981)

[254] M. Matsuzawa: J. Phys. B **17**, 795 (1984)

[255] B.P. Kaulakys: Litov. Fiz. Sb. **22**, 3 (1982)

[256] V.S. Marchenko: Khim. Fiz. **4**, 595 (1985) [Engl. Transl.: Sov. J. Chem. Phys. **4**, 963 (1987)]

[257] A.Z. Devdariani, A.N. Klucharev, A.V. Lazarenko and V.A. Sheverev: Pis'ma Zh. Tekh. Fiz. **4**, 1013 (1978) [Engl. Transl.: Sov. Tech. Phys. Lett. **4**, 408 (1978)]

[258] V.P. Zhdanov and M.I. Chibisov: Zh. Eksp. Teor. Fiz. **74**, 75 (1978)

[259] R.K. Janev and A.A. Mihajlov: Phys. Rev. A **21**, 819 (1980); A.A. Mihajlov and R.K. Janev: J. Phys. B **14**, 1639 (1981)

[260] E.L. Duman and I.P. Shmatov: Zh. Eksp. Teor. Fiz. **78**, 2116 (1980)

[261] V.S. Lebedev and V.S. Marchenko: Khim. Fiz. **3**, 210 (1984) [Engl. Transl.: Sov. J. Chem. Phys. **3**, 311 (1986)]

[262] V.S. Lebedev: J. Phys. B **24**, 1993 (1991)

[263] V.S. Lebedev: in *Dymamics of Elementary Atomic–Molecular Processes in Gas and Plasma*, ed. by V.A. Shcheglov (Nova Science Publishers, New York 1996), p. 255

[264] V.A. Smirnov and A.A. Mihajlov: Opt. Spectrosk. **30**, 984 (1971) [Engl. Transl.: Opt. Spectrosc. **5**, 525 (1971)]

[265] R.K. Janev and A.A. Mihajlov: Phys. Rev. A **20**, 1890 (1979)

[266] A.N. Klucharev and M.L. Janson: *Elementary Processes in the Plasmas of Alkali Metals* (Energoatomizdat, Moscow 1988) (in Russian)

[267] V.A. Ivanov, V.S. Lebedev and V.S. Marchenko: Zh. Eksp. Teor. Fiz. **94**, 86 (1988) [Engl. Transl.: Sov. Phys. - JETP **67**, 2225 (1988)]

[268] V.A. Ivanov, V.S. Lebedev and V.S. Marchenko: Pis'ma Zh. Tekh. Fiz. **14**, 1575 (1988) [Engl. Transl.: Sov. Tech. Phys. Lett. **14**, 686 (1988)]

[269] A.M. Dykhne and G.L. Yudin: Usp. Fiz. Nauk **125**, 377 (1978) [Engl. Transl.: Sov. Phys. Usp. **21**, 549 (1978)]

[270] W.R. Wadt: J. Chem. Phys. **68**, 402 (1978); *ibid.* **73**, 3915 (1980)

[271] D.R. Bates and A. Dalgarno: in *Atomic and Molecular Processes*, ed by D.R. Bates (Academic Press, New York and London 1962)

[272] L.M. Biberman, V.S. Vorob'ev and I.T. Yakubov: *Kinetics of Non-equilibrium Low-Temperature Plasmas* (Consultants Bureau 1987)

[273] C.T. Belyaev and G.I. Budker: in *Physics of Plasma and Problem of Controlled Thermonuclear Reactions*, ed. by M.A Leontovich, Vol. 3 (Academy of Science of the USSR Press, Moscow 1958) p. 41

[274] A.V. Gurevich and L.P. Pitaevskii: Zh. Eksp. Teor. Fiz. **46**, 1281 (1964) [Sov. Phys. JETP **19**, 870 (1964)

[275] A.V. Gurevich: Geomagnetism and Aeronomy **4**, 3 (1964)

[276] D.R. Bates, A.E. Kingston and R.W.P. McWhirter: Proc. Roy. Soc. London. Ser. A **267**, 297 (1963)

[277] D.R. Bates and A.E. Kingston: Planet. Space Sci. **11**, 1 (1963)

[278] V.A. Ivanov: Usp. Fiz. Nauk. **162**, 35 (1992) [Sov. Phys. Usp. **35**, 1 (1992)

[279] L.D. Landau and E.M. Lifshitz: *Statistical Physics* (Pergamon, Oxford, 1980)

[280] N. Davidson: *Statistical Mechanics* (McGraw-Hill Book Co., New York, 1962), pp. 169-210

[281] P.W. Atkins: *Physical Chemistry*, 4th Ed., (Oxford University Press, Oxford, 1990)

[282] *Molecules in the Stellar Environment* edited by U.G. Jrgensen, Lecture Notes in Physics, Vol. 428 (Springer-Verlag, Berlin-Heidelberg, 1994)

[283] V.S. Lebedev, L.P. Presnyakov and I.I. Sobelman: Astron. Zh., **77**, 3090 (2000) [Engl. Transl.: Astron. Rep., **44**, 338 (2000)]

[284] V.S. Lebedev, L.P. Presnyakov and I.I. Sobelman: Pis'ma v ZhETF, **72**, 256 (2000) [Engl. Transl.: JETP Lett., **72**, 178 (2000)]

[285] J.L. Dunham: Phys. Rev. **41**, 721 (1932)

[286] K.P. Huber and G. Herzberg: *Molecular Spectra and Molecular Structure. Constants of Diatomic Molecules*, Vol. IV (Van Nostrand, New York, 1979)

[287] J.E. Mayer and M. Göppert-Mayer: Phys. Rev. **43**, 605 (1933)

[288] *Thermodynamic Properties of Individual Substances*, ed. by L.V. Gurvich *et al.*, Vol. I (Hemisphere Publishing Corporation, New York 1989)

[289] A.J. Sauval and J.B. Tatum: Ap. J. Suppl. Ser. **56**, 193 (1984); J.B. Tatum, Pub. Dom. Ap. Obs. Victoria, **13**, 1 (1966)

[290] D.R. Bates and C.S. McKibbin: J. Phys. B **6**, 2485 (1973)

[291] D.R. Bates and C.S. McKibbin: Proc. Roy. Soc. London, Ser. A **339**, 13 (1974)

[292] D.R. Bates: Proc. Roy. Soc. London, Ser. A **360**, 1 (1978)

[293] M.R. Flannery: in *Recent Studies in Atomic and Molecular Processes*, ed. by A.E. Kingston (Plenum Press 1987), p. 167

[294] M.R. Flannery: J. Chem. Phys. **95**, 8205 (1991)

[295] V.S. Lebedev: to be published

[296] D.R. Bates: Mon. Not. R. Astron. Soc. **111**, 303 (1951); *ibid.* **112**, 40 (1952)

[297] H. Wind: J. Chem. Phys. **42**, 2371 (1965); *ibid.* **43**, 2956 (1965)

[298] D.D. Konowalow, M.E. Rozenkrantz, and M.L. Olson: J. Chem. Phys. **72**, 2612 (1980)

[299] B.M. Smirnov and A.S. Yatsenko: Physics–Uspekhi, **39**, 211 (1996)

[300] R.E. Moss: J. Phys. B **32**, L89 (1999)

INDEX

PHYSICS REVIEWS

Editor·
I M Khalatnikov
L D Landau Institute
of Theoretical Physics
Russian Academy of Sciences
Moscow

Aims and Scope

Physics Reviews reports on significant developments in physics research and presents reviews by scientists from the former Soviet Union and will be of particular interest to research scientists who do not read Russian

World Wide Web Addresses

Additional information is also available through the Publisher's web home page site at http //www cambridgescientificpublishers com

Ordering Information

Each volume consists of an irregular number of parts depending upon extent Issues are available individually as well as by subscription 2004 Volume 21/22.

Orders may be placed with your usual supplier or at the address shown below Journal subscriptions are sold on a per volume basis only Claims for nonreceipt of issues will be honored if made within three months of publication of the issue All issues are dispatched by airmail throughout the world

Subscription Rates

Base list subscription price per volume EUR 120 00 * This price is available only to individuals whose library subscribes to the journal OR who warrant that the journal is for their own use and provide a home address for mailing. Orders must be sent directly to the Publisher and payment must be made by check or credit card

Separate rates apply to academic and corporate/government institutions

'EUR (Euro) The Euro is the worldwide base list currency rate All other currency payments should be made using the current conversion rate set by Publisher Subscribers should contact their agents of the Publisher All prices are subject to change without notice

Orders should be placed through the Publisher at the following address

Cambridge Scientific Publishers Ltd
PO Box 806
Cottenham
Cambridge
CB4 8RT
UK
Tel +44 (0)1954 251283
Fax +44 (0)1954 252517
Email janie wardle@cambridgescientificpublishers.com
Website www.cambridgescientificpublishers.com

Printed in UK

August 2004

Physics Reviews

Notes for Contributors

Manuscripts

Papers should be typed with double spacing and wide margins (3 cm) on good quality paper and submitted in triplicate. Authors may also submit papers on disk in any format. Papers should be sent to I.M. Khalatnikov, L.D Landau Institute of Theoretical Physics, Kosygin Str 2, Moscow, 117940, Russia or to Jane Wardle, Cambridge Scientific Publishers, P.O. Box 806, Cottenham, Cambridge, CB4 8RT, UK. Email. janiewardle@cambridgescientificpublishers.com. Submission of a paper to *Physics Reviews* will be taken to imply that it represents original work not previously published, that is not being considered for publication elsewhere, and that if accepted for publication it will not be published in the same language without the consent of the Editors and Publisher.

Language: The language of publication is English.

Abstract: Each paper requires an abstract of 100-150 words summarizing the significant coverage and findings. It is a condition of acceptance by the Editor of a typescript for publication that the Publishers acquire automatically the copyright in the typescript throughout the world

Figures

All figures should be numbered with consecutive arabic numbers, have descriptive captions and be mentioned in the text. Keep figures separate from the text, but indicate an approximate position for each in the margin.

Preparation: Figures submitted must be of a high enough standard for direct reproduction. Line drawings should be prepared in black (India) ink on white paper or on tracing cloth, with all the lettering and symbols included. Alternatively, good sharp photoprints ("glossies") are acceptable. Photographs intended for half-tone reproduction must be glossy original prints of maximum contrast. Clearly label each figure with author's name and figure number, indicate "top" where this is not obvious. Redrawing or retouching of unusable figures will be charged to the authors

Size: Figures should be planned so that they reduce to 12.7 cm column width. The preferred width of line drawings is 15 to 22 cm with capital lettering 4 mm high, for reduction by one-half. Photographs for half-tone reproduction should be about twice the desired size.

Color plates: Whenever the use of color is an integral part of the research, or where the work is generated in color, the Journal will publish the color illustrations without charge to the author Reprints in color will carry a surcharge Please write to the Publisher for details.

Equations and Formulae

Any mathematical or chemical notation should be clearly marked

Mathematical: Mathematical equations should preferably be typewritten, with subscripts and superscripts clearly shown. It is helpful to identify unusual or ambiguous symbols in the margin when they first occur. To simplify typesetting, please use: 1) the "exp" form of complex exponential functions; 2) fractional exponents instead of root signs; and 3) the solidus (/) to simplify fractions – e.g. $\exp x^{1}$.

Chemical: Ring formulae and other complex chemical matter are extremely difficult to typeset Please. therefore. supply reproducible artwork for equations containing such chemistry

Marking: Where chemistry is straightforward and can be set (e.g. single-line formulae), please help the typesetter by distinguishing between, e.g., double bonds and equal signs, and single bonds and hyphens, where there is ambiguity. The printer finds it extremely difficult to identify which symbols should be set in roman (upright) or italic or bold type, especially where the paper contains both mathematics and chemistry. Therefore, please underline all mathematical symbols to be set in italic and put a wavy line under bold symbols. Other letters not marked will be set in roman type.

Tables

Number tables consecutively with arabic numerals and give a clear descriptive caption at the top. Avoid the use of vertical rules in the tables. Indicate in the margin where the printer should place the tables

References and Notes

References and Notes are indicated in the text by consecutive superior arabic numbers (without parentheses). The full list should be collected and typed at the end of the paper in numerical order. Listed references should be complete in all details but excluding article titles in journals. Authors' initials should follow their names; journal title abbreviations should conform to *Physical Abstracts* style. Examples:

Smith, A.B and Jones, C.D., 1990, *J Appl. Phys.* 34, 296

Brown, R.B *Molecular Spectroscopy*, Gordon and Breach, New York, 1965, 3rd ed., Chap 6, pp. 95-106.

Proofs

Contributors from the former Soviet Union will receive page proofs (including figures) for correction via our internal courier network to Moscow. These must be returned to our Moscow office (Victor Selivanov, Lebedev Physical Institute, 53 Leninsky Prospect, Moscow 117924, Russia Email: victor@vandv.ru) within 48 hours of receipt All other contributors will receive page proofs (including figures) by airmail for correction, which must be returned to the printer within 48 hours of receipt Please ensure that a full postal address if given on the first page of the typescript, so that proofs are not delayed in the post. Author's alterations in excess of 10% of the original composition cost will be charged to authors.

Reprints

Additional reprints may be ordered by completing the appropriate form sent with proofs

Page Charges

There are no page charges to individuals or institutions.

www.ingramcontent.com/pod-product-compliance
Lightning Source LLC
Chambersburg PA
CBHW021030210326

41598CB00016B/975